MW01639580

Using Computers

Programming and Problem Solving

Gerald H. Elgarten
Alfred S. Posamentier

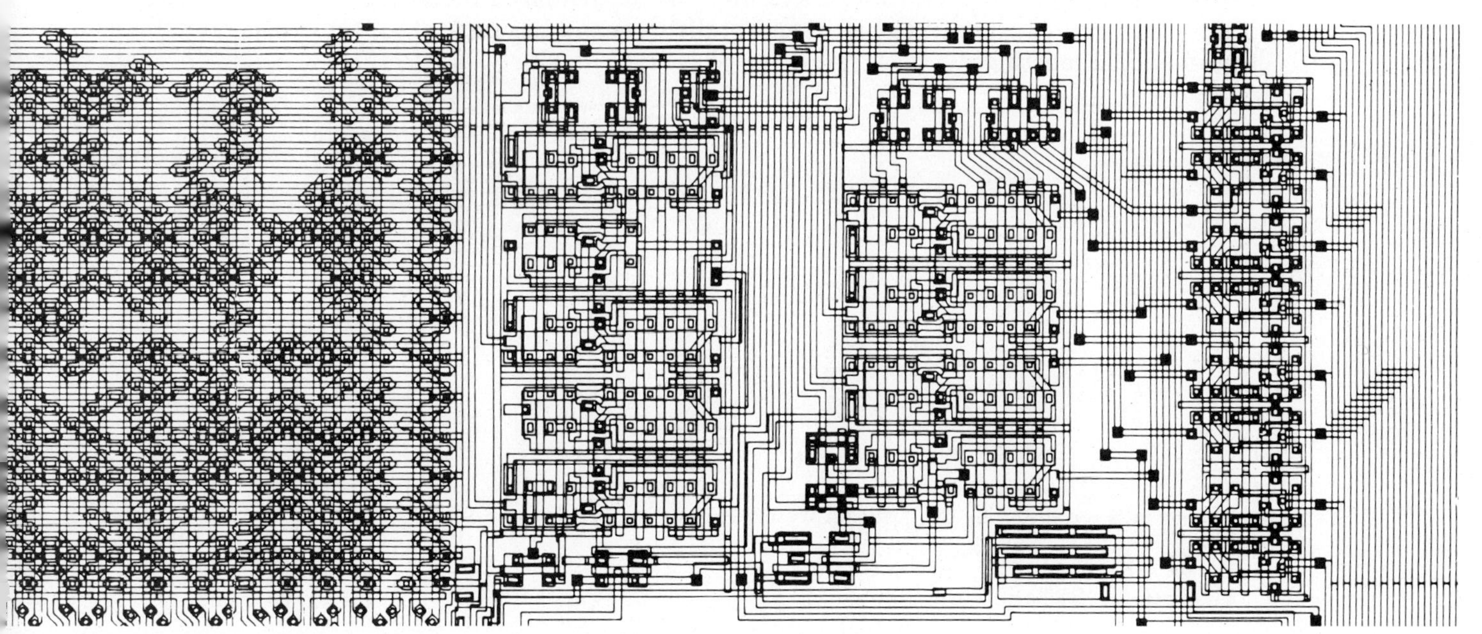

Addison-Wesley Publishing Company
Menlo Park, California · Reading, Massachusetts
London · Amsterdam · Don Mills, Ontario · Sydney

ABOUT THE AUTHORS

Gerald H. Elgarten is Supervisor of the Mathematics and Computer Science Department at Francis Lewis High School, Queens, New York. He has developed courses and taught at The City College of the City University of New York for the past four years to teachers in grades 3 through 12. Mr. Elgarten has been a junior high school mathematics teacher, and a consultant to schools and school districts, assisting them in developing curricula to introduce computer courses and computer components into existing courses. He has been a speaker locally and nationally on the subject of introducing computers in the classroom.

Dr. Alfred S. Posamentier is Professor of Mathematics Education and Chairman of the Department of Secondary and Continuing Education at The City College of the City University of New York. There he is involved with teacher education and secondary school curriculum development. He has served as project director of numerous National Science Foundation in-service institutes. He is currently also a director of the City College Select Program in Science and Engineering (SPISE).

Dr. Posamentier has had extensive experience both as a secondary school curriculum supervisor and mathematics teacher, where he has taught all grade levels and ability levels. In addition to being the author of a number of mathematics books and journal articles, he has served as a consultant to many school districts and has presented lectures at numerous professional meetings.

This book is published by the **Addison-Wesley Innovative Division**.

Design and composition: Suzanne Bonfield
Illustrations: Richard Lujan

ISBN-0-201-20205-0
DEFGHIJ-VH-898765

Foreword

This book is designed to be used either as a text for an introductory computer course or as a text to provide an ongoing component of computer instruction in the pre-senior high school curriculum. In either case, students will be able to use the book to learn how to use the computer as a tool for solving problems with mathematical and business applications and to develop understanding of concepts appropriate to their ability and grade level. By using this approach, we believe that at the same time students are becoming computer literate, they will be increasingly motivated to study mathematics, their ability to conceptualize will improve, and they will be ready for algebra, all of which contribute to an overall improvement in mathematical competency.

Using the *Teacher's Guide* with its 140 detailed lesson plans, and the *Solutions Manual* with complete solutions to all problems, a teacher with little or no computer background will be able to learn the material and simultaneously teach the course.

Because students' mastery occurs at different rates, allowances have been made throughout the book to provide for individual differences. Activities in the text require the use of various levels of skills and techniques, from simple analysis to completely organizing computer solutions to problems. Students are given opportunities to do the following:

1. revise already-written programs to produce different outputs;
2. code programs using a flowchart which has already been developed in the text;
3. develop their own algorithms and flowcharts to solve problems presented in the text.

This book has been successfully field-tested in various middle, junior, and senior high schools. It has also been used as a primary resource for in-service training of teachers.

We hope that you will enjoy working with the text as much as we enjoyed preparing the manuscript.

Gerald H. Elgarten
Alfred S. Posamentier

Acknowledgments

In preparing the manuscript for this book the authors have been fortunate to receive valuable suggestions from computer and mathematics educators at various school levels. These include elementary, middle, and secondary school teachers and supervisors, as well as some college professors. Representatives from industry were also consulted in the preparation of this book.

We would like to thank *Rita Petosa*, Adrien Block Intermediate School (Flushing, N.Y.), for field testing the first draft of the manuscript and for her important suggestions for the revision.

The authors are grateful to *Robert Bluman*, Francis Lewis High School (Queens, N.Y.); *Martin Elgarten*, Evander Childs High School (Bronx, N.Y.); *Marie Fico*, Riverdale Junior High School (Riverdale, N.Y.); and *Regina McCann* and *Lol Prashad*, Harry S. Truman High School (Bronx, N.Y.), for their valuable contributions to the solutions manual.

We also extend our sincere gratitude to the following district mathematics coordinators, mathematics chairpersons, and teachers who field tested the book in their schools: *Jack Berger*, *Lillian Strauss Greenberg*, and *Doris Hirschhorn*, Adrien Block Intermediate School (Flushing, N.Y.); *Janet Beaird*, *George Braia*, *Connie Conrad*, *Donald Maguire*, *Evelyn Perilli*, and *Charles Terry*, Henry H. Wells Middle School (Brewster, N.Y.); *Steven Buchsbaum* and *Fred Remer*, Community School District #6 (Manhattan, N.Y.); *James Burke*, Sayville Schools (Sayville, N.Y.); Eleanor Roosevelt Intermediate School (Manhattan, N.Y.); *Maria De Cesare*, Elizabeth Barrett Browning Intermediate School (Bronx, N.Y.); *L. Materassi-Eaton*, Community School District #5 (Manhattan, N.Y.); *Marie Fico*, *Arthur Jacobs*, and *Peter Uttal*, Riverdale Junior High School (Riverdale, N.Y.); *Constance Goldsby*, *Clarence J. Jones*, *Howard C. Lew*, and *David Simon*, Frederick Douglas Intermediate School (Manhattan, N.Y.); *Gary J. Kava*, Junior High School #117 (Bronx, N.Y.); *Regina Kiczek* and *Ellen Short*, J. J. Ferris High School (Jersey City, N.J.); *Susan Morris*, AMD Middle School (Ossining, N.Y.); *Jesse Reisman*, John Peter Tetard Junior High School (Bronx, N.Y.); *Gayle Rubinstein*, Michelangelo Intermediate School (Bronx, N.Y.); *Marcelino Santiago*, B.B.A.S. Public School #7 (Manhattan, N.Y.); *Gloria Scheskin*, Louis Armstrong Intermediate School (Elmhurst, N.Y.); *Barbara Shearer* and *Richard Woodall*, William L. Dickinson High School (Jersey City, N.J.); *Rose A. Solimene*, Community School District #10 (Bronx, N.Y.); *Murray Spero* and *Seymour Samuels*, Francis Lewis High School (Queens, N.Y.); *Sylvia Stahl*, Pablo Casals Intermediate School (Bronx, N.Y.); *Steven Weiss*, Flushing High School (Flushing, N.Y.).

A special thanks to *Theodore Liebersfeld* for providing valuable assistance in the preparation of the manuscript.

Contents

Chapter 1

Beginning Programming Concepts

Section 1.1 WHAT IS COMPUTER PROGRAMMING?

Before we begin working with a computer we must understand what is meant by computer programming. The following activity will introduce you to what takes place when we program a computer.

Class Exercise 1 Have a sheet of paper along with a pen or pencil in front of you before you begin.

a. Memorize the instructions that you will read.
b. Each time you read the word RUN, follow the instructions that you have memorized.

Instructions:
1. Hold your pen or pencil in hand.
2. On the sheet of paper draw these four figures.

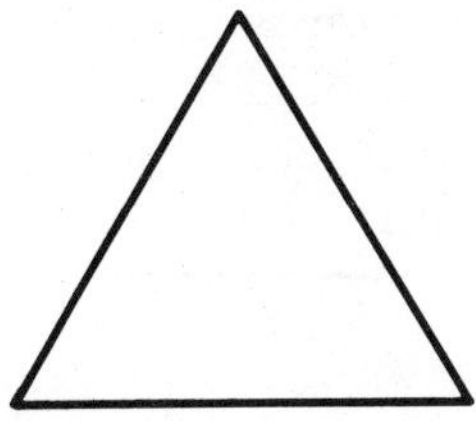 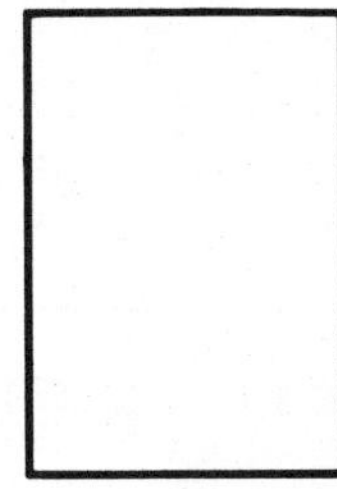  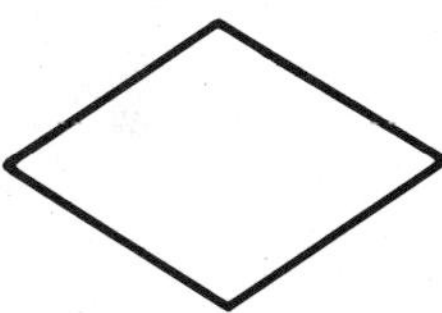

3. Rest the pen or pencil on the sheet of paper.
4. Relax!
5. End of activity.

RUN
RUN
RUN

How many times did you follow the instructions?
Did you remember to wait for the command RUN before following the instructions?

Let us now discuss (analyze) how this activity is similar to what takes place when you program a computer.

In this exercise we were the *programmers* and you were the *computer.*

In Our Exercise	Using a Computer
1. We had a plan. Our plan was to use an activity to show what takes place when programming a computer.	1. Before using a computer you must develop a plan. This plan will outline the steps you wish the computer to follow to solve a problem.
2. We gave you instructions in a language that you understood.	2. When using a computer you will give it instructions to solve a problem. These instructions must be in a language that the computer understands. There are many computer languages. In this book you will be working with the language BASIC (*B*eginner's *A*ll-Purpose *S*ymbolic *I*nstruction *C*ode).
3. While you were reading the instructions you were memorizing them (storing them in your memory).	3. Instructions are stored in a computer's memory. The computer waits for the proper signal before it follows (executes) the instructions.
4. You executed (followed) these instructions as many times as you read the word RUN.	4. A computer will execute the instructions that it has stored in its memory as many times as it receives the appropriate signal (command).

Summary: What Is Computer Programming?

1. Developing a plan to use the computer to solve a problem.

   In the computer field the person who develops the plan is called a *computer scientist* or *systems analyst.*

2. Using the plan to write a sequence of instructions in a computer language.

   This sequence of instructions is called a *program.* The title given to someone who writes programs for a computer is *computer programmer* or *programmer.*

3. Entering the program and having the computer execute the sequence of instructions (run the program).

    If a computer system uses cards to enter a program, the person responsible for typing the information onto the cards is a *keypunch operator. Terminal operator* is the job title for the person who gets the computer to run programs.

In this book you will learn how to use the computer to solve problems. As we go through the book you will do things that are part of all the jobs we just mentioned.

1. You will be the *systems analyst* when you develop a plan to solve a problem.

2. While you translate the plan into the computer language BASIC you will be the *programmer.*

3. You are the *keypunch* and *terminal operator* when you type in and run your program on the computer.

We feel and hope that as you read this book you will find the experience valuable and enjoyable.

Exercise 1.1

a. Describe the role of each in programming a computer.

1. systems analyst
2. computer programmer
3. keypunch operator
4. terminal operator

b. What are the educational requirements for each of the four careers mentioned above. List them.

Section 1.2 PREPARING A PROBLEM FOR A COMPUTER SOLUTION

Let us now consider our first problem for solution by computer.

Problem

Find the perimeter of a square when you are given the length of a side.

Before we can teach (program) the computer to find the perimeter of a square we must be able to find the perimeter ourselves. A brief review of finding the perimeter of a square might be helpful.

Example 1 Find the perimeter of the following square:

5 cm

5 cm 5 cm

5 cm

Recall that *perimeter* is the distance around the figure. Therefore, the perimeter of the square is 5 + 5 + 5 + 5 or 20 cm.

Class Exercise 1 Find the perimeters of the following squares.

1.

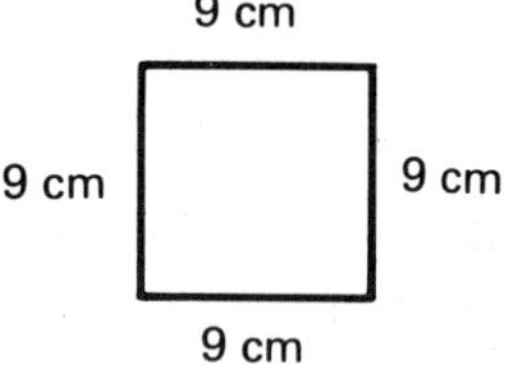

2.

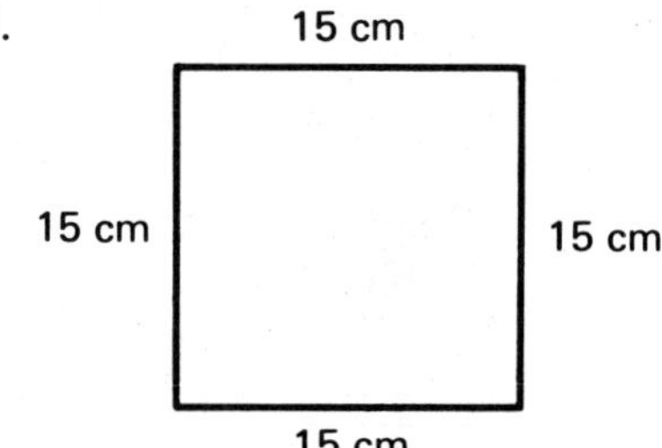

3.

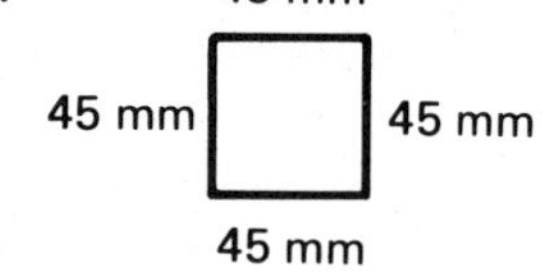

4.

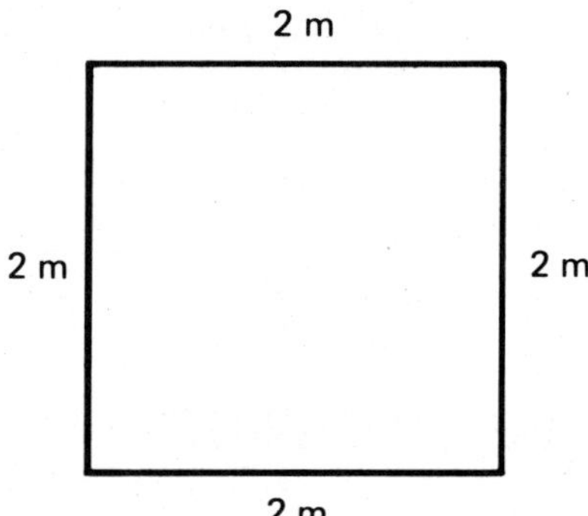

Find the perimeters of squares with sides of length:

5. 8.6 cm 6. 15.7 in. 7. 34 mm 8. 12.36 m

We are now ready to prepare our problem for a computer solution.

Computers can solve any problem that can be broken down into steps which can be translated into a computer language.

> The step-by-step procedure that leads to the solution of a problem is called an *algorithm.*

Developing an algorithm for the perimeter problem:
To find the perimeter of a square we can add the lengths of the sides or multiply the length of one side by 4.

Perimeter (P) = 4 × side (S) or $P = 4 \times S$

Example 2 Use the algorithm $P = 4 \times S$ to find the perimeter of a square with sides of length 6 cm each.

Solution:

$P = 4 \times S$

$P = 4 \times 6$ ← replacing S with 6

$P = 24$ cm

Try the next exercise to practice using the algorithm.

Class Exercise 2 Use the algorithm† $P = 4 \times S$ to find the perimeter of the following squares with sides of length:

1. 19 mm
2. 5.3 m
3. 9.8 cm
4. 5.84 m

Now that we have an algorithm, we are ready to prepare a plan.

When developing the plan *keep in mind how you used the algorithm to solve each of the problems in Class Exercise 2.*

The *plan* for using a computer to find the perimeter of a square given the length of a side:

1. Put in a value for S.
2. Calculate: $P = 4 \times S$.
3. Print the value of P.
4. This ends the instructions.

This plan will be used in the next section when we write our first program in the computer language BASIC (*B*eginner's *A*ll-Purpose *S*ymbolic *I*nstruction *C*ode).

---

†In mathematics we refer to $P = 4 \times S$ as a formula. In work with computers, formulas are usually referred to as algorithms. As we move through the book you will see that not all algorithms are formulas.

Exercise 1.2 Find the perimeters of the following squares.

1.

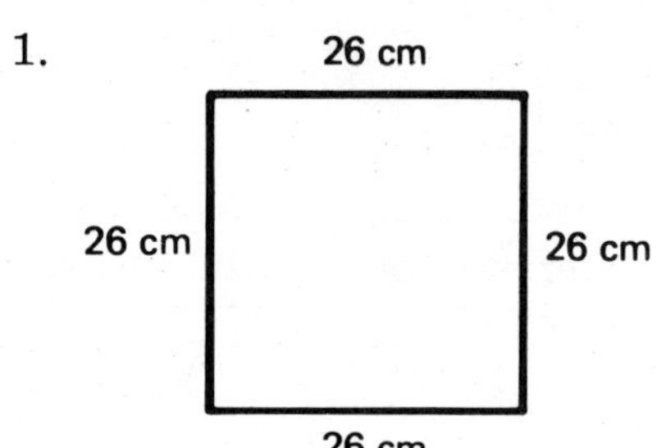

2.

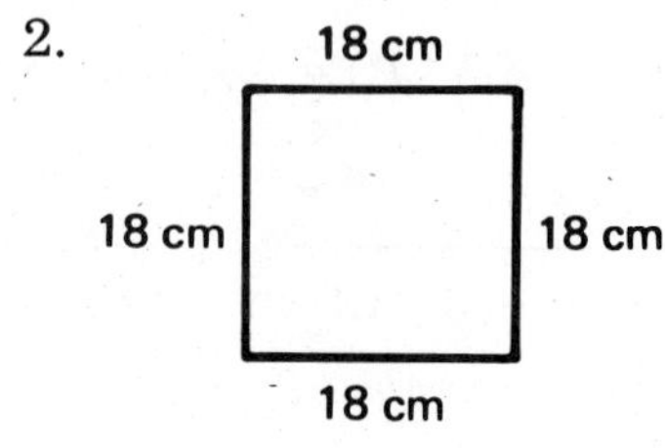

3.

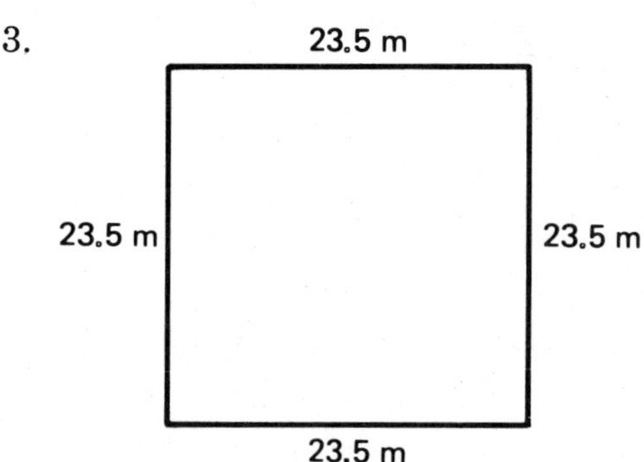

4.

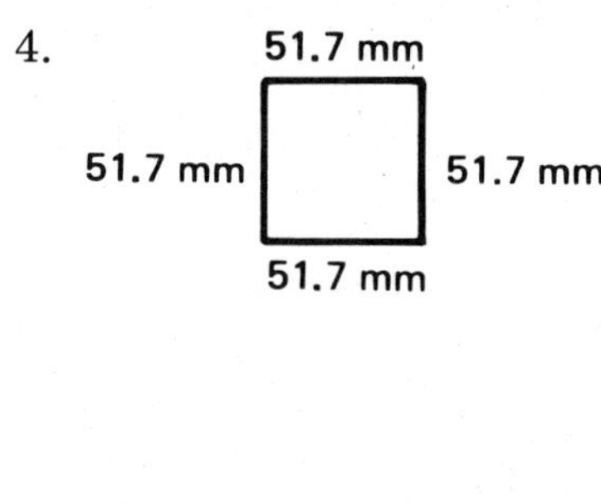

Find the perimeters of the following squares with sides of length:

5. 11.9 mm 6. 38 ft. 7. 29.6 yds. 8. 762.5 mm

Use the algorithm $P = 4 \times S$ to find the perimeter of the following squares with sides of length:

9. 436 mm 10. 43.6 cm 11. 4.36 in. 12. 0.436 m

Section 1.3 WRITING OUR FIRST PROGRAM IN BASIC (INPUT, PRINT, END)

We are now ready to write (code) our first program.

The table on the next page shows words and symbols used in translating our plan (at the end of Section 1.2) into the computer's language, BASIC.

Translating Our Plan into BASIC

Plan	*BASIC Instructions*
1. Put in a value for *S*.	10 INPUT *S*
2. Calculate: $P = 4 \times S$.	20 $P = 4 * S$
3. Print the value of *P*.	30 PRINT *P*
4. This ends the instructions.	40 END

Notice that our BASIC instructions have numbers on each line. We need the line numbers because the computer follows (executes) the instructions in numerical order. (In Section 1.7 you will see why we number the lines 10, 20, 30 and 40.)

Let us now analyze each line of our program.

> An INPUT statement stops the computer. It then allows us to enter (store) information (data) into its memory.

From our program:

```
10  INPUT S
```

S is the name of a memory location (or memory cell).

A memory location (or variable name for such a location) must begin with a letter of the alphabet. It may be followed by

1. another letter, or
2. a numeral from 0 to 9.

Q, *B*, *ET*, *LP*, *Z*3, *A*7, *WF* are all legal (correct) variable names.

Other variable names will be discussed in later chapters.

Class Exercise 1 Which of the following are legal variable names?

1. *C*7	2. !*Y*	3. *R*	4. 6*S*
5. *DD*	6. $*X*	7. *X*	8. *Y*2

Calculations in BASIC

From our program:

```
20  P=4*S
```

This is read *P is assigned the value of 4 times S.*

It uses one symbol showing an assignment (=) and one arithmetic operation (*).

The table below compares arithmetic operation symbols used in mathematics and BASIC.

Arithmetic Operation Symbols

Mathematics	*Operation*	*BASIC*
+	Addition	+
−	Subtraction	−
×	Multiplication	*
÷	Division	/
C^D	Exponentiation†	$C \uparrow D$‡

†When using the exponentiation key we sometimes encounter a rounding error. For example, 5 ↑ 2 might read 25.0000001. To avoid this error, consider calculating 5 * 5 or see Section 6.1 (rounding).
‡Some computers use ∧ for exponentiation ($A \wedge B$). Others use ↑ but display [as the symbol.

See how we use these symbols.

Arithmetic Phrase	$3 \times 5 + 2^3$
BASIC Translation	3 * 5 + 2 ↑ 3
Algebraic Equation	$X = 3E + CD - C \div D$
BASIC Translation	X = 3 * E + C * D − C/D

In algebra

3*E* means 3 times *E* → 3 * E

CD means *C* times *D* → C * D

In BASIC the symbol * must be used whenever you multiply.

Class Exercise 2 Translate the following into BASIC.

1. $Y = 6T + Q \div R$
2. $L = AB - CD$
3. $R = F \div D + ST$
4. $B = C^4$

We shall now return to analyzing our program.

```
30  PRINT P
```

Will print the value stored in memory location *P*. (More about PRINT in Section 1.6.)

```
40  END
```

Signals the computer that there are no other instructions to be executed. The program ends.

In the next section we will learn how to enter our program into the computer and RUN it.

Exercise 1.3 Which of the following are legal variable names?

1. M
2. $C8$
3. $8C$
4. A, B
5. CE
6. $A\uparrow$
7. Z
8. $L\uparrow$

Translate the following into BASIC.

9. $P=4C$
10. $Y=AR-C\div Z$
11. $M=S^5$
12. $X=C\div R+7YZ$

Section 1.4 TRACING, ENTERING AND RUNNING OUR FIRST PROGRAM

Unless you have your own computer you probably will not have much time to run your program. Here is a way of seeing if your program will work before using the computer. It is called *tracing* your program.

Procedure for Tracing a Program

Place a chart of the computer's memory next to your program.

Program

```
10  INPUT S
20  P=4*S
30  PRINT P
40  END
```

Chart of Computer Memory Cells

S	P
5	
	4 * 5 = 20
	[20]

P ← Variables used in the program

[20] ← Shows value displayed.

The number or numbers (in this case, 5) you use to test (trace) a program are called test data. Always choose numbers that you used when the problem was solved with paper and pencil. Since our answer, 20, is the same as we calculated in Section 1.2, our program seems to be correct.

Now we put the program to the real test. We will enter and RUN the program on the computer!

To Enter a Program

1. Type in the word NEW and press [RETURN]. (The TRS-80 has an [ENTER] key instead.) This erases whatever was in the computer's memory and prepares the computer to receive a NEW program.

2. Type in your program. At the end of each line press [RETURN]. This places the instruction in the computer's memory.

3. To see the instructions that are in the memory, type in [L][I][S][T] and press [RETURN]. (We will write LIST [RETURN] in the remainder of the text.)

To Run a Program

Type in RUN [RETURN]. This signals the computer to follow the instructions (program).

Notice what appears on the display of three different computers when you enter, LIST and RUN your program.

TRS-80 Model III

```
CASS?
MEMORY SIZE?
RADIO SHACK MODEL III BASIC
(C)/80 TANDY
READY
  10  INPUT S
  20  P=4*S
  30  PRINT P
  40  END
LIST
  10  INPUT S
  20  P=4*S
  30  PRINT P
  40  END
READY
RUN
?
```

Commodore PET

```
***COMMODORE BASIC 4.0***
15359 BYTES FREE
READY.
NEW
READY.
  10  INPUT S
  20  P=4*S
  30  PRINT P
  40  END
LIST
  10  INPUT S
  20  P=4*S
  30  PRINT P
  40 END
READY.
RUN
?
```

Apple II

```
NEW
  10  INPUT S
  20  P=4*S
  30  PRINT P
  40  END
LIST
  10  INPUT S
  20  P=4*S
  30  PRINT P
  40  END
RUN
?
```

On each display you see a question mark (?). This means the computer is waiting for a value to be put in (INPUT) S. If we now enter a 5 RETURN you will see

```
? 5
20
```

Enter and RUN this program on your computer. See if you get the same result as our display above. Good luck!

The next examples will help you understand how the computer follows instructions.

Example 1 Find the value assigned to A in the following computer instruction if D=2, E=6 and F=10.

70 $A=D+E+F$

Solution: The 70 is the line number of the computer instruction and does not affect the value of A.

By replacing D, E and F by 2, 6 and 10, respectively (in that order) we get:

$A=2+6+10$ or
$A=18$.

Class Exercise 1 Find the value assigned to A in the following computer instructions if D=2, E=6 and F=10.

1. 70 $A=D*E$
2. 30 $A=E*F$
3. 60 $A=F-E$
4. 50 $A=E\uparrow D$
5. 20 $A=F/D$
6. 40 $A=D*E*F$

Example 2 What is the value assigned to T after executing the two computer instructions in order? In each case B=7 and C=2.

20 $A=B+C$
30 $T=A+B$

Solution: In line 20, after replacing B with 7 and C with 2,

$A=B+C$
$A=7+2$
$A=9$

Then in line 30, remember the number 9 is stored in memory cell A (from line 20) and 7 is still stored in B. Therefore,

$T=A+B$
$T=9+7$
$T=16$

Class Exercise 2 Find the value assigned to T after executing the two computer instructions in order. In each case $B=7$ and $C=2$.

1. 20 $F=B-C$
   30 $T=F+B$
2. 40 $R=B*C$
   50 $T=R-6$
3. 70 $S=B\uparrow C$
   80 $T=S-B$
4. 30 $L=5*C$
   40 $T=70/L$
5. 10 $D=3*B*C$
   20 $T=D-B$
6. 60 $P=B/C$
   70 $T=P-2$

Example 3 The listing below shows a program that was just typed into a computer.

```
10  INPUT A,B
20  C=A*B
30  PRINT A
40  PRINT B
50  PRINT C
60  END
```

a. If the command RUN were typed into the computer, what would appear on the screen?
b. If $A=3$ and $B=4$, what will the output look like?

Solution:

a.
```
RUN
?
```
A question mark will appear on the screen because of the INPUT statement on line 10.

b. We will trace the program putting 3 in memory cell A and 4 in memory cell B.

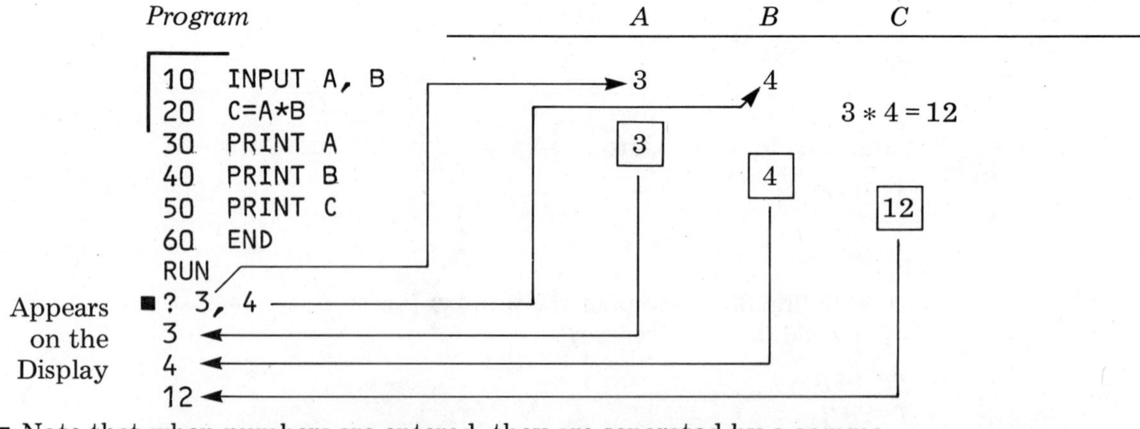

■ Note that when numbers are entered, they are separated by a comma.

Summary of the Solution:

1. What appears on the display when an INPUT instruction is executed?
   A question mark will appear on the display.
2. What appears on the screen when the computer executes line 20?
   Nothing. The calculation is performed in the computer's memory. The result is stored in the computer's memory.
3. When will the computer display what is stored inside its memory?
   Whenever it receives a command to PRINT (as in lines 30, 40 and 50).

Class Exercise 3 In each of the following problems, the listing shows a program that was just typed into a computer. Answer all questions for each program.

1.
```
10  INPUT A, B
20  C=A+B
30  PRINT A
40  PRINT B
50  PRINT C
60  END
```

a. If $A=12$ and $B=7$, show the output.
b. What is the output if $A=17.6$ and $B=19.8$?
c. What is the output if $A=368.43$ and $B=591.79$?

2.
```
10  INPUT R, S
20  PRINT R
30  PRINT S
40  X=R*S
50  PRINT X
60  END
```

a. If $R=9$ and $S=3$, show the output.
b. What is the output if $R=8.5$ and $S=.7$?
c. What is the output if $R=6.38$ and $S=5.3$?

3.
```
10  INPUT C, D, E
20  F=8*C
30  G=D/E
40  PRINT C
50  PRINT D
60  PRINT E
70  PRINT F
80  PRINT G
90  END
```

a. If $C=3$, $D=14$ and $E=7$, show the output.
b. What is the output if $C=4.93$, $D=2.4$ and $E=4$?
c. What is the output if $C=.5$, $D=12$ and $E=24$?

4.
```
10  INPUT X, Y
20  PRINT X
30  C=X/Y
40  PRINT Y
50  S=2*C
60  PRINT C
70  PRINT S
80  END
```

a. If $X=30$ and $Y=5$, show the output.
b. What is the output if $X=36.8$ and $Y=2$?
c. What is the output if $X=83.6$ and $Y=.2$?

Exercise 1.4

Find the value assigned to B in the following computer instructions if $A=8$ and $C=4$.

1. 100 $B=A-C$
2. 90 $B=A/C$
3. 90 $B=A*C$
4. 70 $B=C\uparrow 3$
5. 70 $B=6*A*C$
6. 30 $B=5.2*A$

In each of the following, find the value assigned to Y after executing the two computer instructions in order. In each case $X=12$ and $Z=3$.

7. 80 $S=X+Z$
   90 $Y=S-4$
8. 20 $T=X*Z$
   30 $Y=T/6$
9. 30 $B=X-Z$
   40 $Y=B/Z$
10. 40 $P=X/Z$
    50 $Y=8.3*P$
11. 70 $L=Z\uparrow 2$
    80 $Y=X*L$
12. 60 $M=4.5*Z$
    70 $Y=M-7$

In each of the following problems, the listing shows a program that was just typed into a computer. Answer all questions for each program.

13.
```
10  INPUT K, L
20  B=K-L
30  PRINT K
40  PRINT L
50  PRINT B
60  END
```

a. If $K=16$ and $L=5$, show the output.
b. What is the output if $K=35$ and $L=2.4$?
c. What is the output if $K=394.6$ and $L=39.7$?

14.
```
10  INPUT Q, P
20  PRINT Q
30  G=Q/P
40  PRINT P
50  PRINT G
60  END
```

a. If $Q=16$ and $P=4$, show the output.
b. What is the output if $Q=81.9$ and $P=9$?
c. What is the output if $Q=4.32$ and $P=.3$?

15.
```
10  INPUT Z
20  A=3*Z
30  PRINT Z
40  B=2*A
50  PRINT A
60  PRINT B
70  END
```

a. If $Z=12$, show the output.
b. What is the output if $Z=6.8$?
c. What is the output if $Z=4.62$?

16.
```
10  INPUT N, J
20  P=J↑3
30  Q=N*P
40  PRINT P
50  PRINT Q
60  END
```

a. If $N=26$ and $J=2$, show the output.
b. What is the output if $N=8.24$ and $J=4$?
c. What is the output if $N=.36$ and $J=3$?

Section 1.5 DEVELOPING A PROGRAM FOR FINDING THE PERIMETER OF A TRIANGLE (ORDER OF OPERATIONS IN BASIC)

How are the shapes of the following triangles different?

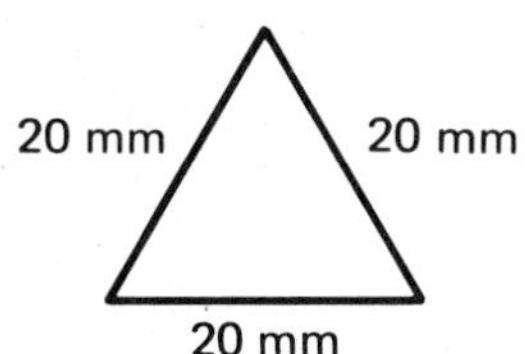

Triangle 1

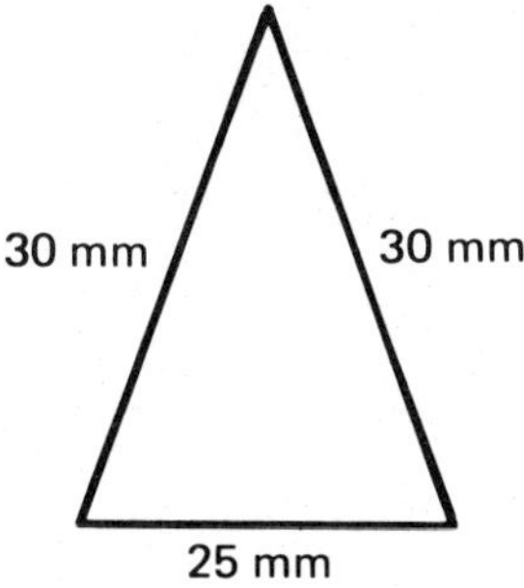

Triangle 2

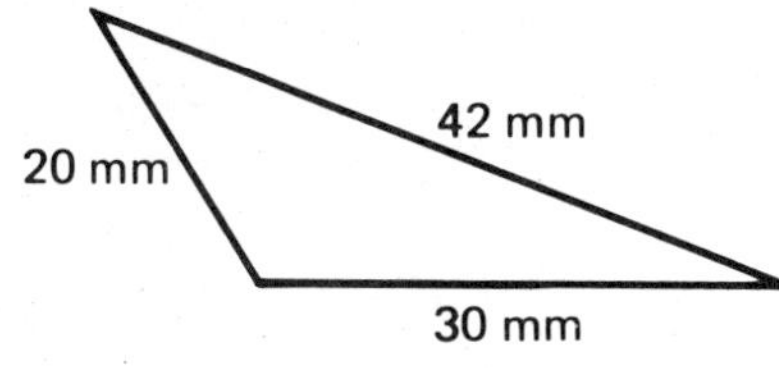

Triangle 3

Triangle 1 has all sides equal in length (or congruent). Triangle 2 has only two sides equal in length and Triangle 3 has no two sides that are the same length.

We can group (classify) triangles by the number of sides that are congruent (or equal in length) as follows:

An *equilateral* triangle is a triangle with three sides equal in length.

An *isosceles* triangle is a triangle with at least two sides equal in length.

A *scalene* triangle is a triangle with no sides equal in length.

Class Exercise 1 Classify the triangles on the previous page as *equilateral*, *isosceles* or *scalene*.

1. Triangle 1
2. Triangle 2
3. Triangle 3

To find the perimeter of a triangle we add the lengths of the sides. For example, the perimeter of this scalene triangle is

40 + 30 + 20 or 90 mm.

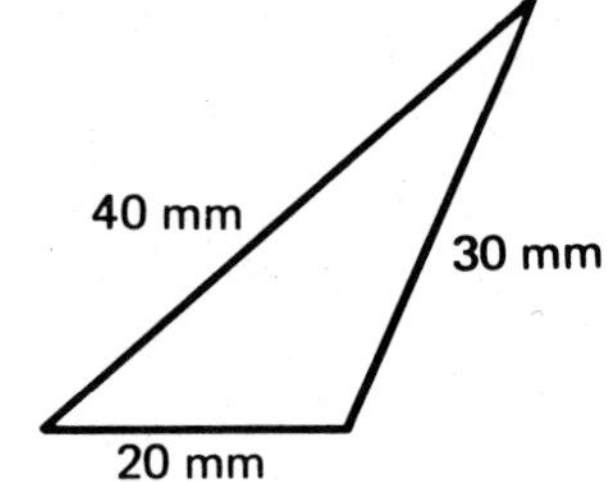

Class Exercise 2 a. Classify each of the following triangles as scalene, isosceles or equilateral.
b. Find the perimeter of each triangle.

1.

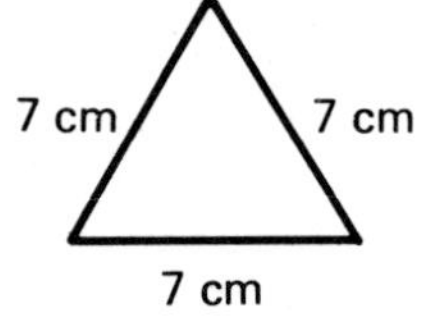

2.

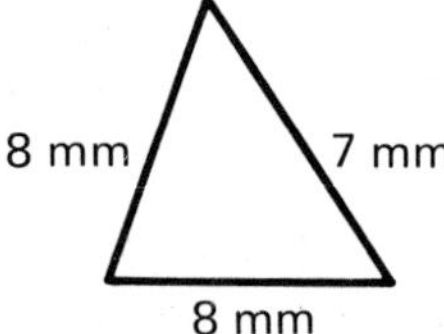

3.

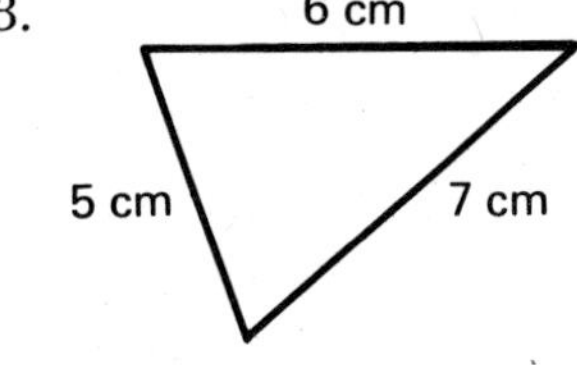

4.

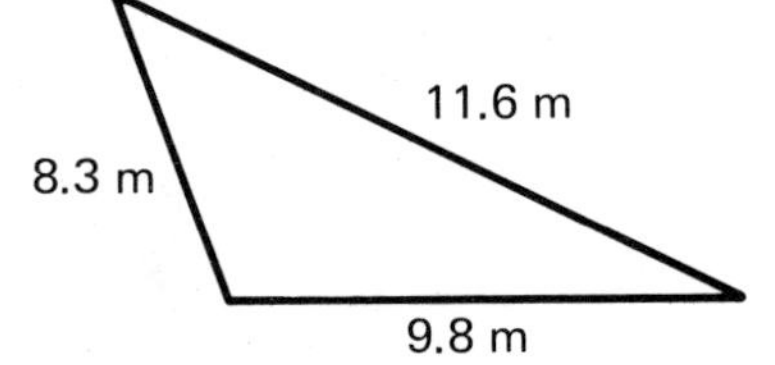

Let us now turn to the computer to solve the following problem.

Problem Find the perimeter of a scalene triangle when we are given the lengths of the sides.

How to find a computer solution to a problem:	
a. *Analysis*	We develop the algorithm.
b. *Plan*	We list the steps we take to use the algorithm to solve the problem.
c. *Program*	We translate the plan into BASIC.

Solution:

a. *Analysis:* How do we find the perimeter of a scalene triangle? The perimeter of a scalene triangle = first side's length + second side's length + third side's length.

How can this be written in a shorter form?

$P=F+S+T$ — Where P is the perimeter,
F is the length of the first side.
S is the length of the second side.
T is the length of the third side.

b. *Plan:*

1. Put in values for F, S, T.
2. Calculate: $P=F+S+T$.
3. Print P.
4. End.

Next we will write the program and show the output when the sides have lengths 20 mm, 30 mm and 42 mm, respectively.

c. *Program*

```
   10  INPUT F, S, T
   20  P=F+S+T
   30  PRINT P
   40  END
   RUN
■  ? 20, 30, 42
■■ 92
```

Discussion of the Program:

■ Computer stops to allow us to enter the values.

■■ Perimeter is printed.

Let us examine an isosceles triangle. See if we can develop an algorithm for finding the perimeter.

[Remember that developing an algorithm involves first solving a problem with numbers and then using letters (variables). When you replace numbers with letters you are *generalizing.*]

Example 1 Find the perimeter of the two isosceles triangles below. (*Note:* We first solve the problem with numbers.)

1.

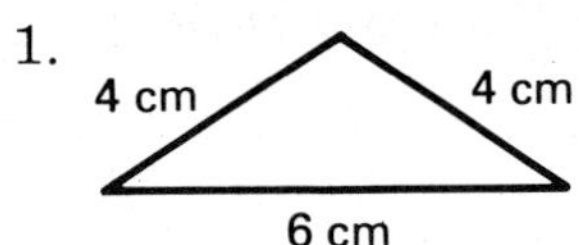

$P=6+4+4$
$P=14$ cm

2.

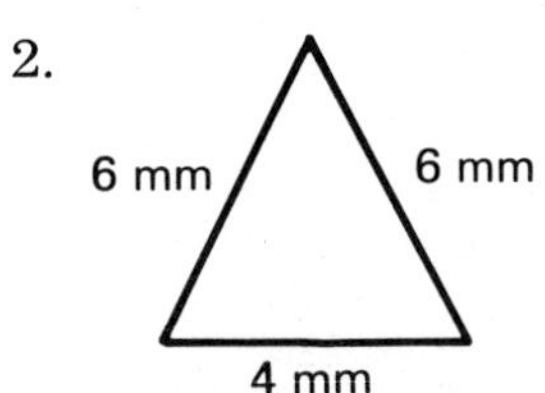

$P=4+6+6$
$P=16$ mm

Solution: To generalize, we replace the numbers with letters.

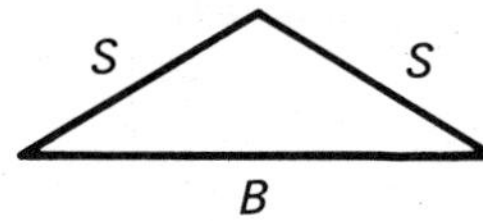

Since an isosceles triangle has two congruent (equal) sides we can use the same letter to represent these sides.

The algorithm is $P=B+S+S$. Since S appears twice we can write:

$P=B+2\times S$ or $P=B+2S$.

Let us use the algorithm to find the perimeter of the isosceles triangle above (Example 1) where $B=6$ cm and $S=4$ cm.

Examine the two possibilities. Which is correct?

1. $P=B+2S$
   $P=6+2\times 4$
   $P=8\times 4$
   $P=32$

2. $P=B+2S$
   $P=6+2\times 4$
   $P=6+8$
   $P=14$

Problem 2 is correct (see Example 1 above). Notice that we multiplied first and then added.

The computer performs arithmetic operations in the same order that we use in mathematics.

Let us review this order.

Order of Performing Arithmetic Operations

Step 1: Do any operations in parentheses.

Step 2: Work out exponents.

Step 3: Perform all multiplications and divisions, in order, from left to right.

Step 4: Perform all additions and subtractions, in order, from left to right.

Look at the two examples below to see how these rules are used.

Example 2 Find the value assigned to R.

$R=7+4\cdot(3+2)$

Note: This dot is used in mathematics as a multiplication sign.

Solution: Following Step 1, we do the operation in parentheses.

$R=7+4(3+2)$ becomes $R=7+4(5)$

Note: A number followed by parentheses means multiply. Therefore, we can omit the dot.

Using Step 3, we perform all multiplications.

$R=7+4(5)$ becomes $R=7+20$

Then we add (Step 4) to get

$R=27$

Example 3 Find the value assigned to W.

$W=12-2^3$

Solution: Since there are no parentheses, we go directly to Step 2 and evaluate the exponent. Since

$2^3 = 2 \cdot 2 \cdot 2$ which is 8,

Note: Here we need the dot for multiplication.

$W = 12 - 2^3$ becomes

$W = 12 - 8$.

By Step 4 we subtract, so

$W = 4$.

Let us practice using the order of performing arithmetic operations by doing the problems in the next class exercise.

Class Exercise 3 Find the values assigned to R.

1. $R = 6 + 5 \cdot (7 + 1)$
2. $R = 11 + 8 \cdot (5 - 2)$
3. $R = 2 + 3^2$
4. $R = 5 + 7 \cdot 3$
5. $R = 8 \cdot 4 - 3 \cdot 6$
6. $R = 16 - 8/2$
7. $R = 5 + 6 \cdot 3 + 4^3$
8. $R = (5 + 6) \cdot 3 + 3^4$

Example 4 Find the value assigned to W in the following computer instruction if $X = 2$, $Y = 8$ and $Z = 5$.

```
40 W=(Y+Z)*X
```

Solution: The 40 is the line number of the computer instruction and does not affect the value of W.

By replacing Y, Z and X by 8, 5 and 2 respectively, we get

$W = (8 + 5) * 2$.

Following Step 1, $W = 13 * 2$.

Following Step 3, $W = 26$.

Class Exercise 4 Find the value assigned to W in each computer instruction if $X=2$, $Y=8$ and $Z=5$.

1. 30 $W=(Y-Z)*X$
2. 40 $W=Y+Z*X$
3. 20 $W=Z+Y/X$
4. 50 $W=Z+Y\uparrow X$
5. 150 $W=3*Y+4*X$
6. 170 $W=X*Y/Y*X$
7. 60 $W=(X*Y)/(Y*X)$
8. 10 $W=Y/X+Z\uparrow X$

Example 5 The listing below shows a program that was just typed into a computer. Show the complete output for the program.

```
10  INPUT A, B, C
20  D=A+B*C
30  PRINT A
40  PRINT B
50  PRINT C
60  PRINT D
70  END
RUN
? 3, 4, 5
```

Solution:

```
   RUN
   ? 3, 4, 5
a. 3
b. 4
c. 5
d. 23
```

3, 4, and 5 are placed into memory cells A, B and C (in that order).
a. 3 is displayed from line 30.
b. 4 is displayed from line 40.
c. 5 is displayed from line 50.
d. From line 20, $D=3+4*5$
$D=3+20$
$D=23$.

This value is not seen on the display until line 60 is executed.

Now try the next class exercise!

Class Exercise 5 Each listing below shows a program that was just typed into a computer. Show the complete output for the program.

1.
```
10  INPUT A, B
20  F=A+B*A
30  PRINT A
40  PRINT B
50  PRINT F
60  END
RUN
? 5, 6
```

2.
```
10  INPUT G, H
20  J=G-H↑3
30  PRINT H
40  PRINT G
50  PRINT J
60  END
RUN
? 16, 2
```

3.
```
10  INPUT C, D
20  PRINT C
30  PRINT D
40  E=C*D+C/D
50  PRINT E
60  END
RUN
? 10, 5
```

4.
```
10  INPUT X, Y
20  PRINT X
30  Z=X+X*Y-Y
40  PRINT Y
50  PRINT Z
60  L=Z/Y
70  PRINT L
80  END
RUN
? 12, 3
```

Exercise 1.5

a. Classify each triangle according to the length of its sides.
b. Find the perimeter of each triangle.

1.
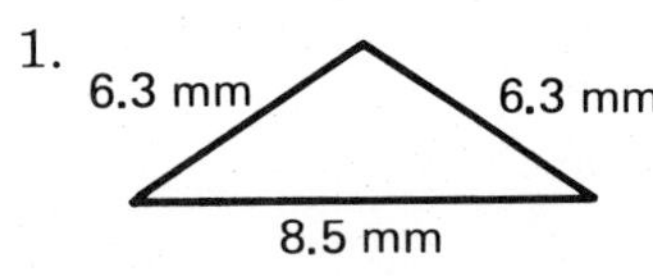

2.
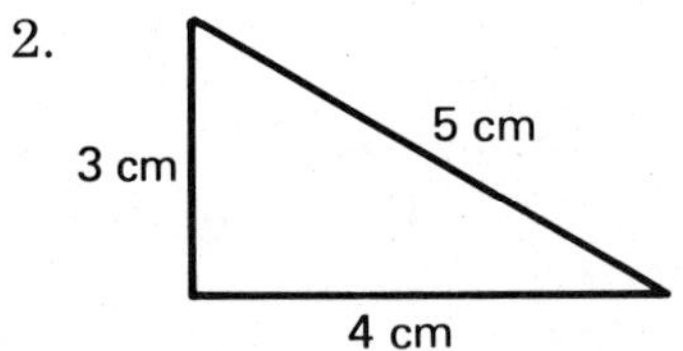

3.
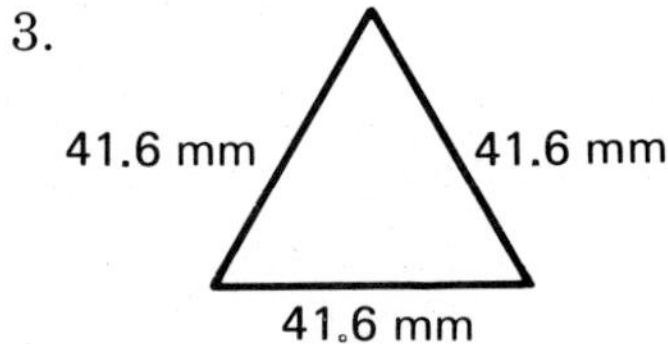

4. The listing below shows a program that was just typed into the computer. (Note that it is the same program we developed to find the perimeter of a scalene triangle.)

```
10  INPUT F, S, T
20  P=F+S+T
30  PRINT P
40  END
```

a. If $F=8$ cm, $S=9$ cm and $T=10$ cm, show the output.
b. What is the output if $F=41.6$ mm, $S=38.9$ mm and $T=43.7$ mm?

5. Revise the program in Problem 4 so that if $F=4$ mm, $S=5$ mm and $T=6$ mm, the output would appear as follows:

```
RUN
? 4, 5, 6
4
5
6
15
```

For the following, find the values assigned to L.

6. $L=9+7\cdot(8+5)$
7. $L=16+9\cdot(7-4)$
8. $L=6+4^3$
9. $L=25-6\cdot 3$
10. $L=7\cdot 5-6\cdot 2$
11. $L=36-12\div 3$
12. $L=9-3\cdot 2+5^2$
13. $L=(9-3)\cdot 2+2^3$

Find the value assigned to D in each of the following computer instructions if $A=12$, $B=3$ and $C=4$.

14. 70 $D=(A-C)/2$
15. 20 $D=A+B*C$
16. 10 $D=A-C/2$
17. 60 $D=A+C\uparrow B$
18. 50 $D=A*B+C*B$
19. 90 $D=A/B*C/2$
20. 80 $D=(A/B)*(C/2)$
21. 10 $D=C*B/A+A$

In the following, find the value assigned to E by executing the two computer instructions in order. In each case A=8 and B=3.

22. 30 $C=A+B$
    40 $E=4*C$
23. 10 $X=A*B$
    20 $E=X-2*B$
24. 70 $Y=A+B*A$
    80 $E=40-Y$
25. 60 $R=B\uparrow 2$
    70 $E=(R-1)/A$

Each listing below shows a program that was just typed into a computer. Show the complete output for the program.

26.
```
10  INPUT K, L
20  N=K+L*K
30  PRINT K
40  PRINT L
50  PRINT N
60  END
RUN
? 7, 12
```

27.
```
10  INPUT L, M, N
20  PRINT L
30  S=L-M/N
40  B=S+L*M
50  PRINT S
60  PRINT B
70  END
RUN
? 10, 12, 6
```

28.
```
10  INPUT P, Q
20  R=P+Q↑2
30  PRINT R
40  S=(P+Q)*P
50  PRINT S
60  END
RUN
? 3, 4
```

29.
```
10  INPUT C, D, E
20  F=C+D/E
30  PRINT F
40  G=(C+D)/E
50  PRINT G
60  END
RUN
? 12, 14, 2
```

Section 1.6 MAKING OUTPUTS PROVIDE MORE INFORMATION (PRINT WITH PUNCTUATIONS)

Look at the following output:

```
? 5
20
```

What problem does the program that produced this output solve?

There are many possible answers to this question. Since there is not enough information given, you may say that the program finds the perimeter of a square when we are given the length of a side. (See Section 1.4.)

In this section we will look at ways of making our output give more information to the person using the program (program user).

Look at this program:

```
10  INPUT N
20  PRINT "I CAN PRINT N"
30  PRINT "OR I CAN PRINT"; N
40  END
```

When we run this program entering 3 for N (that is, typing in RUN [RETURN] [3]

RETURN) our output will appear as

```
   RUN
   ? 3
■  I CAN PRINT N
■■ OR I CAN PRINT 3
```

■ When *N* is inside quotes, the letter *N* is displayed (line 20).

■■ When *N* is outside quotes (line 30), the value in the memory cell is printed.

If we had left out the semicolon in line 30, our result would not be affected. However, in this book we will use the semicolon since it makes the program easier to read.

Let us look at ways to use the semicolon and comma in PRINT statements.

Program #1 (using semicolons)

```
10  PRINT "PRINTING WITH SEMICOLONS"
20  INPUT C, D, E
30  PRINT C; D; E
40  END
RUN
PRINTING WITH SEMICOLONS
? 2, 3, 4
2  3  4
```

This program uses semicolons to separate the list of variables to be printed. (If your computer does not leave a space between the numbers, refer to your manual.)

Now let us examine the next program.

Program #2 (using commas)

```
10  PRINT "PRINTING WITH COMMAS"
20  INPUT C, D, E
30  PRINT C, D, E
40  END
RUN
PRINTING WITH COMMAS
? 2, 3, 4
2          3          4
```

The comma divides the display into 3, 4 or 5 fields depending upon the computer you are using.

Notice that the numbers are printed in the leftmost position (left justified) in each field.

The following exercises are designed to give you practice using punctuations with PRINT statements. After you complete the exercises, we will return to our first program to improve its output.

Example 1 Show the complete output for the following program.

```
10  INPUT A
20  PRINT "A="; A
30  PRINT "HELLO"
40  B=6
50  PRINT B
60  PRINT A*B; A+B
70  END
RUN
? 4
```

Solution:

```
   RUN
a. ? 4
b. A=4
c. HELLO
d. 6
e. 24 10
```

a. 4 is placed into memory cell A.
b. $A=4$ is printed from line 20.
c. HELLO is printed from line 30.
d. The instruction $B=6$ assigns a 6 to memory location B.
6 is printed from line 50.
(*Note:* The variable must be on the *left* side in an assignment statement: $B=6$ as opposed to $6=B$. See line 40.)
e. The computer will also print any arithmetic operation (line 60).

Class Exercise 1 For each program, show the complete output.

1.
```
10  INPUT C
20  PRINT "THE NUMBER IS"
30  PRINT C
40  END
RUN
? 3
```

2.
```
10  INPUT D
20  PRINT "THE NUMBER IS"; D
30  END
RUN
? 5
```

3.
```
10  INPUT Q, R
20  PRINT "Q="; Q
30  PRINT "R="; R
40  END
RUN
? 6, 7
```

4.
```
10  INPUT A, B, C
20  PRINT A, B, C
30  PRINT A; B; C
40  END
RUN
? 7, 8, 9
```

You may find that you want to change your output in order to provide more information to the user. The next example shows how we can make changes in our program to produce a specific output.

Example 2 Here is a planned output and a program. *A* ***planned output*** *is the output you wish the program to produce.* Correct the program so that it will produce the required output.

Planned Output

```
RUN
? 7
Q=7
RUN
? 13
Q=13
```

Program

```
10  INPUT Q
20  PRINT Q
30  END
```

Solution: The ? 7 in our planned output indicates an INPUT statement in our program (line 10). Line 20 in our program will only print what is in memory location *Q* (7). If we change the line to:

```
20  PRINT "Q=7"
```

what would the computer print? The computer would print *Q*=7 just as we wrote in our planned output. Our program would then be:

New Program

```
10  INPUT Q
20  PRINT "Q=7"
30  END
```

What would happen if we then entered a 13 for *Q* as in our planned output?

The computer would print *Q*=7. To have the number on the display change every time a new number is entered, we must use a variable after the equal sign.

Corrected Program

```
10  INPUT Q
20  PRINT "Q="; Q  ◄——  Computer will print Q= followed by whatever value is in
30  END                 memory location Q.
```

Class Exercise 2 Given a planned output and a program, correct the program so that it will produce the required output.

1. *Planned Output*

```
RUN
? 8
X=8
RUN
? 2
X=2
```

Program

```
10  INPUT X
20  PRINT X
30  END
```

2. *Planned Output*

```
RUN
? 10, 11
F=10
G=11
RUN
? 5, 6
F=5
G=6
```

Program

```
10  INPUT F, G
20  PRINT F
30  PRINT "G="
40  END
```

3. *Planned Output*

```
RUN
LARA'S PROGRAM
? 100
TEST SCORE IS 100
RUN
LARA'S PROGRAM
?  99
TEST SCORE IS 99
```

Program

```
10  "LARA'S PROGRAM"
20  INPUT L
30  PRINT "TEST SCORE IS"; 100
40  END
```

4. *Planned Output*

```
RUN
PROGRAM BY JEREMY
ENTER A NUMBER
? 5
THE NUMBER IS 5
RUN
PROGRAM BY JEREMY
ENTER A NUMBER
? 8
THE NUMBER IS 8
```

Program

```
10  PRINT PROGRAM BY JEREMY
20  "ENTER A NUMBER"
30  INPUT H
40  PRINT "THE NUMBER IS"; H
50  END
```

We are now ready to write a program to produce a planned output. After studying the example below, try the exercises that follow.

Example 3 Write a program to produce the planned output.

Planned Output

```
   RUN
a. ? 4, 13
b. A=4
c. B=13
   RUN
   ? 5, 1
   A=5
   B=1
```

Solution:

a. ? 4, 13 indicates an INPUT statement with two variables:

10 INPUT *A*, *B*

b. *A*=4 indicates that *A*= is constant and the number that follows is changing:

20 PRINT "*A*="; *A*

c. *B*=13 is programmed the same way, using *B* as the variable:

30 PRINT "*B*="; *B*

The following program will also produce the remainder of the output (when 5 and 1 are input).

```
10  INPUT A, B
20  PRINT "A="; A
30  PRINT "B="; B
40  END
```

Class Exercise 3 Write a program to produce the planned output.

1. *Planned Output*

```
RUN
? 2, 3
2 3
RUN
? 4, 5
4 5
```

2. *Planned Output*

```
RUN
? 15, 16
Q=15
R=16
RUN
? 23, 6
Q=23
R=6
```

3. *Planned Output*

```
RUN
LYNDA'S PROGRAM
WHAT IS THE NUMBER?
? 32
CORRECT!
RUN
LYNDA'S PROGRAM
WHAT IS THE NUMBER?
? 30
CORRECT!
```

4. *Planned Output*

```
RUN
USING PUNCTUATIONS
? 7, 8, 9
7 8 9
7             8             9
RUN
USING PUNCTUATIONS
? 1, 2, 3
1 2 3
1             2             3
```

Exercise 1.6 For each program, show the complete output.

1.
```
10  INPUT L, M, N
20  PRINT L, M, N
30  END
RUN
? 3, 2, 4
```

2.
```
10  PRINT "WHAT IS YOUR NUMBER?"
20  INPUT V
30  PRINT "V="; V
40  END
RUN
WHAT IS YOUR NUMBER?
? 17
```

3.
```
10  R=7
20  S=3
30  A=R+S*R
40  PRINT R; S; A
50  END
RUN
```

4.
```
10  A=4
20  B=6
30  PRINT "THE NUMBERS ARE"
40  PRINT "A="; A; "B="; B
50  END
RUN
```

Given a planned output and a program for each of the following, correct the program so that it will produce the required output.

5. *Planned Output*

```
RUN
? 37
Y=37
RUN
? 14
Y=14
```

Program

```
10  INPUT Y
20  PRINT Y
30  END
```

6. *Planned Output*

```
RUN
? 1, 2
D=1
E=2
RUN
? 2, 1
D=2
E=1
```

Program

```
10  INPUT D, E
20  PRINT D=1
30  PRINT "E="
40  END
```

7. *Planned Output*

```
RUN
? 4, 5
F=4       G=5
RUN
? 3, 47
F=3       G=47
```

Program

```
10  INPUT A, B
20  PRINT "F="; F, "G="; G
30  END
```

8. *Planned Output*

```
RUN
NO INPUT
3
4
RUN
NO INPUT
3
4
```

Program

```
10  "NO INPUT"
20  X=3
30  Y=4
40  PRINT "X="; X
50  PRINT "Y"
60  END
```

For each of the following, write a program to produce the planned output.

9. *Planned Output*

```
RUN
? 9
THE ANSWER IS 9
RUN
? 18
THE ANSWER IS 18
```

10. *Planned Output*

```
RUN
? 31
THE ANSWER IS
31
RUN
? 19
THE ANSWER IS
19
```

11. *Planned Output*

```
RUN
WHAT IS YOUR NUMBER LISA?
? 4
YOU CHOSE 4
RUN
WHAT IS YOUR NUMBER LISA?
? 19
YOU CHOSE 19
```

12. *Planned Output*

```
RUN
ENTER A NUMBER DAVID
? 21
21 IS YOUR NUMBER
RUN
ENTER A NUMBER DAVID
? 42
42 IS YOUR NUMBER
```

Section 1.7 CODING A PROGRAM AFTER DESIGNING A PLANNED OUTPUT

Now that we know more about the PRINT statement, let us improve our first program from Section 1.3,

```
10  INPUT S
20  P=4*S
30  PRINT P
40  END
```

to give us the planned output that follows.

Planned Output

```
    RUN
a.  FINDING THE PERIMETER OF A SQUARE
b.  ENTER THE LENGTH OF A SIDE
c.  ? 6
d.  THE PERIMETER IS 24
```

Without looking ahead, see if you can rewrite our program to produce the planned output. After writing the program, RUN it on the computer. Good luck!

Improving the Output

In order to get the computer to print statements like those appearing at (a) and (b) above, we must use quotes with our PRINT statement:

```
10 PRINT "FINDING THE PERIMETER OF A SQUARE"
20 PRINT "ENTER THE LENGTH OF A SIDE"
```

The question mark at (c) indicates an INPUT statement. Here, we need a variable:

```
30 INPUT S
```

We must now ask the computer to perform the calculation (from line 20 of the first program) to find *P*:

```
40 P=4*S
```

and then print the answer as shown in the planned output:

```
50 PRINT "THE PERIMETER IS"; P
60 END
```

Instead of redoing the entire program, as we did, let us see if we can make the revision while keeping as many of the old lines as possible.

Compare the original program to our revised one.

Original Program

```
10  INPUT S
20  P=4*S
30  PRINT P
40  END
```

Revised Program

```
10  PRINT "FINDING THE PERIMETER OF A SQUARE"
20  PRINT "ENTER THE LENGTH OF A SIDE"
30  INPUT S
40  P=4*S
50  PRINT "THE PERIMETER IS"; P
60  END
```

Which instructions are the same in the original program as in the revised program? (For the moment, disregard the line numbers.)

There are three instructions that are the same.

10 INPUT *S*
20 *P*=4**S*
40 END

But notice that in the revised program these instructions have different line numbers: 30, 40 and 60, respectively. Let us try to revise the program by keeping the original three instructions and inserting the three new ones into the original program.

The revision would be made as follows.

Original Program

```
10  INPUT S
20  P=4*S
30  PRINT P
```

New Instructions

```
5   PRINT "FINDING THE PERIMETER OF A SQUARE"
7   PRINT "ENTER THE LENGTH OF A SIDE"
30  PRINT "THE PERIMETER IS"; P
```

Note that we retyped line 30. This erases the previous line 30 and places the new line in memory.

Since we numbered our original lines in multiples of 10, we left room to insert new instructions between these numbers.

When we type RUN at this point, *the computer will execute the instructions in numerical order, regardless of the order in which each line was typed.*

To see the program printed in numerical order, we can type in LIST RETURN and the following will appear on the display.

Original Program Plus New Lines

```
10  INPUT S
20  P=4*S
30  PRINT P
40  END
5   PRINT "FINDING THE PERIMETER OF A SQUARE"
7   PRINT "ENTER THE LENGTH OF A SIDE"
30  PRINT "THE PERIMETER IS"; P
LIST
```

Revised Program Listed with Lines in Numerical Order

```
5   PRINT "FINDING THE PERIMETER OF A SQUARE"
7   PRINT "ENTER THE LENGTH OF A SIDE"
10  INPUT S
20  P=4*S
30  PRINT "THE PERIMETER IS"; P
40  END
```

We call this procedure of revising programs by inserting and retyping lines *editing*.

Example 1

a. Show the output of the old program.
b. Edit the old program to produce the new output. Try to keep as much of the old program as possible.

Old Program

```
10  INPUT C, B
20  D=C-B
30  PRINT D
40  END
```

New Output

```
ENTER TWO NUMBERS
? 8, 2
THE DIFFERENCE IS
6
```

Solution:
a. Using the values from the new output we run the old program.

```
10  INPUT C, B
20  D=C-B
30  PRINT D
40  END
RUN
? 8, 2
6
```

This program prints the difference of two numbers. But looking at our output, all we see is the number 6.

We will now revise this program so that when it is used it will display the information in the *new output.*

b. The following lines should be added to the program:

```
5    PRINT "ENTER TWO NUMBERS"
25   PRINT "THE DIFFERENCE IS"
```

For the first line, which we labeled 5, we could have chosen any whole number from 0 to 9. For the line called 25, we could have used any integer from 11 to 19, or 21 to 29.

Revised Program

```
10   INPUT C, B
20   D=C-B
30   PRINT D
40   END
5    PRINT "ENTER TWO NUMBERS"
25   PRINT "THE DIFFERENCE IS"
```

The following exercises will give you practice in editing and writing programs.

Class Exercise 1 a. Show the output for the old program.
b. Edit the old program to produce the new output. Try to keep as much of the old program as possible.

1. *Old Program*

```
10   INPUT R, T
20   S=R+T
30   PRINT S
40   END
```

New Output

```
ENTER TWO NUMBERS
? 7, 1
THE SUM IS 8
```

2. *Old Program*

```
10   INPUT A, B
20   P=A*B
30   PRINT P
40   END
```

New Output

```
FINDING A PRODUCT
ENTER TWO NUMBERS
? 3, 4
THE PRODUCT IS
12
```

3. *Old Program*

```
10  INPUT C, D, E
20  A=C+D+E
30  PRINT A
40  END
```

New Output

```
ADDING THREE NUMBERS
PUT IN THREE NUMBERS
? 5, 6, 7
THE RESULT IS 18
```

4. *Old Program*

```
10  INPUT Q, R, S
20  L=Q+R*S
30  PRINT L
40  END
```

New Output

```
MULTIPLY AND THEN ADD
ENTER THREE NUMBERS
? 2, 3, 6
THE ANSWER IS 20
```

Now we can put these skills to work writing computer programs.

Example 2

For the following problem:
a. Develop an algorithm to solve the problem.
b. Write a plan showing how you use the algorithm to solve the problem.
c. Plan an output.
d. Code the program.

Problem

Find the perimeter of an equilateral triangle given the length of a side.

Solution:
a. *Analysis:* The perimeter of an equilateral triangle is the sum of the lengths of the three sides. Since the lengths of the sides are the same, $P=3*S$ is our algorithm.

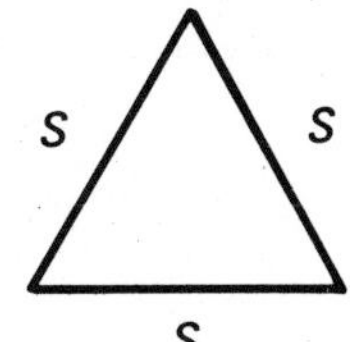

Next we show how we use the algorithm to solve the problem.

b. *Plan:*
1. Put in the length of a side of an equilateral triangle: S.
2. Calculate: $P=3*S$.
3. Print the perimeter, P.
4. End.

Notice that we use the algorithm *after* we enter the length of a side.

c. *Planned Output:* We will give enough information in our planned output so that the program user will understand the problem being solved.

```
PERIMETER OF AN EQUILATERAL TRIANGLE
ENTER THE LENGTH OF A SIDE
? 8
THE PERIMETER IS 24
```

d. *The Program*

```
10  PRINT "PERIMETER OF AN EQUILATERAL TRIANGLE"
20  PRINT "ENTER THE LENGTH OF A SIDE"
30  INPUT S
40  P=3*S
50  PRINT "THE PERIMETER IS"; P
60  END
```

Class Exercise 2 For each of the following problems:

a. Develop an algorithm to solve the problem.
b. Write a plan showing how you use the algorithm to solve the problem.
c. Plan an output.
d. Code the program.

1. **Problem:** Find the perimeter of a regular pentagon given the length of a side. (A regular pentagon is a five-sided polygon with all sides the same length and all angles equal in measure.)

2. **Problem:** Enter two numbers. Find the product of the numbers.

3. **Problem:** Enter two nonzero integers. Find the quotient of the integers.

4. **Problem:** Find the average (arithmetic mean) of two numbers.

Exercise 1.7

a. Show the output of the old program.
b. Edit the old program to produce the new output. Try to keep as much of the old program as possible.

1. *Old Program*

```
10  INPUT X
20  F=X+5
30  PRINT F
40  END
```

New Output

```
ENTER A NUMBER
? 4
FIVE MORE IS 9
```

2. *Old Program*

```
10  INPUT A
20  R=A-2
30  PRINT R
40  END
```

New Output

```
ENTER A NUMBER
? 6
TWO LESS IS 4
```

3. *Old Program*

```
10  INPUT W
20  S=W↑2
30  PRINT S
40  END
```

New Output

```
PUT IN A NUMBER
? 3
SQUARED IS 9
```

4. *Old Program*

```
10  INPUT E, F, G
20  H=E-F*G
30  PRINT H
40  END
```

New Output

```
MULTIPLY AND THEN SUBTRACT
ENTER THREE NUMBERS
? 16, 3, 4
THE RESULT IS 4
```

For each of the following problems:
a. Develop an algorithm to solve the problem.
b. Write a plan showing how you use the algorithm to solve the problem.
c. Plan an output.
d. Code the program.

5. **Problem:** Enter a number. Print 7 more than the number. (*Hint:* See Exercise 1 above.)

6. **Problem:** Enter a number greater than 6. Print 6 less than the number.

7. **Problem:** Find the mean (average) of three numbers.

8. **Problem:** Enter two numbers. Find the sum of the numbers. Then find the product of the numbers.

END OF CHAPTER EXERCISES

Section 1.2 Find the perimeters of the following squares.

1. 8.4 cm

8.4 cm 8.4 cm

8.4 cm

2. 3.52 mm

3.52 mm 3.52 mm

3.52 mm

3.

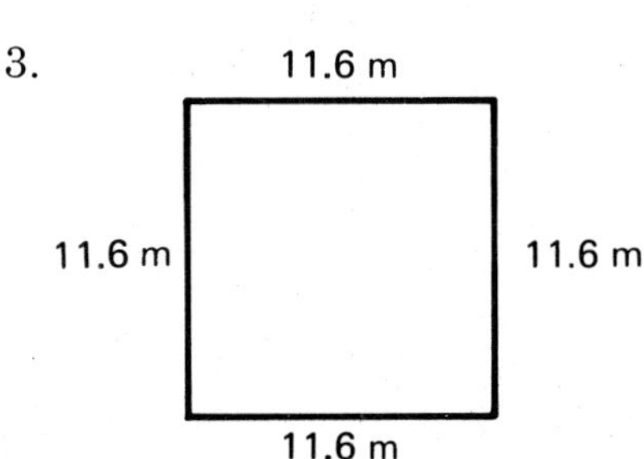

4. 5.8 cm

5.8 cm 5.8 cm

5.8 cm

Find the perimeters of squares with sides of length:

5. 3.9 cm 6. 15.4 m 7. 91.65 mm 8. 4.57 m

Use the algorithm $P=4\times S$ to find the perimeter of the following squares with sides of length:

9. 642 mm 10. 76.3 m 11. 8.49 cm 12. 7.5 m

Section 1.3 Which of the following are legal variable names?

13. D 14. $\$Q$ 15. $9E$ 16. $E9$

Translate the following into BASIC:

17. $S=4R$

18. $T=5C-3D$

19. $Y=\frac{C+E}{D}$

20. $L=A+BC$

Section 1.4

Find the value assigned to C in the following computer instructions if $A=16$ and $B=4$.

21. 70 $C=A*B$
22. 40 $C=A/B$
23. 50 $C=B\uparrow 2$
24. 10 $C=5*A$
25. 90 $C=3*B*A$
26. 60 $C=8.3*B$

Find the values assigned to Y after executing the two computer instructions in order. In each case $T=15$ and $R=3$.

27. 80 $L=R+3$
    90 $Y=L+T$
28. 10 $P=T*R$
    20 $Y=P/5$
29. 20 $D=T-R$
    30 $Y=D/2$
30. 40 $L=6*R$
    50 $Y=L-T$
31. 50 $U=R\uparrow 2$
    60 $Y=U/R$
32. 30 $N=2\uparrow R$
    40 $Y=4.5*N$

In each of the following problems, the listing shows a program that was just typed into a computer. Answer all questions for each program.

33.
```
10  INPUT B, C
20  S=B+C
30  PRINT B
40  PRINT C
50  PRINT S
60  END
```
a. If $B=13$ and $C=6$, show the output.
b. What is the output if $B=188$ and $C=117$?
c. What is the output if $B=23.58$ and $C=741.94$?

34.
```
10  INPUT C
20  PRINT C
30  Y=4*C
40  PRINT Y
50  END
```
a. If $C=14$, show the output.
b. What is the output if $C=6.4$?
c. What is the output if $C=27.9$?

35.
```
10  INPUT R, S
20  E=R-S
30  F=3*E
40  PRINT E
50  PRINT F
60  END
```
a. If $R=23$ and $S=17$, show the output.
b. What is the output if $R=16.7$ and $S=3.6$?
c. What is the output if $R=29.8$ and $S=10.9$?

36.
```
10  INPUT P
20  C=P*P
30  S=C/2
40  PRINT C
50  PRINT S
60  END
```

a. If $P=5$, show the output.
b. What is the output if $P=8$?
c. What is the output if $P=7$?

Section 1.5

a. Classify each triangle according to the lengths of its sides.
b. Find the perimeter of each triangle.

37.

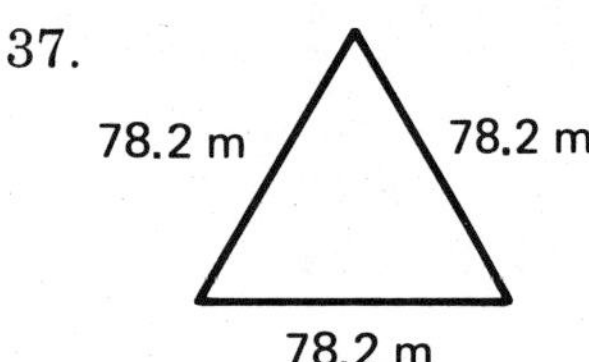

38.

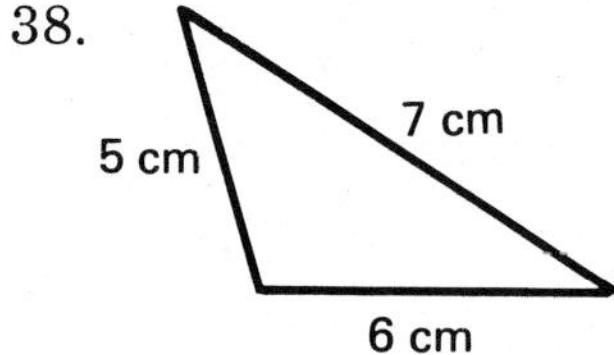

39.

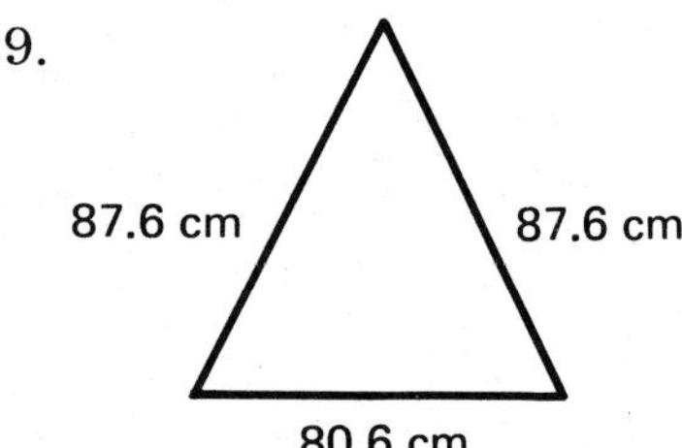

Find the values assigned to M:

40. $M=8+6\cdot3$
41. $M=(8+6)\cdot3$
42. $M=16-6\div2$
43. $M=(16-6)\div2$
44. $M=4+3^2$
45. $M=18-2^3$
46. $M=9\cdot3-8\div2$
47. $M=(8+2)^2+7\cdot3^2$

Find the value assigned to E in each of the following computer instructions if $A=18$, $B=2$ and $C=3$.

48. 30 $E=A-4*B$
49. 70 $E=A+B*C$
50. 20 $E=(A+B)*C$
51. 80 $E=A*B+B*C$
52. 90 $E=C\uparrow B$
53. 60 $E=A/B-C*B$
54. 40 $E=A/B*C$
55. 10 $E=A/(B*C)$

For the following, find the value assigned to F by executing the two computer instructions in order. In each case $G=16$ and $H=2$.

56. 10 $Q=G+H/2$
    20 $F=Q/H$

57. 20 $R=G-3*H$
    30 $F=R+2*H$

58. 60 $S=10\uparrow H$
    70 $F=S/(G+2*H)$

59. 40 $L=H+G/H$
    50 $F=L\uparrow2$

Each listing shows a program that was just typed into a computer. Show the complete output for the program.

60.
```
10  INPUT B, C
20  A=B+B*C
30  PRINT B
40  PRINT C
50  PRINT A
60  END
RUN
? 5, 6
```

61.
```
10  INPUT K, L, M
20  PRINT K
30  PRINT L
40  PRINT M
50  M=L+K*K
60  PRINT M
70  END
RUN
? 3, 5, 19
```

62.
```
10  INPUT A, B
20  S=4*B
30  R=S+A*B
40  PRINT S
50  PRINT R
60  END
RUN
? 8, 3
```

63.
```
10  INPUT V, W, Y
20  PRINT V
30  G=V+W*Y
40  H=(V+W)*Y
50  I=G+H
60  PRINT G
70  PRINT I
80  END
RUN
? 3, 4, 5
```

Section 1.6

For each program, show the complete output.

64.
```
10  INPUT A, B, C
20  PRINT A; B; C
30  PRINT A, B, C
40  END
RUN
? 4, 8, 2
```

65.
```
10  PRINT "ENTER A VALUE"
20  INPUT J
30  T=6+3*J
40  PRINT "J="; J, "T="; T
50  END
RUN
ENTER A VALUE
? 17
```

66.

```
10  PRINT "WHAT IS YOUR NUMBER?"
20  INPUT X
30  PRINT "THREE MORE IS"
40  K=X+3
50  PRINT K
60  END
RUN
WHAT IS YOUR NUMBER?
? 7
```

67.

```
10  J=12
20  K=8
30  PRINT "FIRST NUMBER"; J
40  PRINT "SECOND NUMBER"; K
50  S=K+J
60  PRINT "SUM IS"; S
70  END
RUN
```

Given a planned output and a program for each of the following, correct the program so that it will produce the required output.

68. *Planned Output*

```
RUN
? 4
S=4
RUN
? 11
S=11
```

Program

```
10  INPUT S
20  PRINT S
30  END
```

69. *Planned Output*

```
RUN
? 8, 9
X=8
Y=9
RUN
? 7, 2
X=7
Y=2
```

Program

```
10  INPUT A, B
20  PRINT "X="; A
30  PRINT "Y=B"
40  END
```

70. *Planned Output*

```
RUN
ENTER A NUMBER
? 7
NUMBER IS 7
RUN
ENTER A NUMBER
? 31
NUMBER IS 31
```

Program

```
10  PRINT ENTER A NUMBER
20  INPUT N
30  PRINT N
40  END
```

71. *Planned Output*

```
RUN
PUT IN A VALUE
? 11
YOUR NUMBER IS
11
RUN
PUT IN A VALUE
? 100
YOUR NUMBER IS
100
```

Program

```
10  PUT IN A VALUE
20  INPUT 11
30  PRINT "YOUR NUMBER IS"
40  PRINT 11
50  END
```

For the following, write a program to produce the planned output.

72. *Planned Output*

```
RUN
? 3
THE NUMBER IS 3
RUN
? 7
THE NUMBER IS 7
```

73. *Planned Output*

```
RUN
? 19
THE VALUE IS
19
RUN
? 39
THE VALUE IS
39
```

74. *Planned Output*

```
RUN
CHOOSE A NUMBER
? 71
71 IS YOUR CHOICE
RUN
CHOOSE A NUMBER
? 76
76 IS YOUR CHOICE
```

75. *Planned Output*

```
RUN
ENTER TWO NUMBERS
? 63, 28
NUMBERS ARE 63 AND 28
RUN
ENTER TWO NUMBERS
? 54, 75
NUMBERS ARE 54 AND 75
```

Section 1.7

a. Show the output of the old program.
b. Edit the old program to produce the new output. Try to keep as much of the old program as possible.

76. *Old Program*

```
10  INPUT N
20  D=2*N
30  PRINT D
40  END
```

New Output

```
PUT IN A NUMBER
? 6
DOUBLE IS 12
```

77. *Old Program*

```
10  INPUT J
20  E=J+8
30  PRINT E
40  END
```

New Output

```
ADDING TO A NUMBER
ENTER A NUMBER
? 3
EIGHT MORE IS 11
```

78. *Old Program*

```
10  INPUT Y
20  T=3*Y
30  PRINT T
40  END
```

New Output

```
ENTER A NUMBER
? 14
NUMBER IS 14
TRIPLE IS 42
```

79. *Old Program*

```
10  INPUT A, B, C, D
20  R=A*B+C*D
30  PRINT R
40  END
```

New Output

```
MULTIPLY BEFORE ADDING
ENTER 4 NUMBERS
? 3, 4, 5, 6
ANSWER IS 42
```

For each of the following problems:
a. Develop an algorithm to solve the problem.
b. Write a plan showing how you use the algorithm to solve the problem.
c. Plan an output.
d. Code the program.

80. **Problem:** Enter a number. Print the number and print double the number.

81. **Problem:** Enter a number. Print the number and then print 9 more than the number.

82. **Problem:** Find the arithmetic mean (average) of four numbers.

83. **Problem:** Enter three numbers. Print the product of the first and second. Then, print the product of the second and third.

Chapter 2

Applications Using Beginning Programming Concepts

Section 2.1 CONVERTING FROM OTHER BASES TO BASE TEN

How could 1101 not be one thousand one hundred one?

When it is written in a base other than ten. Let us examine more closely what this means.

Look at the group of asterisks (*) below.

****

We will group these symbols by tens, fives, and eights.

Grouping by tens:

**********	***
1 ten	3 ones

A system that groups by tens is called a base ten numeration system.

The numeral 13 in our base ten system shows the grouping of 1 ten and 3 ones. The base ten numeration system uses the following ten digits to write numbers: 0, 1, 2, 3, 4, 5, 6, 7, 8, 9.

Grouping by fives:

*****	*****	***
2 fives		3 ones

A system that groups by fives is called a base five numeration system. The numeral 23_{five} in a base five system shows the grouping of 2 fives and 3 ones. A base five numeration system uses the following five digits to write numbers: 0, 1, 2, 3, 4.

Grouping by eights:

********	*****
1 eight	5 ones

A system that groups by eights is called a base eight numeration system. The numeral 15_{eight} in a base eight numeration system shows the grouping of 1 eight and 5 ones. A base eight numeration system uses the following eight digits to write numbers: 0, 1, 2, 3, 4, 5, 6, 7.

The numerals 23_{five} and 15_{eight} each name the number 13 in the base ten system (which is our number system).

Example 1 What base ten numeral does each name?

1. 42_{five} 2. 34_{eight} 3. 23_{six}

Solutions:

1. $42_{five} = 4$ fives + 2 ones
$= 4 \cdot 5 + 2 \cdot 1$
$= 20 + 2$
$42_{five} = 22$

2. $34_{eight} = 3$ eights + 4 ones
$= 3 \cdot 8 + 4 \cdot 1$
$= 24 + 4$
$34_{eight} = 28$

3. $23_{six} = 2$ sixes + 3 ones
$= 2 \cdot 6 + 3 \cdot 1$
$= 12 + 3$
$23_{six} = 15$

Class Exercise 1 What base ten numeral does each name?

1. 34_{five} 2. 56_{eight} 3. 43_{six} 4. 67_{nine}
5. 23_{seven} 6. 23_{four} 7. 32_{five} 8. 11_{two}

So far we have only looked at two-digit numerals. How do you change (convert) numerals that have more than two digits in base ten?

To answer the question, we must first understand the place values in base ten.

Examine the numeral 3412.

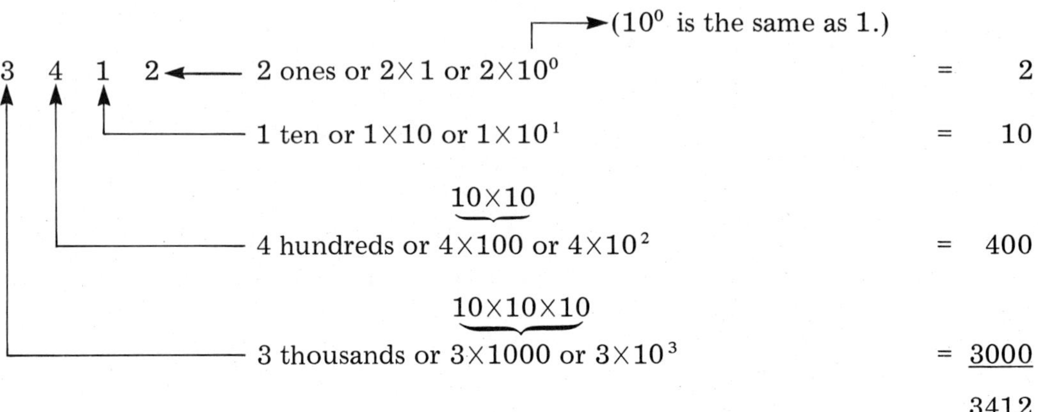

In the base ten numeration system, as you move from right to left, each place has a value that is ten times the value of the place to its immediate right.

In a base five numeration system, as you move from right to left, each place has a value that is five times the value of the place to its immediate right.

Example 2 Convert 324_{five} to base ten.

Solution: The numeral 4 is in the units (or ones) place. Each numeral to the left has a place value that is five times the place value of the place to its right.

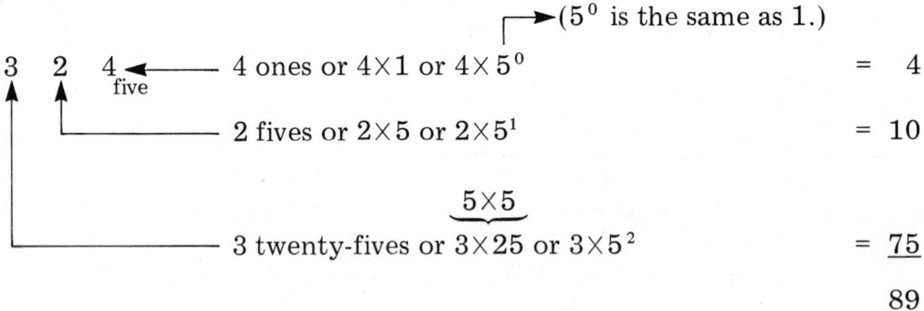

Therefore, 324_{five} is 89 in base ten.

Class Exercise 2 Convert each to base ten.

1. 243_{five}
2. 243_{six}
3. 243_{eight}
4. 101_{two}
5. 107_{eight}
6. 111_{two}
7. 1312_{five}
8. 4231_{six}

The computer can be used to convert numbers from one base to another.

Problem

Enter a three-digit numeral, in base five, one digit at a time. The left most digit should be entered first. Convert the numeral to base ten.

Solution:

a. *Analysis:* In base five the only digits are 0, 1, 2, 3 and 4. After the units digit each place to the left has five times the value of the place to its immediate right.

For example, if the base five numeral to be converted is

234_{five}

the digits 2, 3, 4 are entered in that order. The base ten equivalent will be

$$4\times 1 \quad + \quad 3\times 5 \quad + \quad 2\times 5\times 5$$
$$4 \quad + \quad 15 \quad + \quad 50 \quad = \quad 69.$$

Now we use this procedure in our plan.

b. *Plan:*

1. Put in values for A, B, C.
2. Calculate: $N = C\times 1 + B\times 5 + A\times 5\times 5$.
3. Print N.
4. End.

c. *Planned Output*

```
CONVERTING FROM BASE FIVE TO BASE TEN
ENTER A BASE FIVE NUMERAL WITH 3 DIGITS
ONE DIGIT AT A TIME
BEGIN WITH THE LEFT MOST DIGIT
? 2, 3, 4
IN BASE TEN YOU GET 69
```

d. *The Program*

```
10  PRINT "CONVERTING FROM BASE FIVE TO BASE TEN"
20  PRINT "ENTER A BASE FIVE NUMERAL WITH 3 DIGITS"
30  PRINT "ONE DIGIT AT A TIME"
40  PRINT "BEGIN WITH THE LEFT MOST DIGIT"
50  INPUT A, B, C
60  N=C*1+B*5+A*5*5
70  PRINT "IN BASE TEN YOU GET"; N
80  END
```

Exercise 2.1

What base ten numeral does each name?

1. 43_{five}
2. 65_{eight}
3. 34_{six}
4. 76_{nine}
5. 32_{seven}
6. 32_{four}
7. 23_{five}
8. 10_{two}

Convert each of the following to base ten.

9. 423_{five}
10. 324_{six}
11. 716_{eight}
12. 100_{two}
13. 724_{eight}
14. 1001_{two}
15. 4312_{five}
16. 3120_{seven}

Revise the program that was written in this section to produce the following outputs.

17.

```
BASE FIVE TO BASE TEN
ENTER A BASE FIVE 3 DIGIT NUMERAL
BEGIN WITH THE LEFT MOST DIGIT
ENTER ONE DIGIT AT A TIME
? 2, 3, 4
FOR 2 3 4 BASE FIVE
BASE TEN IS 69
```

18.

```
CONVERTING BASE FIVE TO BASE TEN
PUT IN A 3 DIGIT BASE FIVE NUMERAL
BEGIN WITH THE LEFT MOST DIGIT
ONE DIGIT AT A TIME PLEASE!
? 2, 3, 4
2 3 4 BASE FIVE IS 69 BASE TEN
```

For the following problems:
a. Write an analysis of the problem that leads to an algorithm.
b. Write a plan that outlines how you use the algorithm to solve the problem.
c. Design a planned output.
d. Code the program and RUN it on the computer.

19. **Problem:** Enter a three-digit numeral, in base eight, one digit at a time. The left most digit should be entered first. Convert the numeral to base ten.

20. **Problem:** Enter a four-digit numeral, in base five, one digit at a time. Begin with the left most digit. Convert the numeral to base ten.

21. **Problem:** Enter a three-digit numeral, one digit at a time. Begin with the left most digit. Enter a base. Convert the numeral to base ten.

Section 2.2 FINDING AREAS OF QUADRILATERALS AND TRIANGLES

Formulas are useful in finding the areas of different figures.
We can find the area of a rectangle by using the formula:

Area = base × height[†] or $A = b \cdot h$.

Example 1 Find the area of a rectangle whose base is 5 and whose height is 2.

5 UNITS
(BASE)
2 UNITS
(HEIGHT)

Solution: Using the formula $A = b \cdot h$ we get:

$A = 5 \cdot 2 = 10$ square units.

(You can count the square units in the rectangle.)

[†]Sometimes the base is called length and the height is called width. The formula is then $A = \ell \cdot w$.

Class Exercise 1 Find the area of the following rectangles.

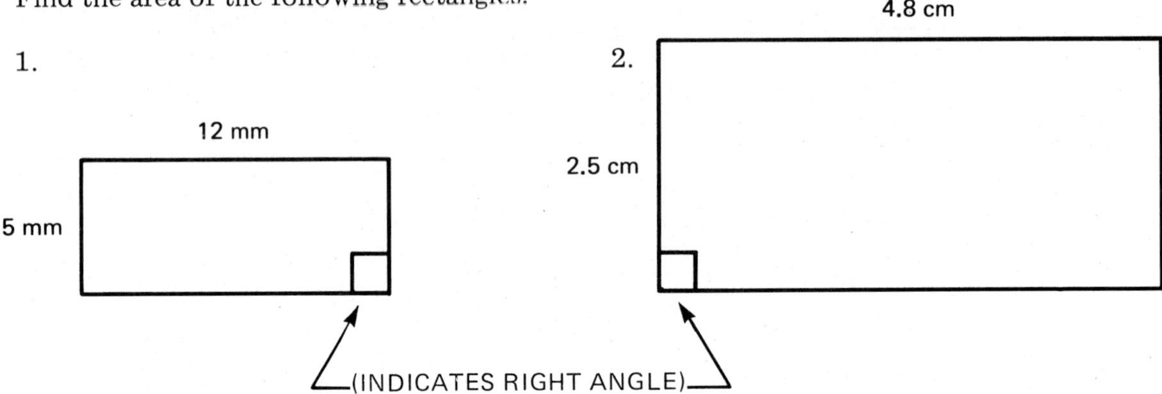

3. Find the area of a rectangle whose base is 14.6 m and whose height is 6.2 m.

4. What is the area of a rectangle whose base is 7.8 cm and whose height is 48 mm?

By drawing a diagonal of the rectangle below we form two right triangles of equal area.

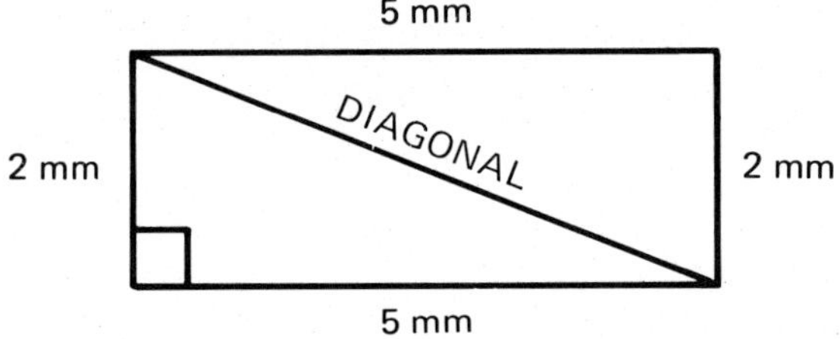

How does the area of one of these triangles compare to the area of the rectangle?

Since the area of a right triangle is one-half the area of the rectangle, we have a new formula:

Area of a right triangle = one half the area of a rectangle or $A = \frac{1}{2}(b \cdot h)$

Example 2

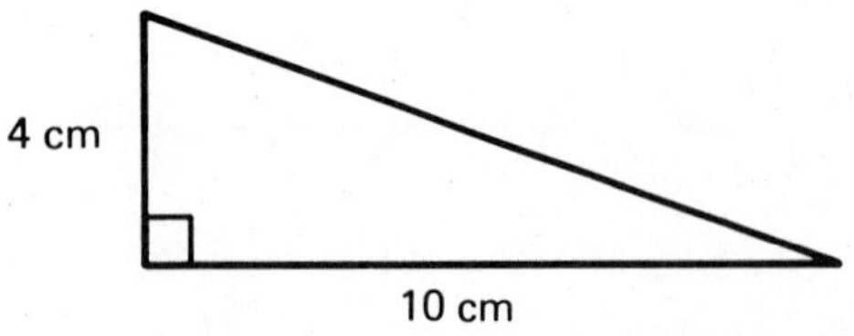

Find the area of a right triangle whose base is 10 cm and whose height is 4 cm.

Solution:

$A=\frac{1}{2}\cdot b\cdot h$ or $A=\frac{1}{2}bh$

$A=\frac{1}{2}\cdot 10\cdot 4$

$A=20$ square centimeters or $20\ cm^2$

Class Exercise 2 Find the area of the following right triangles.

1.

2.

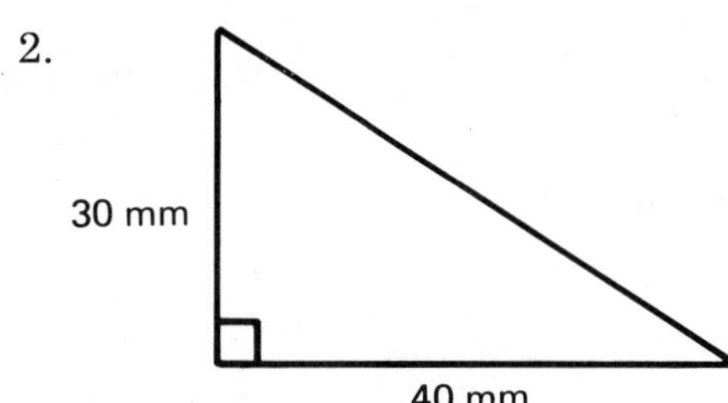

3. Find the area of a right triangle whose base is 7 in. and whose height is 5 in.

4. What is the area of a right triangle whose base is 13 yards and whose height is 10 yards?

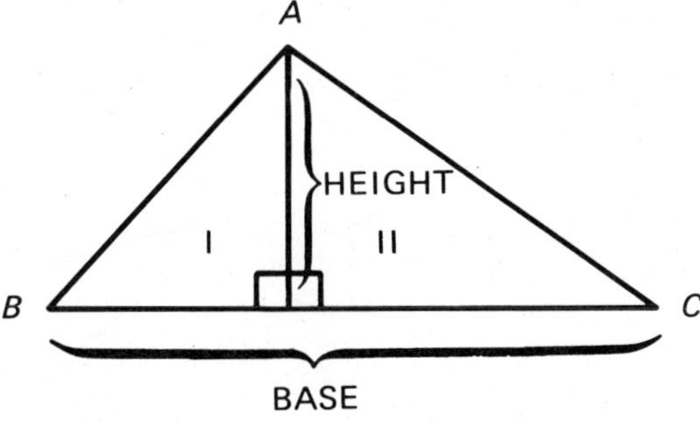

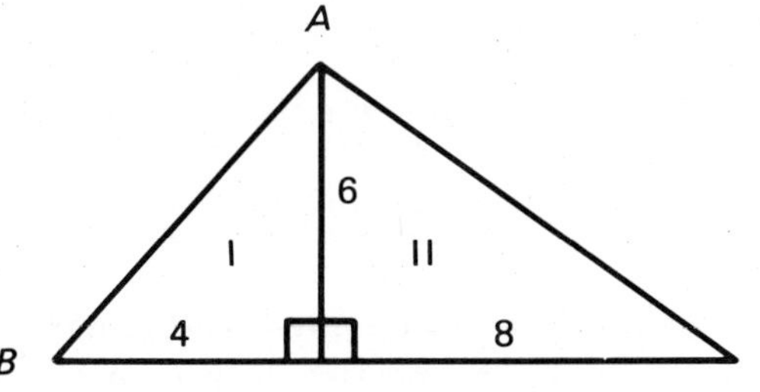

Let us now consider a triangle ($\triangle ABC$) which is not a right triangle. Notice that the height separates the triangle into two right angles. To develop a formula for finding the area of this triangle we will combine the formulas for the areas of the two right triangles (I and II).

We will put numbers in the figure to make the discussion easier to follow.

Area of $\triangle$ I $=\frac{1}{2}\times 4\times 6$

Area of $\triangle$ II $=\frac{1}{2}\times 8\times 6$

Area of $\triangle ABC$ = Area of $\triangle$ I + Area of $\triangle$ II or

Area of $\triangle ABC=\frac{1}{2}(4\cdot 6)+\frac{1}{2}(8\cdot 6)$

Area of $\triangle ABC = 12 + 24$

Area of $\triangle ABC$ = 36 square units.

To get the formula for the area of $\triangle ABC$ let us again look at the line that says

$$\text{Area of } \triangle ABC = \frac{1}{2}(4 \cdot 6) + \frac{1}{2}(8 \cdot 6).$$

By using the distributive property of multiplication over addition, we can write this as:

$$\text{Area of } \triangle ABC = \frac{1}{2} \cdot 6\ (4+8).$$

Height of $\triangle ABC$ (6)   Base of $\triangle ABC$ (4+8)

Therefore the area of $\triangle ABC = \frac{1}{2}(h \cdot b)$ or $A = \frac{1}{2}(b \cdot h)$ where b represents the length of the base of the triangle and h represents the length of the height.

Example 3 Find the area of a triangle if the base is 8 m and the height is 5 m.

Solution:

$A = \frac{1}{2}(b \cdot h)$

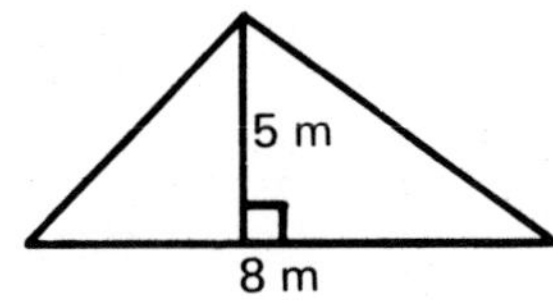

$A = \frac{1}{2} \cdot 8 \cdot 5$

$A = 20\ \text{m}^2$

Class Exercise 3 Find the area of the following triangles.

1.

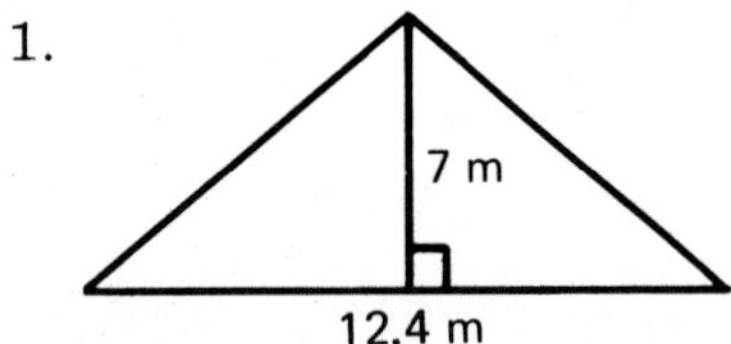

2.

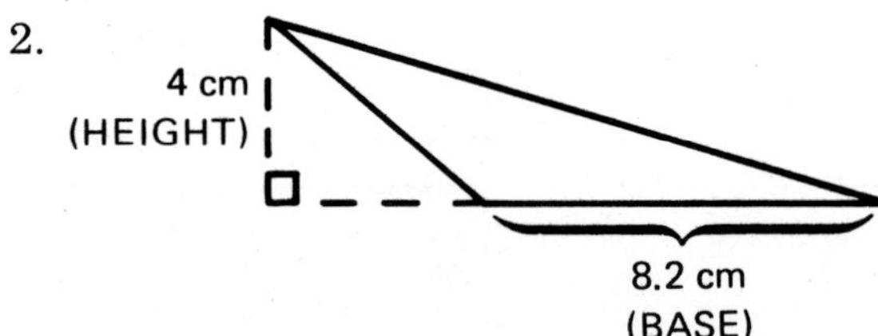

3. Find the area of a triangle if the base is 78 mm and the height is 6 cm.

4. What is the area of a triangle if the base is 39 yards and the height is 60 feet?

A quadrilateral (4-sided polygon) with opposite sides parallel and congruent is called a *parallelogram*. A parallelogram has many useful properties. For example, since its opposite sides have the same length, either diagonal will separate it into two triangles of equal area. This allows us to derive another area formula.

Look at the parallelogram below with one diagonal.

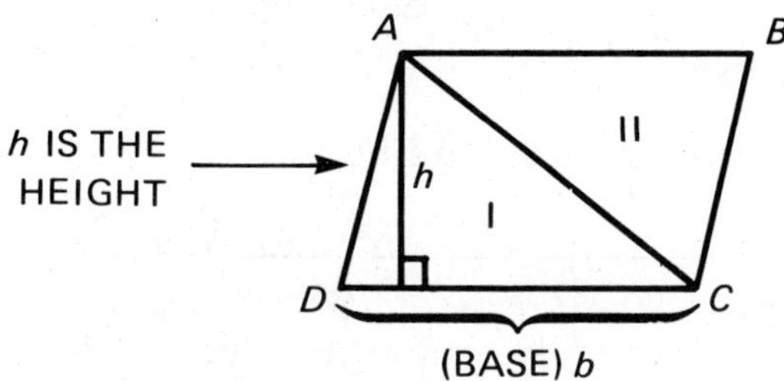

Area of $\triangle$I = Area of $\triangle$II $= \frac{1}{2} \cdot b \cdot h$.

The area of parallelogram $ABCD$ $= 2 \times$ area of either $\triangle$ I or $\triangle$ II.

Therefore, area of parallelogram $ABCD = 2 \cdot (\frac{1}{2} \cdot b \cdot h)$ or

$$A = b \cdot h.$$

Example 4 Find the area of a parallelogram whose base has length 9 cm and whose height is 7 cm long.

Solution:

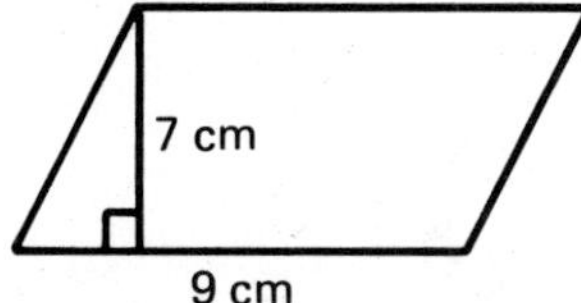

$A = b \cdot h$

$A = 9 \cdot 7$

$A = 63 \text{ cm}^2$

Class Exercise 4 Find the area of the following parallelograms.

1.

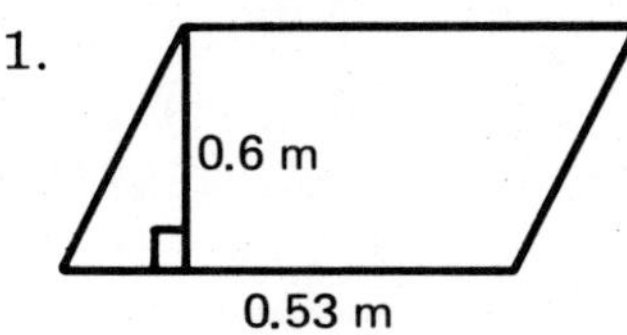

2.

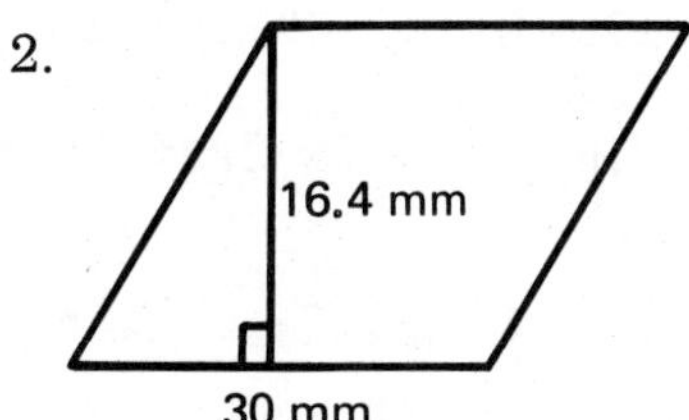

3. Find the area of a parallelogram whose base has length 12.4 cm and whose height is 4.8 cm.

4. What is the area of a parallelogram whose base has length 71.6 cm and whose height is 680 mm?

Let us now use the computer to find the area of various figures.

Problem

Find the area of a parallelogram when we are given the base (B) measured in centimeters and the height (H) measured in millimeters. The area should be given in square centimeters.

Solution:

a. *Analysis:*

1. To find the area of a parallelogram we use the formula: $A = b \cdot h$.
2. To get an area in square centimeters we must have the base and height given to us in centimeters. Since the height is given in millimeters, we must divide it by 10. This converts it to centimeters.

b. *Plan:*

1. Put in values for B, H.
2. Convert height (H) to centimeters: New Height (NH)=$H/10$.
3. Calculate: $A=B \cdot NH$.
4. Print A.
5. End.

c. *Planned Output*

```
FIND THE AREA OF THE PARALLELOGRAM
ENTER THE BASE IN CENTIMETERS
? 5.6
ENTER THE HEIGHT IN MILLIMETERS
? 40
THE AREA IS 22.4 SQUARE CENTIMETERS
```

d. *The Program*

```
   10  PRINT "FIND THE AREA OF THE PARALLELOGRAM"
   20  PRINT "ENTER THE BASE IN CENTIMETERS"
   30  INPUT B
   40  PRINT "ENTER THE HEIGHT IN MILLIMETERS"
   50  INPUT H
   60  NH=H/10
   70  A=B*NH
■  80  PRINT "THE AREA IS"; A; "ƀSQUARE  CENTIMETERS"
   90  END
```

Discussion of the Program:
■ ƀ indicates a *b*lank space. Press the space bar when typing the program. (Some computers space automatically, therefore you would not have to insert the blank.)

Exercise 2.2

Find the area of the following rectangles.

1.

40 mm

3 cm

2.

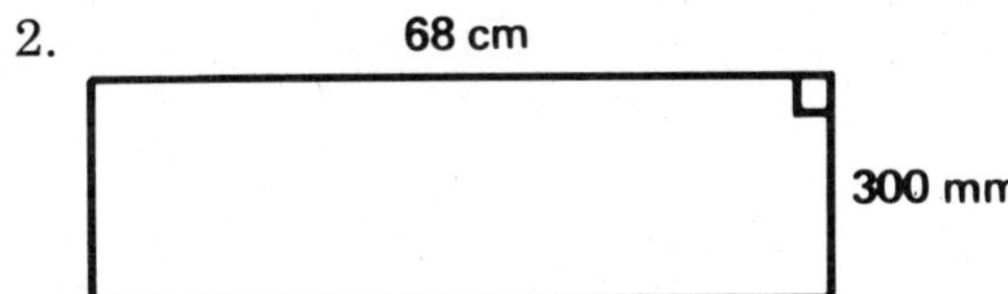

3. Find the area of a rectangle whose base is 17.3 feet and whose height is 8.7 feet.

4. What is the area of a rectangle whose base is 8.7 cm and whose height is 51 mm?

Find the area of the following right triangles.

5.

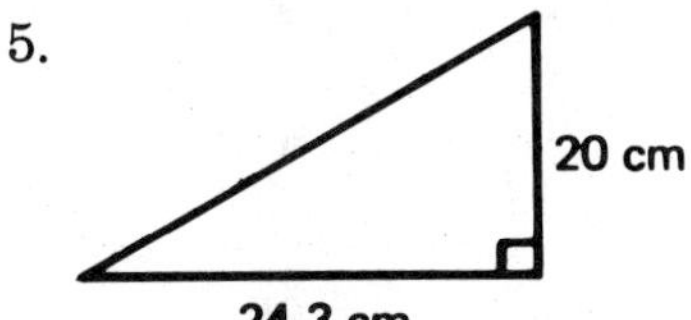

6.

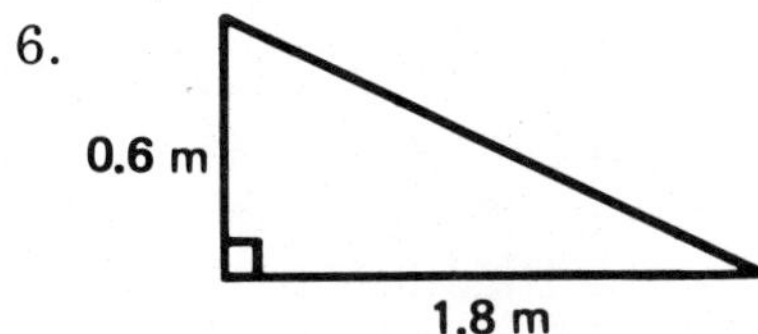

7. Find the area of a right triangle whose base is 8 feet and whose height is 2 yards.

8. What is the area of a right triangle whose base is 7 yards and whose height is 15 feet?

Find the area of the following triangles.

9.

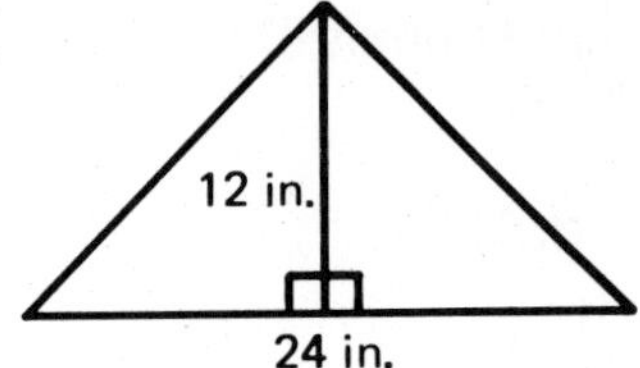

10.

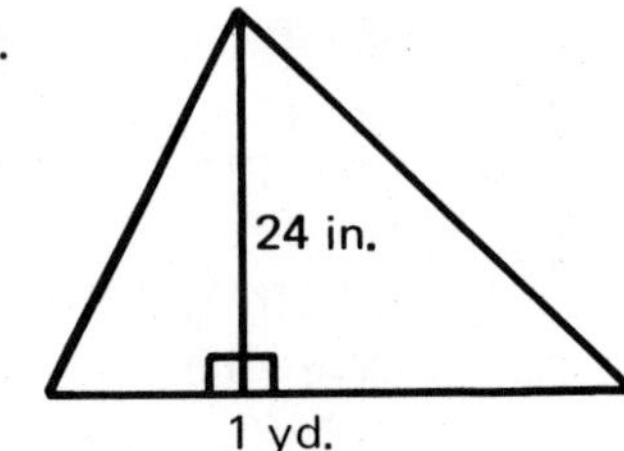

11. Find the area of a triangle if the base is 60 mm and the height is 4.3 cm.

12. What is the area of a triangle if the base is 40 yards and the height is 39 feet?

Find the area of the following parallelograms.

13.

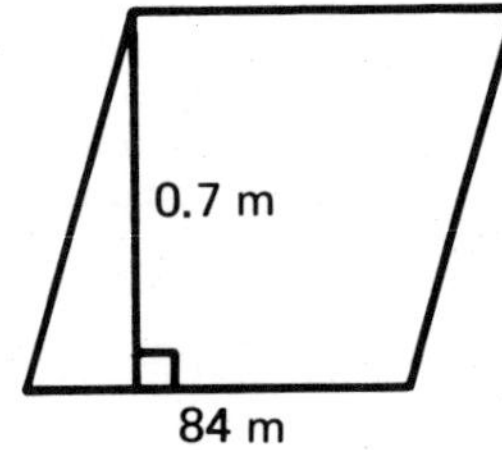

14.

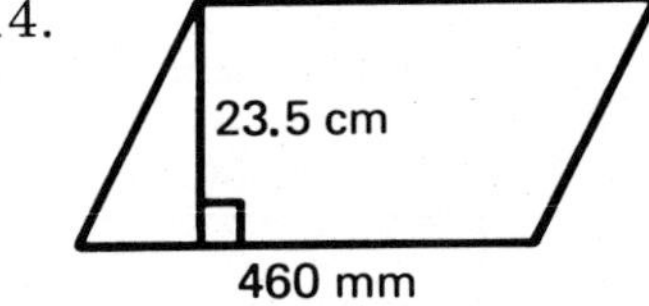

15. Find the area of a parallelogram whose base has length 1.48 m and whose height has length 0.9 m.

16. What is the area of a parallelogram with base 502 mm and height 43 cm?

Revise the program that was written in this section to produce the following outputs.

17.

```
AREA OF A PARALLELOGRAM
ENTER THE BASE AND HEIGHT
THE BASE IN CENTIMETERS
THE HEIGHT IN MILLIMETERS
? 5.6, 40
THE AREA IS 22.4 SQUARE CENTIMETERS
```

18.
```
FINDING THE AREA OF A PARALLELOGRAM
ENTER BASE IN CENTIMETERS
? 5.6
ENTER THE HEIGHT IN MILLIMETERS
? 40
THE HEIGHT IS 40 MM
THE NEW HEIGHT IS 4 CM
THE AREA = 22.4 SQUARE CENTIMETERS
```

For the following problems:
a. Write an analysis of the problem that leads to an algorithm.
b. Write a plan that outlines how you use the algorithm to solve the problem.
c. Design a planned output.
d. Code the program and RUN it on the computer.

19. **Problem:** Find the area of a parallelogram when we are given the base (B) measured in millimeters and the height (H) measured in centimeters. The area should be given in square centimeters.

20. **Problem:** Find the area of a triangle given the lengths of the base (B) and height (H).

21. **Problem:** Find the area of a triangle given the length of the base (B) in feet and the length of the height (H) in yards. The area should be given in square yards.

Section 2.3 CONVERTING FAHRENHEIT AND CELSIUS TEMPERATURES

If someone told you that the temperature was 35° would you know if it was hot or cold? No!

You would need to know on which scale this temperature was measured. Is it hot or cold if this were 35° Celsius?

To convert 35° Celsius (which is temperature in the metric system) to Fahrenheit (which is temperature in the English system) we use the formula:

$$F = \frac{9}{5}C + 32.$$

C — Temperature in Celsius degrees.

F — Temperature in Fahrenheit degrees.

From our problem above, we replace C with 35:

$$F = \frac{9}{5}(35) + 32$$

$$F = 63 + 32$$

$F = 95°$, quite a hot temperature!

Class Exercise 1 Convert each of the following Celsius temperatures to Fahrenheit.

1. 25°   2. 0°   3. 100°   4. 45°

If we need to convert Fahrenheit to Celsius, we use the formula:

$$C=\frac{5}{9}(F-32).$$

Example 1 Convert 50° Fahrenheit to Celsius.

Solution: To convert Fahrenheit to Celsius, we use the formula:

$$C=\frac{5}{9}(F-32).$$

We replace F with 50:

$$C=\frac{5}{9}(50-32)$$

$$C=\frac{5}{9}(18)$$

$$C=10°.$$

Class Exercise 2 Convert each of the following Fahrenheit temperatures to Celsius.

1. 41°   2. 32°   3. 59°   4. 56°

Now let us use the computer to convert temperatures.

Problem Enter a Fahrenheit temperature. Convert it to Celsius degrees.

Solution:

a. *Analysis:* To convert Fahrenheit to Celsius we use the formula:

$$C=\frac{5}{9}(F-32).$$

b. *Plan:*
1. Put in a value for F.
2. Calculate: $C=\frac{5}{9}(F-32)$.
3. Print C.
4. End.

c. *Planned Output*

```
CONVERTING FAHRENHEIT TO CELSIUS
ENTER FAHRENHEIT TEMPERATURE
? 50
CELSIUS TEMPERATURE IS 10 DEGREES
```

d. *The Program*

```
  10  PRINT "CONVERTING FAHRENHEIT TO CELSIUS"
  20  PRINT "ENTER FAHRENHEIT TEMPERATURE"
  30  INPUT F
  40  C=5/9*(F-32)
■ 50  PRINT "CELSIUS TEMPERATURE IS"; C; "ƀDEGREES"
  60  END
```

Discussion of the Program:
■ ƀ indicates a blank space. Press the space bar when typing the program. (This may not be needed with some computers.)

Exercise 2.3 Convert each of the following Celsius temperatures to Fahrenheit degrees.

1. 10° 2. 5° 3. 14° 4. 27°

Convert each of the following Fahrenheit temperatures to Celsius degrees.

5. 109° 6. 82° 7. 68° 8. 56°

Revise the program that was written in this section to produce the following outputs.

9.
```
FAHRENHEIT TO CELSIUS
PUT IN FAHRENHEIT TEMPERATURE
? 50
10 DEGREES CELSIUS
```

10.
```
CONVERTING FAHRENHEIT TO CELSIUS
ENTER FAHRENHEIT TEMPERATURE
? 50
FAHRENHEIT IS 50 DEGREES
CELSIUS TEMPERATURE IS 10 DEGREES
```

For the following problems:
a. Write an analysis of the problem that leads to an algorithm.
b. Write a plan that outlines how you use the algorithm to solve the problem.
c. Design a planned output.
d. Code the program and RUN it on the computer.

11. **Problem:** Enter a Celsius temperature. Convert it to Fahrenheit degrees.

12. **Problem:** Enter two Fahrenheit temperatures (largest and then smallest). Convert them to Celsius degrees. Find the difference between the two Celsius temperatures.

13. **Problem:** Enter two Celsius temperatures. Find the average of their Fahrenheit temperatures.

Section 2.4 ANGLES (SUPPLEMENTARY AND COMPLEMENTARY) AND THE SUM OF THE ANGLES OF A TRIANGLE

When two angles are placed next to each other and form a straight line (as in the figure below), they have a sum of 180°. Since a straight line may be considered a straight angle, a straight angle measures 180°.

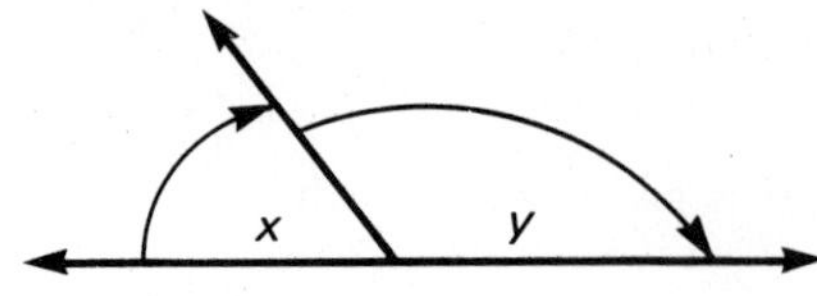

$m\angle x + m\angle y = 180°$

If the sum of the measures of two angles is 180°, we call them *supplementary* angles.

$\angle x$ and $\angle y$ are supplementary angles.

Example 1 What is the supplement of 75°?

Solution: The supplement of 75° is $180-75=105°$.

Class Exercise 1 Find the supplement of each of the following angle measures.

1. 100° 2. 150° 3. 20° 4. 78°

What is the sum of the measures of the following angles?

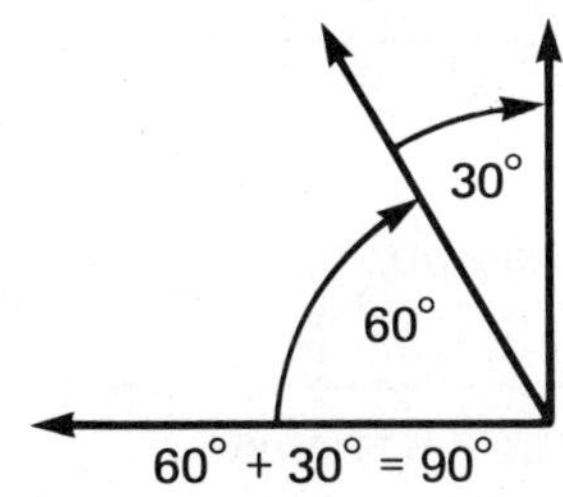

When the sum of the measures of two angles is 90°, we say they are *complementary* angles.

Example 2 Which angle is complementary to an angle of measure 40°?

Solution: The measure of the complement of 40° is $90-40=50°$.

Class Exercise 2 State the measure of the complement of each of the following angle measures.

1. 45° 2. 50° 3. 89° 4. 26°

Let us now look at the angles of a triangle.

Take a piece of paper and cut out a triangle of any shape (see Figure 1). Then tear off each vertex and place them at a common point (see Figure 2).

Notice that the three angles of the triangle can be situated to form a straight angle (straight line). Therefore, the sum of these measurements is 180°.

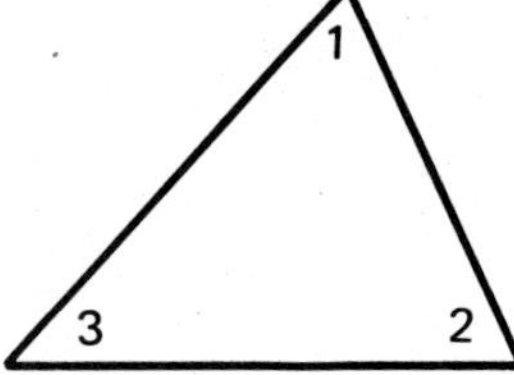

Figure 1

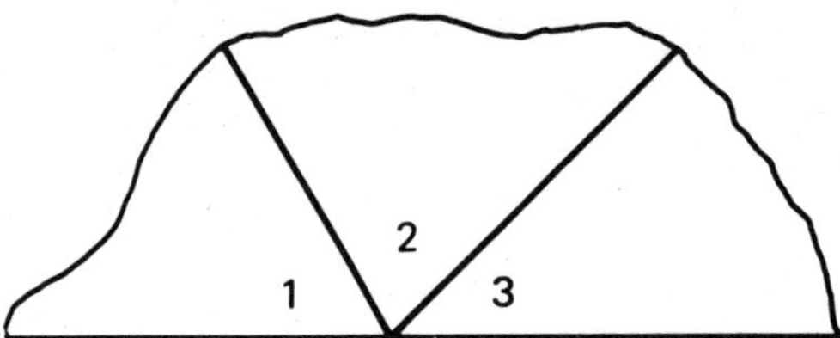

Figure 2

We can say that the *sum of the measures of the angles of a triangle is 180°.*

Example 3 Find the value of x in the triangle.

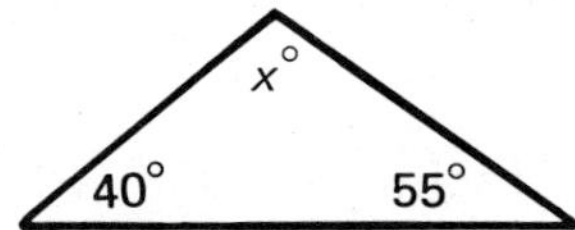

Solution: Since the sum of the measures of the angles of a triangle is 180°, we add $40 + 55 = 95$ and then subtract the result from 180:

$180-(40+55)$ or $180-95=85$.

To check our answer, $40+55+85$ must equal 180. It does!

Class Exercise 3 Find the value of x in each triangle.

1.

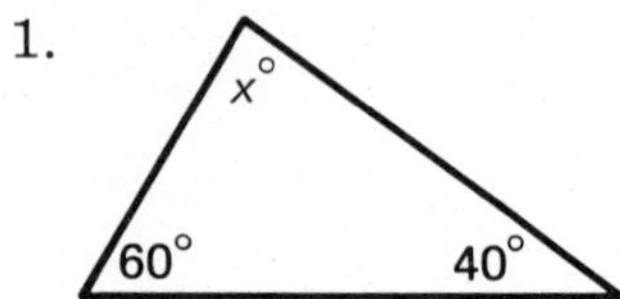

2.

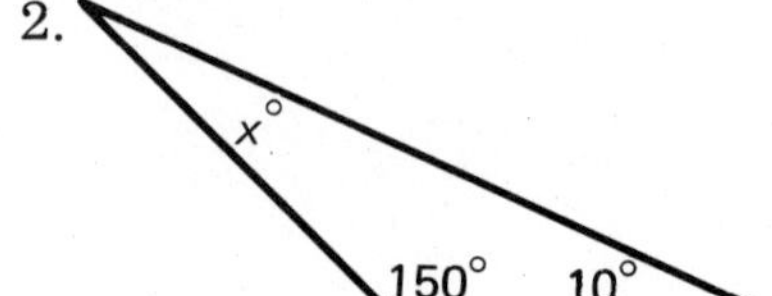

3. (Triangle with angles 42°, 48°, x°)

4. (Triangle with angles 35°, 58°, x°)

Now let us turn to the computer to solve the following problem.

Problem

Enter the measure of an angle (from 0° to 180°). Find the supplement of the angle.

Solution:

a. *Analysis:* Two angles are supplementary if the sum of their measures is 180°.
Let us review how we find the supplement.
For an angle of 50°, the supplement is $180-50=130$.
If the angle is 70°, the supplement is $180-70=110$.
Therefore, for any angle A the supplement is $\boxed{180-A}$.

We will now use $180-A$ to find the supplement of an angle A.

b. *Plan:*
1. Put in a value for A.
2. Supplement $(S)=180-A$ or $S=180-A$.
3. Print S.
4. End.

c. *Planned Output*

```
FINDING THE SUPPLEMENT OF AN ANGLE
ENTER THE MEASURE OF THE ANGLE
? 70
THE SUPPLEMENT IS 110
```

Before writing the program we will make a chart (legend) to show what each variable represents.

Legend

Variable	Represents
A	Angle
S	Supplement of angle A

d. *The Program*

```
10  PRINT "FINDING THE SUPPLEMENT OF AN ANGLE"
20  PRINT "ENTER THE MEASURE OF THE ANGLE"
30  INPUT A
40  S=180-A
50  PRINT "THE SUPPLEMENT IS"; S
60  END
```

Exercise 2.4 Find the supplement of each of the following angle measures.

1. 25° 2. 115° 3. 90° 4. 50°

State the complement of each of the following angle measures.

5. 48° 6. 27° 7. 80° 8. 9°

Find the value of x in each of the following triangles.

9.
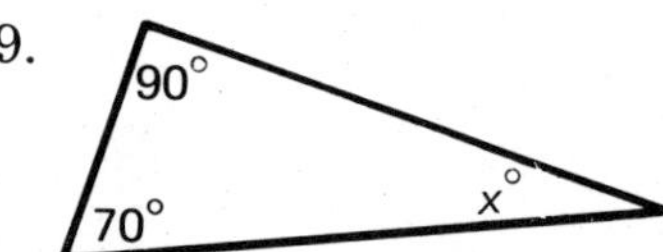

10.
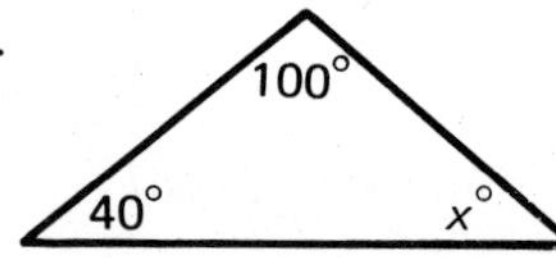

11.
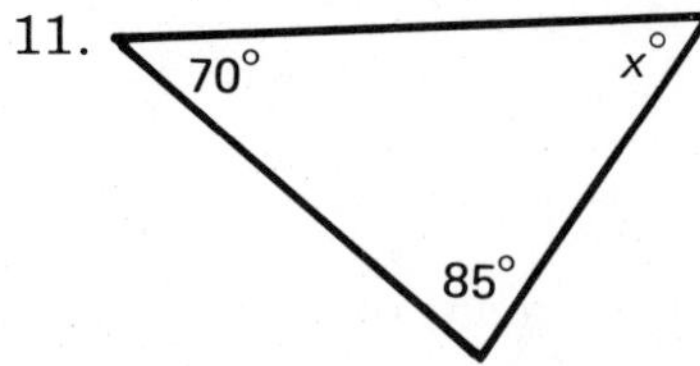

12.
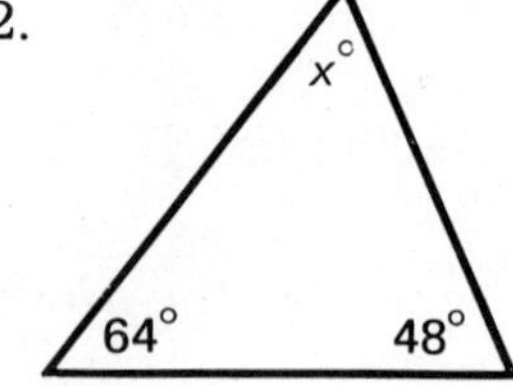

Revise the program that was written in this section so that it will produce each of the following outputs.

13.

```
LET US FIND THE SUPPLEMENT
PUT IN THE MEASURE OF THE ANGLE
? 70
THE SUPPLEMENT IS
110
```

14.

```
WHAT IS THE SUPPLEMENT?
ENTER THE MEASURE OF THE ANGLE
? 70
FOR 70 DEGREES
110 DEGREES IS THE SUPPLEMENT
```

For the following problems:
a. Write an analysis of the problem that leads to an algorithm.
b. Write a plan that outlines how you use the algorithm to solve the problem.
c. Design a planned output.
d. Code the program and RUN it on the computer.

15. **Problem:** Enter the measure of an angle (from 0° to 90°). Find the complement of the angle.

16. **Problem:** Enter the measure of an angle (from 0° to 90°). Find the complement and the supplement of the angle.

17. **Problem:** Enter the measures of two angles of a triangle. Find the measure of the third angle.

Section 2.5 HOURLY WAGE AND STRAIGHT COMMISSION

There are many ways to get paid for a job. We will first discuss being paid by the hour (hourly wage).

Example 1 Diane works after school. She earns $5 per hour. If Diane works 4 hours, how much does she earn?

Solution: The total earned (T) can be found by multiplying the hourly wage (W) by the number of hours (N) worked.

Total (T) = hourly wage (W) × number of hours (N) or

$T = W \cdot N$

$T = 5 \times 4$

$T = 20.$

She earns $20.

Class Exercise 1 Use the formula $T = W \cdot N$ to find the total wages earned when working at:

1. $4 per hour for 5 hours.
2. $4.25 per hour for 3 hours.
3. $3.54 per hour for 7 hours.
4. $4.63 per hour for 10 hours.

Salespersons often are paid a straight commission.

Straight commission means being paid a part (percent) of how much you sell.

Example 2 Charles is paid a straight commission for selling computers. His rate of commission is 8% of his sales. What commission does he earn if his sales are $8000?

Solution: We must find 8% of $8000. First we change the percent to a decimal.

> *To change a percent to a decimal:*
> 1. Remove the percent sign (%).
> 2. Move the decimal point two places to the left.

For example, to change 8% to a decimal

1. Remove the percent sign: 8%
2. Move the decimal point two places to the left: .08

We put a zero if there is no number in that position.

So the commission is now .08 × 8000 = $640.

Before trying the next problem do Class Exercise 2. This will give you practice in changing percents to decimals.

Class Exercise 2 Change the following to decimals.

1. 45% 2. 9% 3. 83% 4. 10%

Now try the problems!

Class Exercise 3 Find the commission when:

1. The rate of commission is 7% and the sales are $5000.
2. The rate of commission is 6% and the sales are $8500.
3. The rate of commission is 8% and the sales are $7200.
4. The rate of commission is 9% and the sales are $9450.

Let us see how the computer would find the commission by solving the following problem.

Problem Enter the total sales (S) and the rate of commission written as a decimal (R). Print the commission.

Solution:
a. *Analysis:* Commission = rate of commission × total sales or $C = R \cdot S$.

We will now use $C = R \cdot S$ to solve the problem.

b. *Plan:*
1. Put in values for R and S. (Remember, R must be written as a decimal.)
2. Calculate: $C = R \cdot S$.
3. Print C.
4. End.

c. *Planned Output*

```
FINDING THE COMMISSION
ENTER THE RATE OF COMMISSION AND TOTAL SALES
TYPE IN THE RATE AS A DECIMAL
? .08, 8000
THE COMMISSION IS $ 640
```

Before writing the program we will make a chart (legend) to show what each variable represents.

Legend

Variable	*Represents*
R	Rate of Commission
S	Total Sales
C	Commission

d. *The Program*

```
10  PRINT "FINDING THE COMMISSION"
20  PRINT "ENTER THE RATE OF COMMISSION AND TOTAL SALES"
30  PRINT "TYPE IN THE RATE AS A DECIMAL"
40  INPUT R, S
50  C=R*S
60  PRINT "THE COMMISSION IS $"; C
70  END
```

Exercise 2.5 Use the formula $T = W \cdot N$ to find the total wages earned when working at:

1. $3.56 per hour for 7 hours.
2. $4.08 per hour for 25 hours.
3. $4.73 per hour for 15 hours.
4. $5.18 per hour for 40 hours.

Change the following to decimals:

5. 32%
6. 3%
7. 92%
8. 19%

Find the commission for each of the following when:

9. The rate of commission is 4% and the sales are $4650.

10. The rate of commission is 2% and the sales are $7068.

11. The rate of commission is 14% and the sales are $3400.

12. The rate of commission is 6% and the sales are $5962.

Revise the program that was written in this section to produce the following outputs.

13.
```
WHAT IS THE COMMISSION?
ENTER RATE OF COMMISSION AND TOTAL SALES
RATE SHOULD BE IN DECIMAL FORM
? .08, 8000
RATE OF COMMISSION IS .08
SALES $ 8000
COMMISSION IS $ 640
```

14.
```
CALCULATING COMMISSIONS
ENTER THE RATE OF COMMISSION
AS A DECIMAL PLEASE!
? .08
ENTER THE SALES
? 8000
$ 640 IS THE COMMISSION
```

For the following problems:
a. Write an analysis of the problem that leads to an algorithm.
b. Write a plan that outlines how you use the algorithm to solve the problem.
c. Design a planned output.
d. Code the program and RUN it on the computer.

15. **Problem:** Enter the hourly wage (W) and the number of hours worked (N). Print the total earned (T).

16. **Problem:** Enter the price of an item (P). If the rate of sales tax is 8%, print the sales tax on the item. (*Example:* If an item costs $21, and the sales tax rate is 8%, the sales tax is $21 \times .08 = \$1.68$.)

17. **Problem:** A bank pays its customers interest on the money they deposit (principal) into savings accounts. Enter a rate of interest (R) and an amount of money deposited (P). Print the annual interest (I). (*Example:* If the rate of interest is 5% and the amount deposited is \$100, then the interest is $.05 \times 100 = \$5$. Note that in this problem the rate of interest and the principal can vary.)

Section 2.6 DISCOUNTS AND SALE PRICES

What would be your discount (amount off) if you bought the TV on sale?

Discount = rate of discount × list price

Discount = 25% of 120

$= .25 \times 120$ ← Change 25% to a decimal.

Discount = \$30

You save \$30 when you buy the TV set on sale.

Class Exercise 1 Find the discount for each item below.

	Item	*List Price*	*Rate of Discount*	*Discount*
1.	Tape Recorder	\$75	20%	
2.	AM/FM Radio	\$60	15%	
3.	Calculator	\$42	50%	
4.	Computer	\$950	12%	

Now that we can find the discount, let us calculate the sale price of our TV set. (Remember that the list price is \$120 and the discount is \$30.)

Sale price = list price − discount

= 120 − 30 ← 30: The discount on \$120 at 25%. ← 120: List price of the TV set.

= \$90 ← Sale price.

Example 1 Find the sale price of the AM/FM Cassette Recorder.

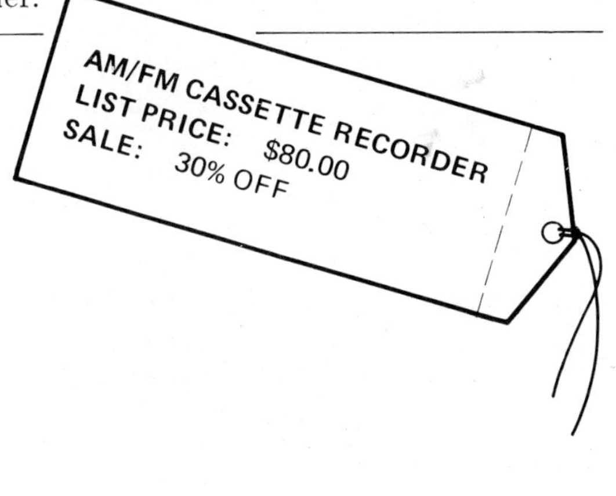

Solution: To use the formula:

Sale price = list price − discount,

we must first find the discount.

Discount = rate of discount × list price

= 30% × 80

= .30 × 80 ← Change 30% to a decimal.

= 24.

Now we can use the formula

Sale price = list price − discount

= 80 − 24 ← (24 from the discount found above)

= 56.

The sale price is \$56.

Class Exercise 2 Find the sale price for each item below.

	Item	*List Price*	*Rate of Discount*	*Discount*	*Sale Price*
1.	TV (Color)	\$560	15%		
2.	Stereo	\$425	20%		
3.	Tape Recorder	\$140	35%		
4.	Calculator	\$45	10%		

We now turn to the computer to solve the following problem.

Problem

Enter the list price and rate of discount (as a decimal). Have the computer find the discount and sale price.

Solution:

a. *Analysis:*
1. To find the discount use the formula: discount = rate of discount × list price.
2. To find the sale price, take the value for the discount from 1 above and place it in the formula: sale price = list price − discount.

b. *Plan:*
1. Put in values for the list price (P) and rate of discount (R). (Rate of discount should be entered as a decimal.)
2. Calculate discount: discount = rate of discount × list price or $D = R \times P$.
3. Calculate sale price: sale price = list price − discount or $S = P - D$.
4. Print D, S.
5. End.

c. *Planned Output*

```
FINDING THE SALE PRICE
ENTER THE LIST PRICE AND RATE OF DISCOUNT
RATE IN DECIMAL FORM PLEASE!
? 120, .25
THE LIST PRICE IS $ 120
THE RATE OF DISCOUNT IS .25
THE DISCOUNT IS $ 30
THE SALE PRICE IS $ 90
```

Before we write the program we will make a chart (legend) to show what each variable will represent.

Legend

Variable	Represents
P	List Price
R	Rate of Discount
D	Discount
S	Sale Price

d. *The Program*

```
10   PRINT "FINDING THE SALE PRICE"
20   PRINT "ENTER THE LIST PRICE AND RATE OF DISCOUNT"
30   PRINT "RATE IN DECIMAL FORM PLEASE!"
40   INPUT P, R
50   PRINT "THE LIST PRICE IS $"; P
60   PRINT "THE RATE OF DISCOUNT IS"; R
70   D=R*P
80   S=P-D
90   PRINT "THE DISCOUNT IS $"; D
100  PRINT "THE SALE PRICE IS $"; S
110  END
```

Exercise 2.6 Find the discount for each item below.

	Item	*List Price*	*Rate of Discount*	*Discount*
1.	Refrigerator	\$730	30%	
2.	TV (Black & White)	\$145	6%	
3.	Dishwasher	\$470	15%	
4.	Bicycle	\$160	22%	

Find the sale price for each item in the chart.

	Item	*List Price*	*Rate of Discount*	*Discount*	*Sale Price*
5.	Typewriter	$460	27%		
6.	Camera	$76	50%		
7.	Clock Radio	$54	30%		
8.	Watch	$68	25%		

Revise the program that was written in this section to produce the following outputs.

9.
```
WHAT IS THE SALE PRICE?
PUT IN THE LIST PRICE AND RATE OF DISCOUNT
DECIMAL FORM FOR RATE OF DISCOUNT PLEASE
? 120, .25
THE DISCOUNT IS $ 30
THE SALE PRICE IS $ 90
```

10.
```
DISCOUNT AND SALE PRICE
ENTER THE LIST PRICE
? 120
ENTER THE RATE OF DISCOUNT (DECIMAL FORM)
? .25
120 DOLLARS IS THE LIST PRICE
.25 IS THE RATE OF DISCOUNT
30 DOLLARS IS THE DISCOUNT
THE SALE PRICE IS $ 90
```

For the following problems:
a. Write an analysis of the problem that leads to an algorithm.
b. Write a plan that outlines how you use the algorithm to solve the problem.
c. Design a planned output.
d. Code the program and RUN it on the computer.

11. **Problem:** A store is having a sale offering 25% off every item. Write a program to enter the list price of an item (*P*). Have the computer print the discount. (Remember the rate of discount for every item is 25%.)

12. **Problem:** Enter the list price of an item (P) and the rate of discount (R) written as a decimal. Have the computer find the discount (D). (As a bonus, revise the program so that R can be entered as a percent.)

13. **Problem:** When you buy an item, a sales tax is usually added on to the price. Write a program to enter the price of an item (P) and the rate of sales tax (R). Have the computer print:
a. The amount of sales tax.
b. The final cost (price) of the item including the sales tax.

Section 2.7 SOLVING EQUATIONS BY ADDITION, SUBTRACTION, MULTIPLICATION OR DIVISION

What value for b will solve the equation (make the open sentence a true statement):

$b + 4 = 9$?

Since $5 + 4$ is 9, by replacing b with 5 the statement is true.

Class Exercise 1 Solve each equation for b.

1. $b + 7 = 10$ 2. $5 + b = 11$ 3. $8 + b = 12$ 4. $b + 2 = 16$

Solve the equation for s:

$s + 73 = 1610$.

We can see that as the numbers get larger it is not so easy to mentally find the unknown value.

Let us develop a procedure to find the value of s that solves the equation.

Consider an equation a balance scale where the left and the right sides are the same. To keep the balance, whatever we do to one side of the equation we must also do to the other side.

To solve an equation we must get the variable by itself. To do this we follow the procedure below.

We do the opposite of addition and *subtract* 73 from both sides:

$$\begin{aligned} s + 73 &= 1610 \\ -73 &= -73 \\ \hline s &= 1537 \end{aligned}$$ is the solution.

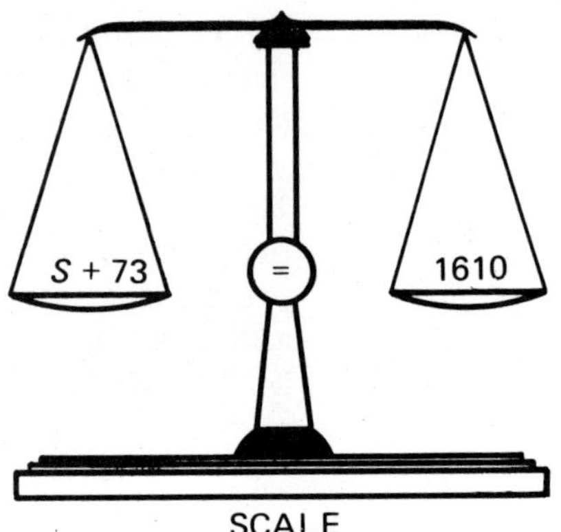

To check the result:

1. Write the original equation again.
2. Replace the variable with the solution.
3. Since both sides are the same (equal), the solution is correct.

Check:

$s + 73 = 1610$

$1537 + 73 = 1610$

$1610 = 1610$

Class Exercise 2 Solve the following equations. Check each result.

1. $C + 431 = 760$
2. $X + 98 = 234$
3. $581 + d = 870$
4. $M + 41.3 = 87.2$

Example 1 Solve this equation. Check your result.

$n - 431 = 689$

Solution: We must get the variable by itself.

We do the opposite of subtraction and *add* 431 to both sides:

$$\begin{aligned} n - 431 &= 689 \\ +431 &= +431 \\ \hline n &= 1120 \end{aligned}$$ is the solution.

Check:

1. We write the original equation again.
2. Replace n with the solution.
3. Since both sides are the same (equal), the solution is correct.

Check:

$n - 431 = 689$

$1120 - 431 = 689$

$689 = 689$

Class Exercise 3 Solve the equations. Check your results.

1. $a - 745 = 682$
2. $d - 983 = 27$
3. $e - 42.3 = 71.9$
4. $f - 325 = 46.7$

By using the opposite operation, we can also solve

$$3C = 432 \quad \text{and} \quad \frac{e}{2} = 53$$

Since $3C$ means 3 times C, the opposite of multiplication is division. Solve the equation and check the result.

$$3C = 432$$

Solution: Divide both sides by 3:

$$\frac{3C}{3} = \frac{432}{3}$$

$1C = 144$; or $C = 144$.

Check:

1. Write the equation: $3C = 432$.
2. Replace C with the solution:

$$3(144) = 432$$
$$432 = 432.$$

Since $\frac{e}{2}$ means e divided by 2, the opposite of division is multiplication. Solve the equation and check the result.

$$\frac{e}{2} = 53$$

Solution: Multiply both sides by 2:

$$2 \cdot \frac{e}{2} = 53 \cdot 2$$

$1e = 106$; or $e = 106$.

Check:

1. Write the equation: $\frac{e}{2} = 53$.
2. Replace e with the solution:

$$\frac{106}{2} = 53$$
$$53 = 53.$$

Both sides are equal, therefore the answers are correct.

Try the next exercise to see if you can solve equations by multiplication or division.

Class Exercise 4 Solve the equations. Check the results.

1. $7f = 4816$
2. $\frac{f}{7} = 93$
3. $8h = 9.616$
4. $\frac{h}{8} = 4.502$
5. $9a = 87.21$
6. $\frac{a}{9} = 63.7$
7. $12b = 1212$
8. $\frac{b}{12} = 5.01$

Writing a program takes quite a bit of time. Therefore, we would *not* write a program to just solve $s + 73 = 1610$ or $C + 431 = 760$. We would write a program to solve *all* equations that look like those.

How can we represent *all* equations that look like $s + 73 = 1610$? Since the numbers may vary, we replace them with variables: $s + a = b$. If $a = 73$ and $b = 1610$, we have our original equation.

$s + a = b$ is called the *general form* of the equation $s + 73 = 1610$.

Example 2 Using the general form of an equation $s + a = b$, find the values a and b for each of the following.

1. $s + 36 = 482$
2. $x + 71 = 580$

Solutions:

1. $a = 36$ and $b = 482$.
2. $a = 71$ and $b = 580$.

Class Exercise 5 Using the general form of an equation $s + a = b$, find the values a and b for each of the following.

1. $c + 43.1 = 350.8$
2. $d + 4.8 = 96.2$
3. $134 + s = 486$
4. $z + 20.7 = 46.9$

Now we turn to the computer to solve the following problem.

Problem Enter values for A and B. Solve for S; all equations of the form $S+A=B$.

Solution:

a. *Analysis:* We must solve for S the equation $S+A=B$. We will follow the same procedure that we used for $s+73=1610$.

Original Equation	*General Form*
Subtract 73 from both sides:	Subtract A from both sides:
$s + 73 = 1610$	$S + A = B$
$-73 = -73$	$-A = -A$
$s = 1610 - 73$	$S = B - A$
$s = 1537.$	

$S=B-A$ is called the *general solution*. Therefore, to solve $s+73=1610$, since $B=1610$ and $A=73$ we need only subtract:

$1610-73=1537$.

We now write how we use the algorithm $S=B-A$ to solve the problem.

b. *Plan:*
1. Put in values for A and B.
2. Calculate: $S=B-A$.
3. Print S.
4. End.

c. *Planned Output*

```
SOLVING EQUATIONS BY SUBTRACTION
ENTER VALUES FOR A AND B
? 73, 1610
THE SOLUTION IS
S= 1537
```

d. *The Program*

```
10  PRINT "SOLVING EQUATIONS BY SUBTRACTION"
20  PRINT "ENTER VALUES FOR A AND B"
30  INPUT A, B
40  S=B-A
50  PRINT "THE SOLUTION IS"
60  PRINT "S="; S
70  END
```

Exercise 2.7

Solve each equation for x.

1. $x+8=9$
2. $x+7=19$
3. $x+2=12$
4. $x+18=23$

Solve the following equations. Check each result.

5. $a+871=983$
6. $485+c=1038$
7. $x+49=83.6$
8. $y+72.4=91$

Solve the following equations. Check the result.

9. $x-11.6=13.5$
10. $b-71.4=36.7$
11. $c-14=6.8$
12. $d-4.5=7$

Solve the following equations. Check each result.

13. $3e = 4182$ 14. $\frac{e}{3} = 713$ 15. $7x = 1414$ 16. $\frac{x}{7} = 5.2$

17. $0.2b = 8$ 18. $\frac{b}{0.2} = 8$ 19. $1.1c = 121$ 20. $\frac{c}{0.6} = 4.3$

Revise the program that was written in this section to produce the following outputs.

21.

```
SOLVING S+A=B
TYPE IN VALUES FOR A AND B
? 73, 1610
WHEN A= 73
AND B= 1610
S= 1537
```

22.

```
SOLVING EQUATIONS OF THE FORM S+A=B
ENTER VALUES FOR A AND B
? 73, 1610
THE SOLUTION TO
S+73= 1610 IS
S= 1537
```

For the following problems:
a. Write an analysis of the problem that leads to an algorithm.
b. Write a plan that outlines how you use the algorithm to solve the problem.
c. Design a planned output.
d. Code the program and RUN it on the computer.

23. **Problem:** Enter values for A and B. Solve for X, all equations of the form $X - A = B$.

24. **Problem:** Enter values for A and B. Solve for X, all equations of the form $AX = B$.

25. **Problem:** Enter values for A and B. Solve for X, all equations of the form $\frac{X}{A} = B$.

END OF CHAPTER EXERCISES

Section 2.1

What base ten numeral does each name?

1. 24_{five} 2. 36_{eight} 3. 34_{nine} 4. 64_{seven}
5. 83_{nine} 6. 23_{six} 7. 55_{eight} 8. 14_{five}

Convert each of the following to base ten.

9. 142_{five} 10. 152_{six} 11. 1111_{two} 12. 600_{seven}
13. 163_{eight} 14. 1234_{five} 15. 3102_{four} 16. 20002_{five}

For the following problems:
a. Write an analysis of the problem that leads to an algorithm.
b. Write a plan that outlines how you use the algorithm to solve the problem.
c. Design a planned output.
d. Code the program and RUN it on the computer.

17. **Problem:** Beginning with the left most position, enter two two-digit numerals in base seven, one digit at a time. Find the sum of the numbers. Write the sum as a base ten numeral.

18. **Problem:** Enter two two-digit numerals in base six, one digit at a time. Begin with the left most position. Find the product of the numbers. Write the product as a base ten numeral.

19. **Problem:** Enter three two-digit numerals, one digit at a time. Begin with the left most position. Enter the base for these numerals. Find the mean of the numbers. Write the mean as a base ten numeral.

Section 2.2

Find the area of the following rectangles.

20. 9 cm; 80 mm

21. 7.3 cm; 65 mm

22. Find the area of a rectangle whose base is 28.4 yards and whose height is 9.3 yards.

23. What is the area of a rectangle whose base is 7.8 cm and whose height is 63 mm?

Find the area of the following right triangles.

24.

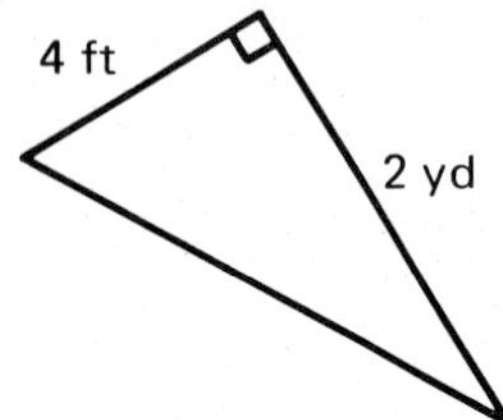

25.

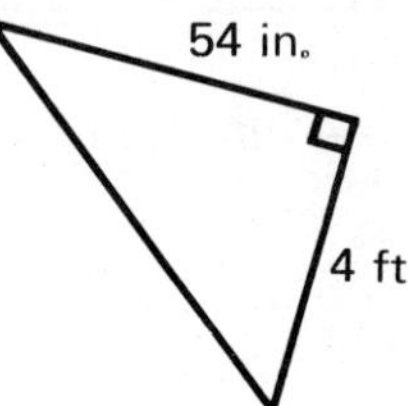

26. Find the area of a right triangle whose base is 15 feet and whose height is 11 feet.

27. What is the area of a right triangle whose base is 5.7 cm and whose height is 438 mm?

Find the area of the following triangles.

28.

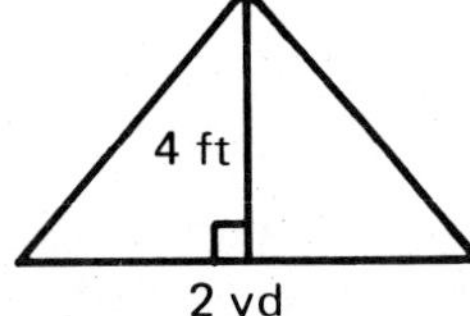

29.

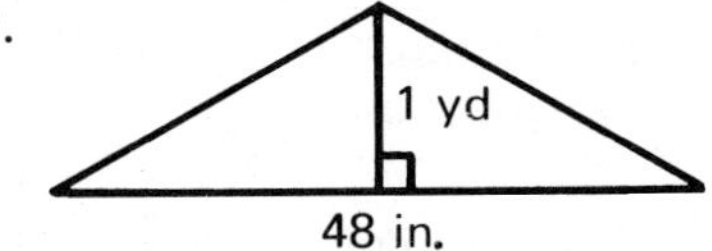

30. Find the area of a triangle if the base is $3\frac{1}{3}$ yards and the height is 7 feet.

31. What is the area of a triangle if the base is 1.03 m and the height is .86 m?

Find the area of the following parallelograms.

32.

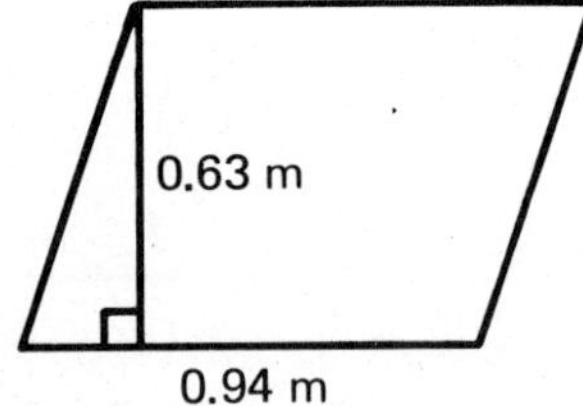

33.

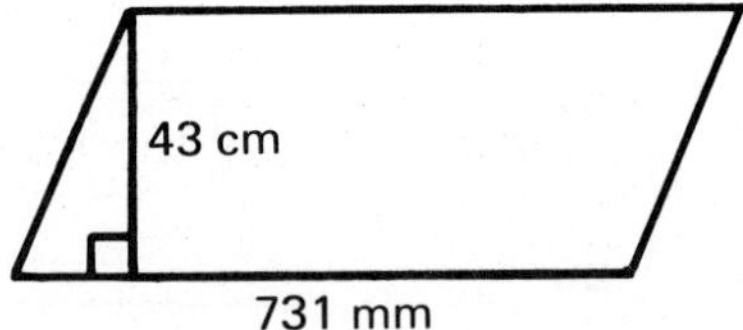

34. Find the area of a parallelogram whose base has length 3.76 m and whose height has length 0.7 m.

35. What is the area of a parallelogram with base 8.2 cm and height 453 mm?

For the following problems:
a. Write an analysis of the problem that leads to an algorithm.
b. Write a plan that outlines how you use the algorithm to solve the problem.
c. Design a planned output.
d. Code the program and RUN it on the computer.

36. **Problem:** Find the area of a rectangle when we are given the base in feet and the height in yards. The area should be given in square feet.

37. **Problem:** Find the area of a square when we are given the length of a side. (*Hint:* A square is a rectangle with all sides congruent.)

38. **Problem:** Carpeting costs $21.00 per square yard. Enter the length and width (in yards) of a rectangular room. Print the area and the cost to carpet the room.

Section 2.3

Convert each of the following Celsius temperatures to Fahrenheit degrees.

39. 55° 40. 30° 41. 20° 42. 32°

Convert each of the following Fahrenheit temperatures to Celsius degrees.

43. 212° 44. 115° 45. 21° 46. 75°

For the following problems:
a. Write an analysis of the problem that leads to an algorithm.
b. Write a plan that outlines how you use the algorithm to solve the problem.
c. Design a planned output.
d. Code the program and RUN it on the computer.

47. **Problem:** Enter a Celsius temperature. After increasing it by 15 degrees, convert this new temperature to Fahrenheit degrees.

48. **Problem:** Enter a Fahrenheit temperature. Increase the temperature by 20 degrees. Find the increase in Celsius temperature.

49. **Problem:** Enter the base and height of a parallelogram. Find the area. Then double the base and take half of the height. Find the area of this new parallelogram. (How do these areas compare?)

Section 2.4

Find the supplement of each of the following angle measures.

50. 134° 51. 27° 52. 107° 53. 9°

State the complement of each of the following angle measures.

54. 84° 55. 72° 56. 3° 57. 11°

Find the value of c in each of the following triangles.

58.

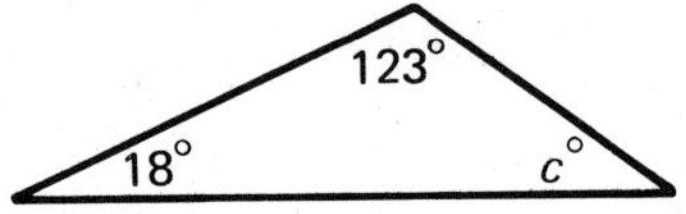

59.

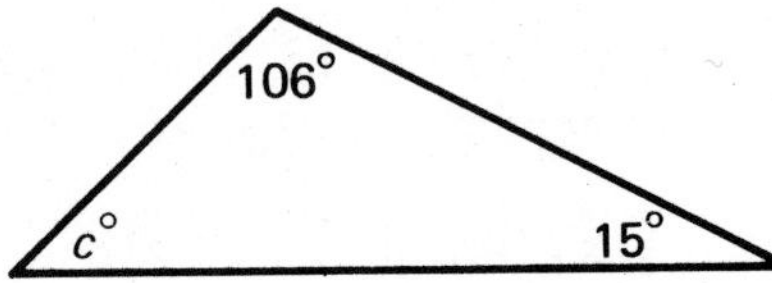

60.

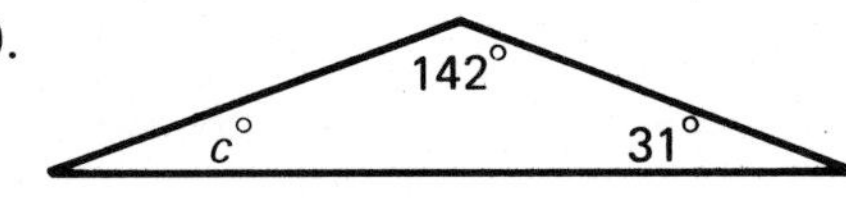

61.

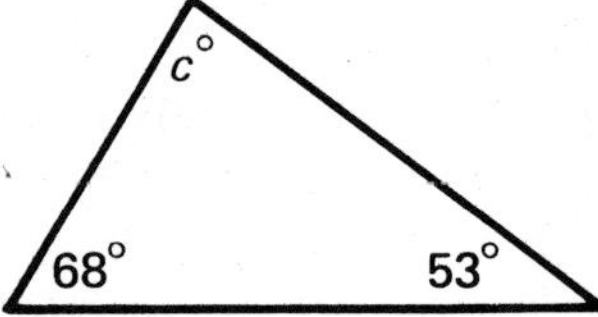

For the following problems:
a. Write an analysis of the problem that leads to an algorithm.
b. Write a plan that outlines how you use the algorithm to solve the problem.
c. Design a planned output.
d. Code the program and RUN it on the computer.

62. **Problem:** A complete rotation is 360°. Enter the measurement of an angle (between 0° and 360°). Find the number of degrees that remain to make the angle a complete rotation.

63. **Problem:** Enter the measure of an angle (between 0 and 90). Display the measure of the angle, double the angle and then display the supplement of double the angle.

64. **Problem:** Enter the measure of an angle (between 0 and 180). Display the measure of the angle, one-half the angle and then display the complement of one-half the angle.

Section 2.5

For the following, use the formula $T = W \cdot N$ to find the total wages (T) earned when working for (Remember: W = hourly wage and N = number of hours worked.):

65. $3.28 per hour for 6 hours.

66. $4.63 per hour for 12 hours.

67. $3.91 per hour for 7.5 hours.

68. $5.47 per hour for 20 hours.

Change the following to decimals.

69. 61% 70. 54% 71. 5% 72. 2%

Find the commission when:

73. the rate of commission is 6% and the total sales are $5218;

74. the rate of commission is 1% and the total sales are $20436;

75. the rate of commission is 25% and the total sales are $1430;

76. the rate of commission is 12% and the total sales are $6700.

77. **Problem:** Enter the price of an item (P). Using the rate of sales tax as 9%, display the sales tax on the item and the total cost of the item including the sales tax.

78. **Problem:** Nancy earns $200 per week plus a 3% commission on her sales. Enter her total sales for the week. Display her commission and the total amount of money she earns for that week.

79. **Problem:** George earns $2.25 per hour babysitting for his next door neighbors. He is paid every four weeks. Enter four numbers representing the number of hours he worked each week. Display the amount of money that he was paid.

Section 2.6

Find the discount for each item below.

	Item	*Price*	*Rate of Discount*	*Discount*
80.	Microwave Oven	$550	18%	
81.	AM/FM Stereo Receiver	$150	25%	
82.	Vacuum Cleaner	$250	27%	
83.	AM/FM Pocket Radio	$22	15%	

Find the sale price for each item in the chart below.

	Item	*Price*	*Rate of Discount*	*Discount*	*Sale Price*
84.	Reversible Ceiling Fan	$290	12%		
85.	Business Computer	$4500	32%		
86.	Video Cassette Recorder	$950	18%		
87.	Coffee Maker	$105	25%		

For the following problems:
a. Write an analysis of the problem that leads to an algorithm.
b. Write a plan that outlines how you use the algorithm to solve the problem.
c. Design a planned output.
d. Code the program and RUN it on the computer.

88. **Problem:** Enter the price of an item (P) and the rate of discount (R). If the rate of sales tax is 9%, display the final cost of the item.

89. **Problem:** Enter the price of an item (P), the rate of discount (R) and the rate of sales tax (RS) written as a percent. Display the discount, sales tax and final cost of the item.

90. **Problem:** If an item in a store is not selling, it is often discounted. If it still does not sell, it may be discounted again. Enter the price of an item (P), the first rate of discount (FR) and the second rate of discount (SR). Display the cost after the first discount and then the cost after the second discount.

Section 2.7

Solve each equation for c.

91. $c + 12 = 15$ 92. $c + 2 = 21$ 93. $c + 3 = 9$ 94. $c + 7 = 15$

Solve the following equations. Check each result.

95. $x + 719 = 1410$ 96. $d + 41.7 = 86.4$ 97. $c + 53 = 100$ 98. $e + 19.3 = 457$

Solve the following equations. Check the result.

99. $x - 41.3 = 16.7$ 100. $b - 5.3 = 18.9$

101. $f - 20 = 38.9$ 102. $h - 6.04 = 25$

Solve the following equations. Check each result.

103. $9x = 3186$ 104. $\frac{c}{5} = 8.3$

105. $7a = 2121$ 106. $\frac{a}{2} = 18.73$

107. $0.4c = 16$ 108. $\frac{c}{0.3} = .72$

109. $8d = 2.48$ 110. $\frac{d}{1.3} = 4.56$

For the following problems:

a. Write an analysis of the problem that leads to an algorithm.
b. Write a plan that outlines how you use the algorithm to solve the problem.
c. Design a planned output.
d. Code the program and RUN it on the computer.

111. **Problem:** Enter values for C and D. Solve for Y, equations of the form:

a. $Y+C=D$ b. $Y-C=D$

112. **Problem:** Enter the area (A) and the base (B) of a parallelogram. (The area must be numerically greater than the base.) Find the height.

113. **Problem:** Enter the area (A) and the height (H) of a triangle. (The area must be numerically greater than the height.) Find the base.

Chapter 3

Intermediate Programming Concepts

Section 3.1 DETERMINING IF A FLOWCHART SOLVES A PROBLEM

In Chapter 1 we used the algorithm $P=4\times S$ to find the perimeter of a square. A side of a square never has a length that is zero or less than zero (negative). How can we make sure that if someone enters a value for S that is less than or equal to zero, we do not have the computer calculate the perimeter?

We must revise our plan from Section 1.2 so that our program includes a check for a side of length less than or equal to zero.

Revised Plan to Find the Perimeter of a Square

1. Put in a value for S.
2. Check to see if $S\leqslant 0$.
   a) If $S\leqslant 0$, print "side's length must be greater than zero." Then go to 1.
   b) If S is not less than or equal to zero, continue.
3. $P=4\times S$.
4. PRINT P.
5. END.

The symbol "=" in line 3 is used here as an assignment symbol. The less than or equal sign ($\leqslant$) in line 2 is a relation symbol.

The following table shows various relation symbols used in mathematics and BASIC.

Relation Symbols

Mathematics	*Meaning*	*BASIC*
$=$	Is equal to	=
$<$	Is less than	<
$>$	Is greater than	>
$\neq$	Is not equal to	<>
$\leqslant$	Is less than or equal	<=
$\geqslant$	Is greater than or equal	>=

Example 1 Show whether the following statement is true or false.

$8+6>5+3\cdot 2$

Solution: We find the value on each side of the relation symbol.

$8+6>5+3\cdot 2$

We multiply first. (arrow to $\cdot$)

We add. (arrow to $+$ in $8+6$)

$14>5+6$

We add. (arrow to $+$)

$14>11$

Since 14 *is* greater than ($>$) 11, the statement is true.

Class Exercise 1 Show whether the following statements are true or false. (The relation and arithmetic symbols are written in BASIC.)

1. $12+3>8+6/2$
2. $3\uparrow 2<2\uparrow 3$
3. $16-4*3=36$
4. $(8+3)*5>=60-5$
5. $3+2*3+2<=2+1\uparrow 3$
6. $25-2*4<>(25-2)*4$

When plans involve checks (as in Step 2 of our revised plan), they become harder to translate into programs. It is helpful to be able to write these plans in another form called a *flowchart*. We will use the following symbols when writing a flowchart.

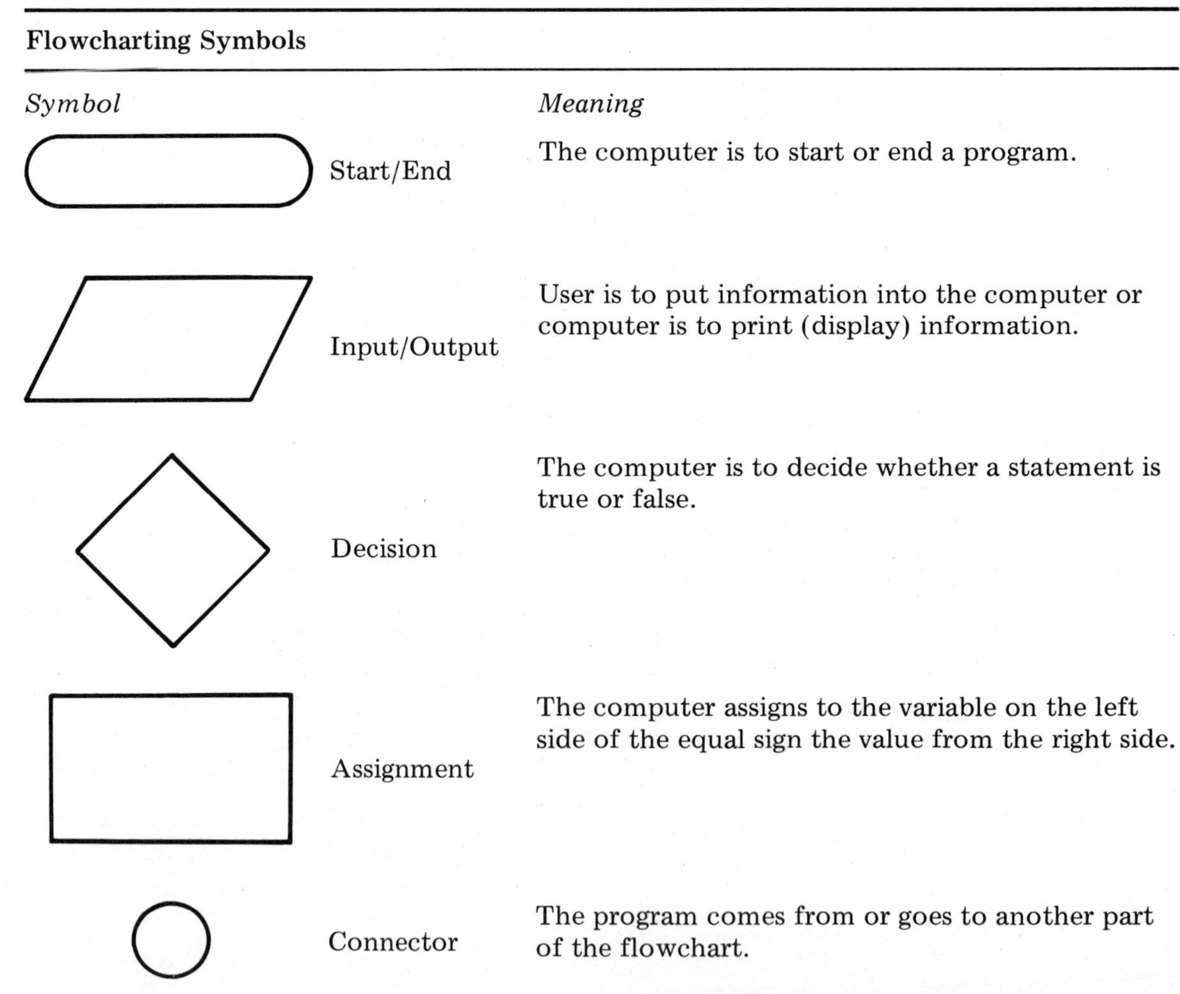

Flowcharting Symbols

Symbol		*Meaning*
	Start/End	The computer is to start or end a program.
	Input/Output	User is to put information into the computer or computer is to print (display) information.
	Decision	The computer is to decide whether a statement is true or false.
	Assignment	The computer assigns to the variable on the left side of the equal sign the value from the right side.
	Connector	The program comes from or goes to another part of the flowchart.

Example 2 Write each using flowchart symbols.

1. INPUT X
2. GO TO 1
3. IS $X = 4 * Y$?

Solution:

1. 

2. 

3. 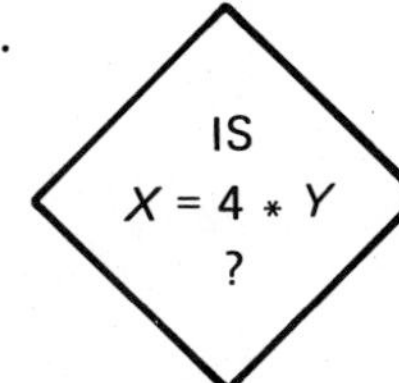

Class Exercise 2 Write each using flowchart symbols.

1. $C = 4 * E$
2. PRINT Y
3. GO TO 3
4. IS $L = 6$?
5. $B = C \uparrow F$
6. INPUT C, D, E

We will now use our symbols to flowchart our revised plan.

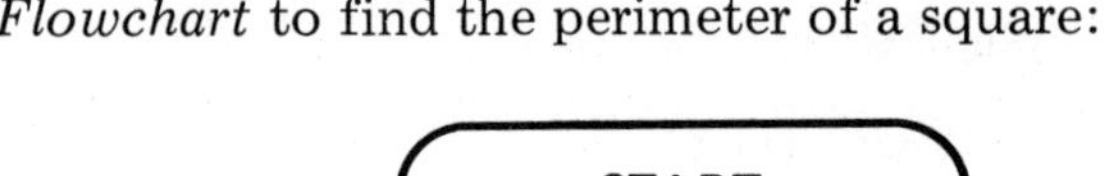

Flowchart to find the perimeter of a square:

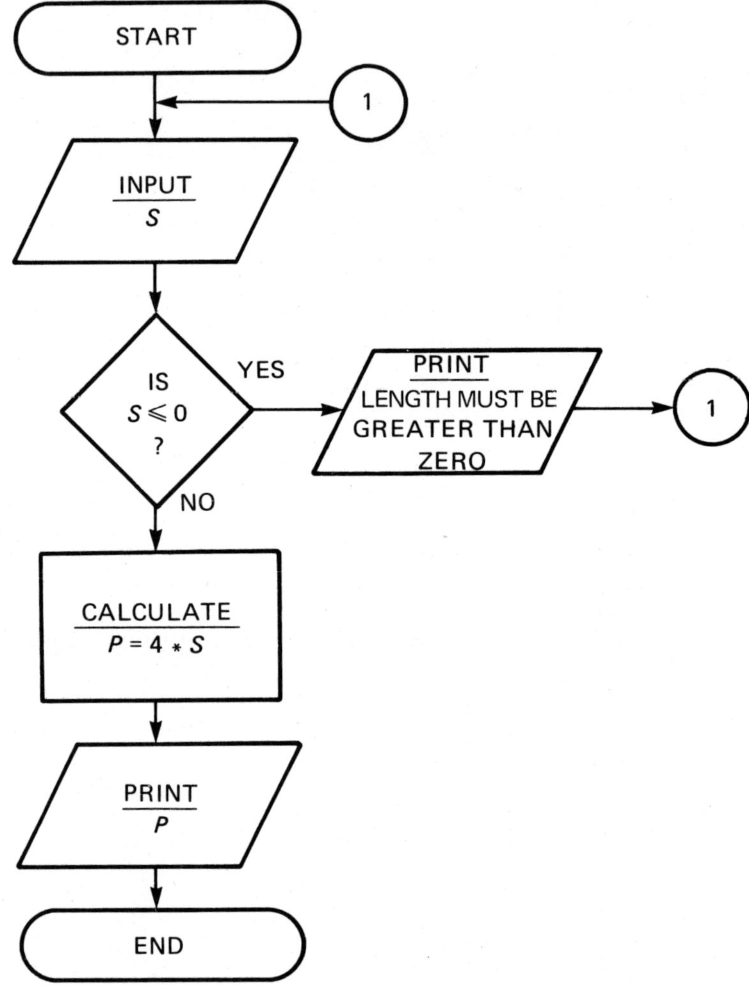

Before coding a program from a flowchart you must make sure that the flowchart solves the problem. The following shows you how to *trace a flowchart*.

Procedure for Tracing a Flowchart

---

1. List in table form all the variables used in the flowchart.
2. a. Follow the arrows in the flowchart.
   b. Perform all tasks.
   c. Record all the information in the table.

We are going to see if the flowchart finds the perimeter of a square with side of length 6. (It should give us an answer of 24.)

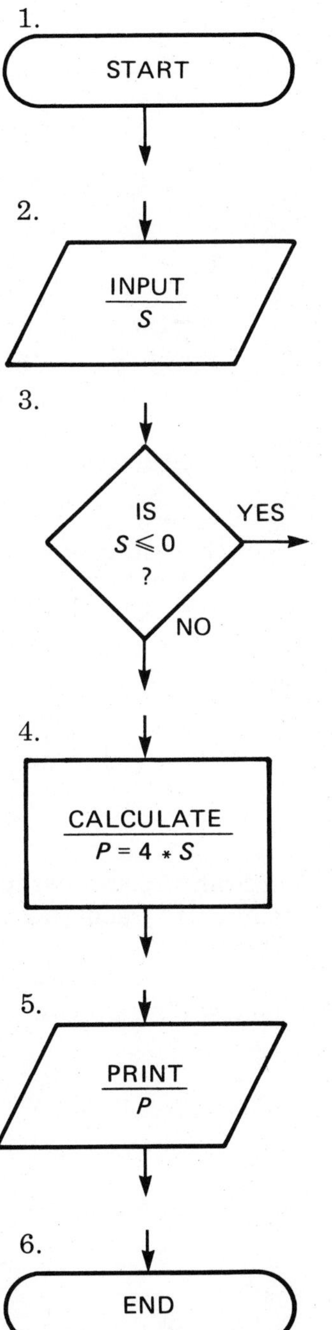

S	P

S	P
6	

Check S. Is $S \leqslant 0$? No; $S=6$. Then follow the arrow down.

S	P
	$4 * 6$
	24

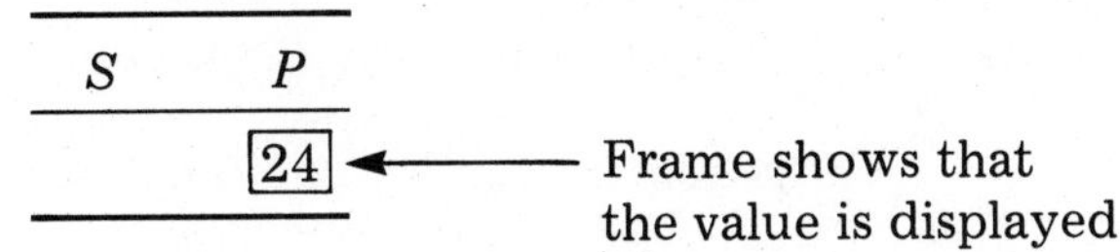

The program ends.

Since 24 is the perimeter, the flowchart solves the problem. We must now see if the flowchart handles the special case when $S \leqslant 0$.

We go back to the INPUT statement and enter a zero for S.

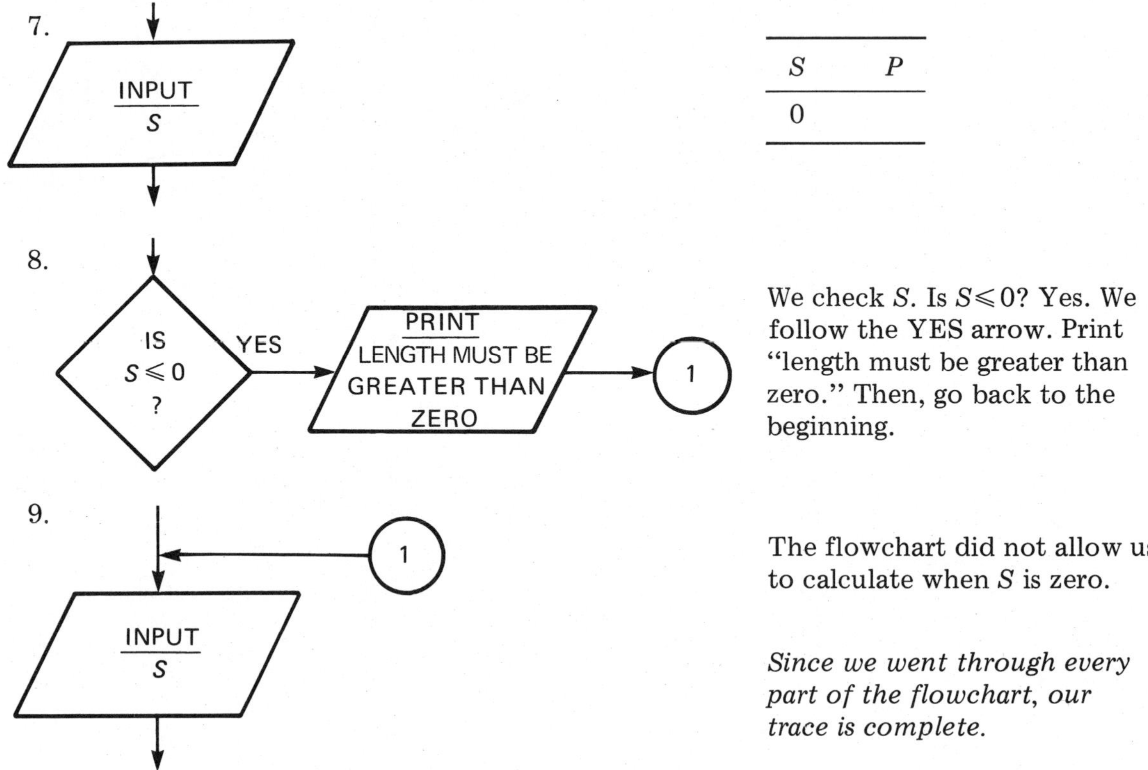

S	P
0	

We check S. Is $S \leqslant 0$? Yes. We follow the YES arrow. Print "length must be greater than zero." Then, go back to the beginning.

The flowchart did not allow us to calculate when S is zero.

Since we went through every part of the flowchart, our trace is complete.

After tracing a flowchart you should have a good plan from which to write a program.

The following example will give you practice in reading and tracing a flowchart.

Example 3 Determine if the flowchart solves the problem. Use the test data along with the procedure previously outlined for tracing a flowchart.

Problem: Enter two numbers. Determine if their product is less than 25.

Test Data: a. 8, 3 b. 6, 8

Flowchart

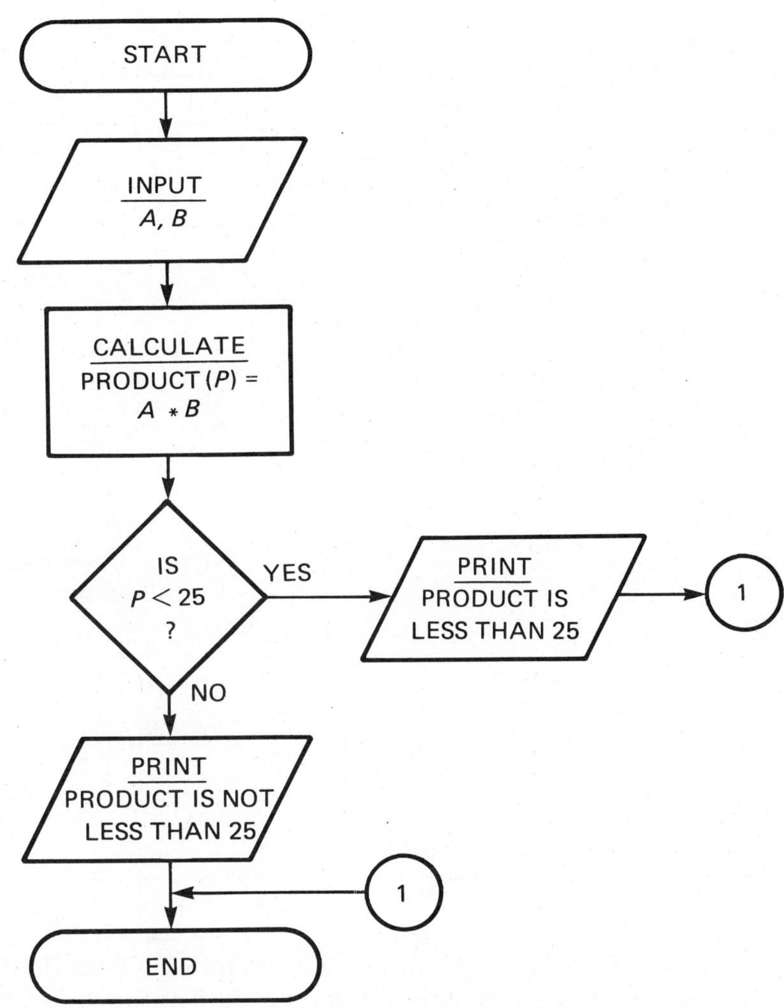

Solution: To see if the flowchart solves the problem, trace the flowchart, using the data 8, 3 and 6, 8.

1. We list the variables and enter the data.

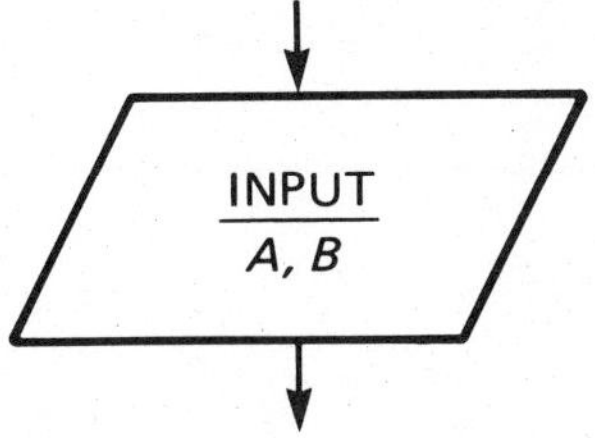

Test #1 (8, 3)

A	*B*	*P*
8	3	

Test #2 (6, 8)

A	*B*	*P*
6	8	

2. We calculate the product (P).

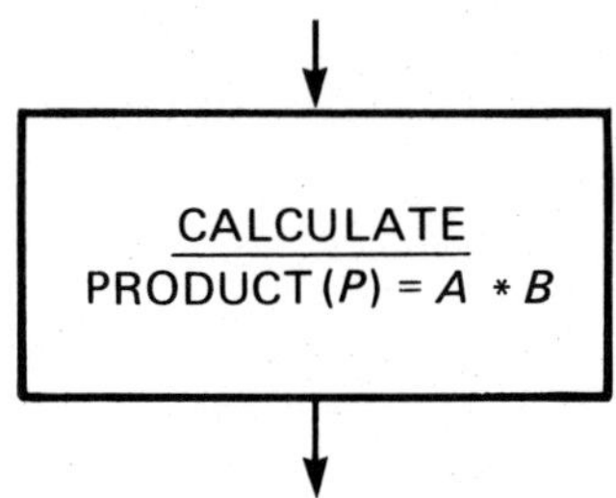

Test #1

A	B	P
		$8 * 3$
		24

Test #2

A	B	P
		$6 * 8$
		48

3. We check P.

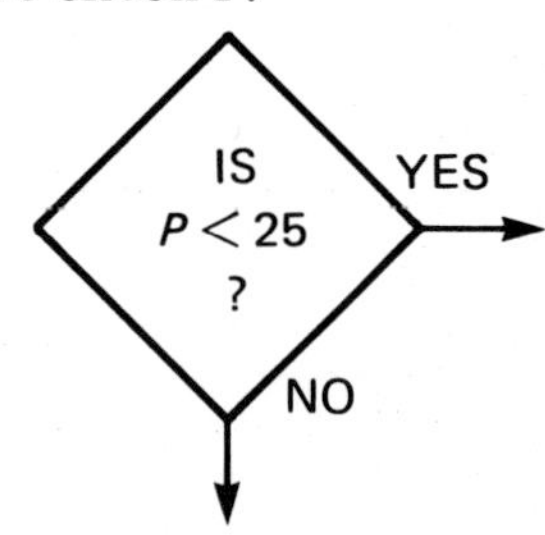

Test #1

A	B	P
8	3	24

Since P is less than 25 we follow the YES.

Test #2

A	B	P
6	8	48

Since P is not less than 25 we follow the NO.

4. For the data in Test #1 we print: PRODUCT IS LESS THAN 25

   For the data in Test #2 we print: PRODUCT IS NOT LESS THAN 25

Since we have executed all parts of the chart (with the correct result), we can now say that the flowchart solves the problem.

Class Exercise 3 In Problems 1 through 4 which follow, determine if the flowchart solves the problem. Use the sets of test data along with the procedure outlined for tracing a flowchart.

1. **Problem:** Enter two numbers A, B. Determine if their sum is less than 15.

Test Data: a. 6, 3 b. 8, 10

Flowchart

START
1
INPUT
A, B
CALCULATE
S = A + B
IS
S < 15
?
YES
NO
PRINT
SUM IS LESS
THAN 15
1
PRINT
SUM IS NOT
LESS THAN 15
1

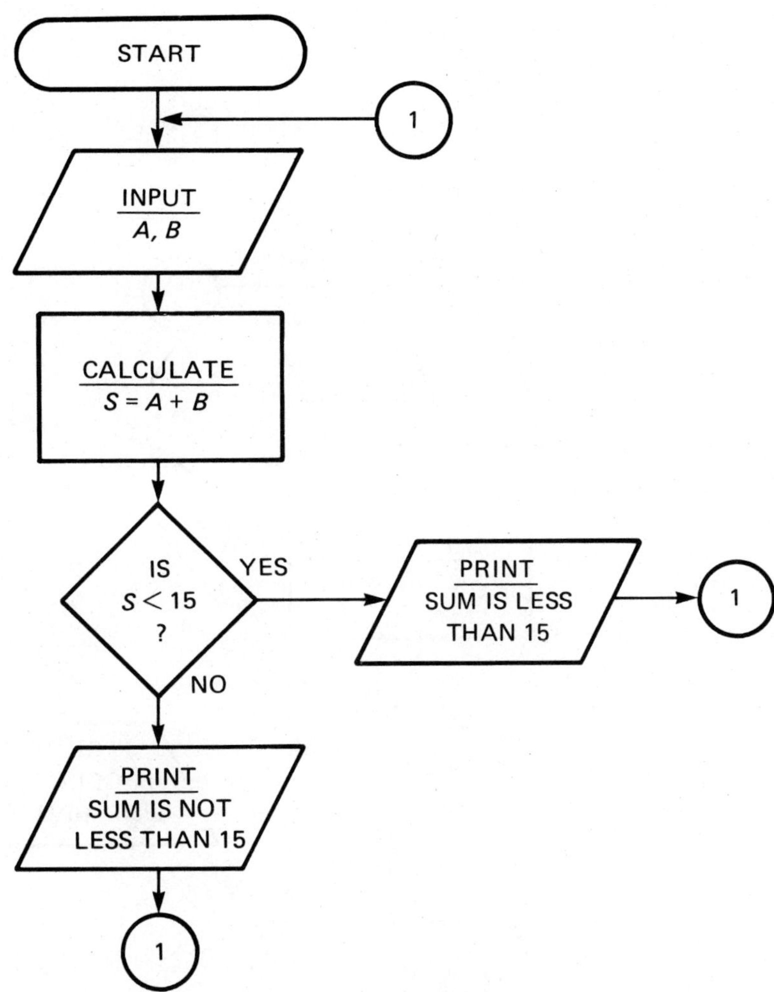

2. **Problem:** Enter two numbers G, H. Determine if the product is greater than or equal to 100.

Test Data: a. 25, 4 b. 7, 3

Flowchart

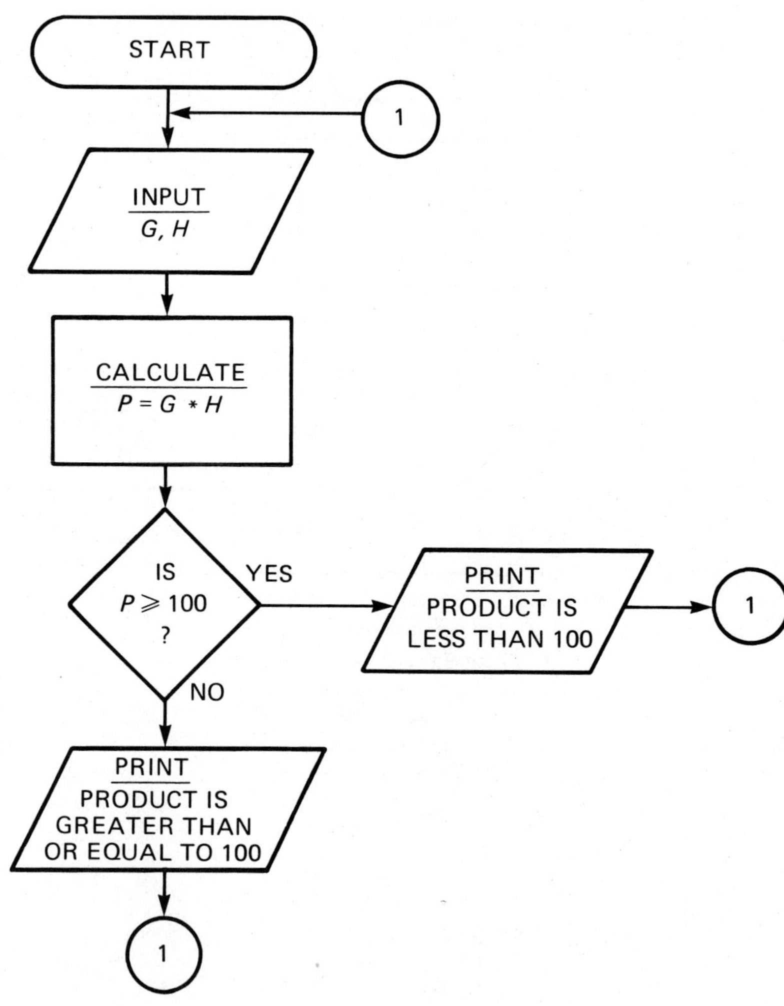

3. **Problem:** Enter three numbers R, S, T. Determine if the sum of the first two is less than the third.

Test Data: a. 6, 8, 10 b. 3, 5, 12

Flowchart

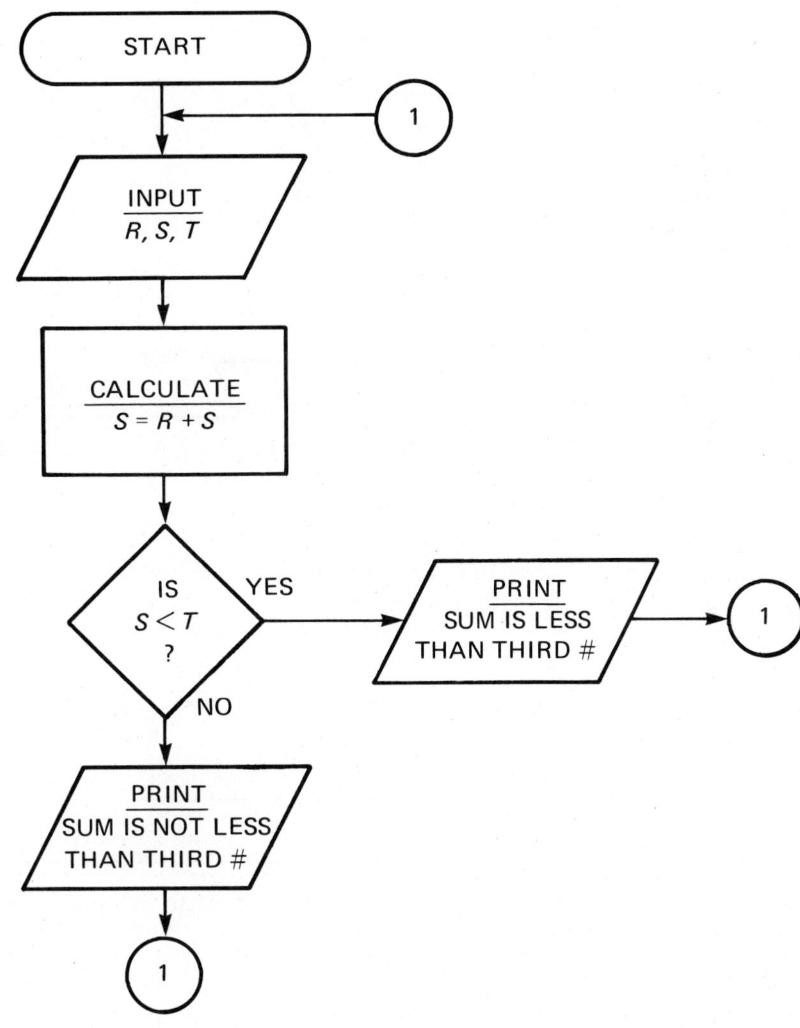

4. **Problem:** Enter two numbers. Determine if the numbers are the same.

Test Data: a. 6, 6 b. 8, 1

Flowchart

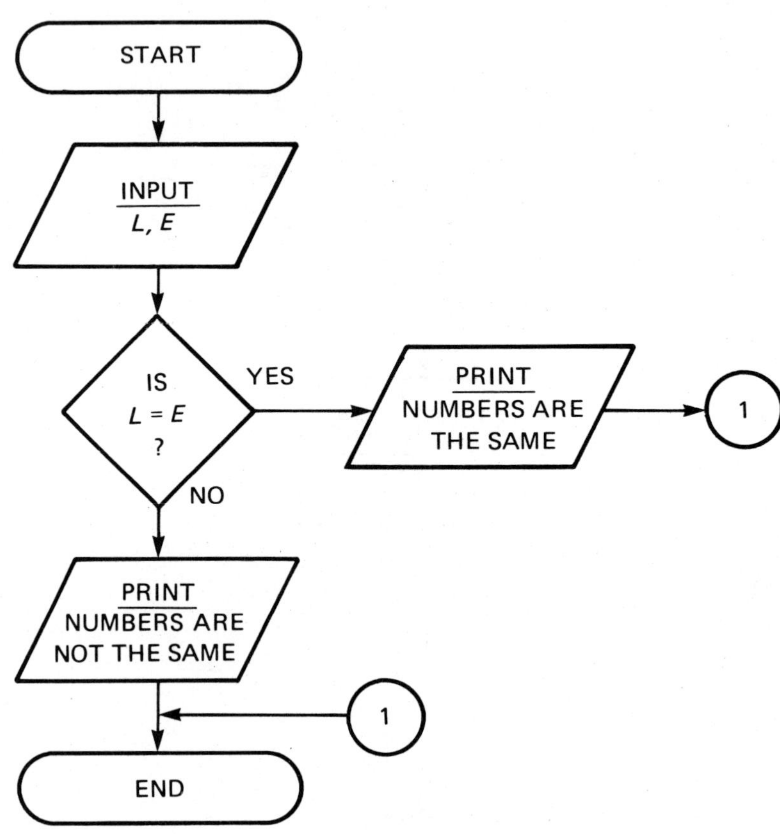

Exercise 3.1 Show whether the following statements are true or false. (The relation symbols and the arithmetic symbols are written in BASIC.)

1. 8 + 6 < 4 * 3
2. 16 − 8 > 2 ↑ 3
3. 16 + 12 >= 5 * 5
4. 2 + 3 * 4 = 2 + (3 * 4)
5. 25 − 6 < 7 + 3 * 3
6. 4 ↑ 2 = 2 ↑ 4
7. 12 + 3 * 8 > (12 + 3) * 8
8. 6 * 2 + 6 * 3 <> 6 * (2 + 3)
9. 2 + 3 ↑ 2 >= 25
10. 2 * 3 + 1 <= 2 * (3 + 1)

Write each of the following using flowchart symbols.

11. INPUT J, R
12. IS $M < S$?
13. $S = 2 * Y$
14. GO TO 4
15. PRINT S
16. $C = 4 \uparrow N$

In Problems 17 through 20 which follow, determine if the flowchart solves the problem. Use the sets of test data along with the procedure outlined for tracing a flowchart.

17. **Problem:** Enter two numbers. Determine if their product is greater than 38.

Test Data: a. 7, 3 b. 9, 8

Flowchart

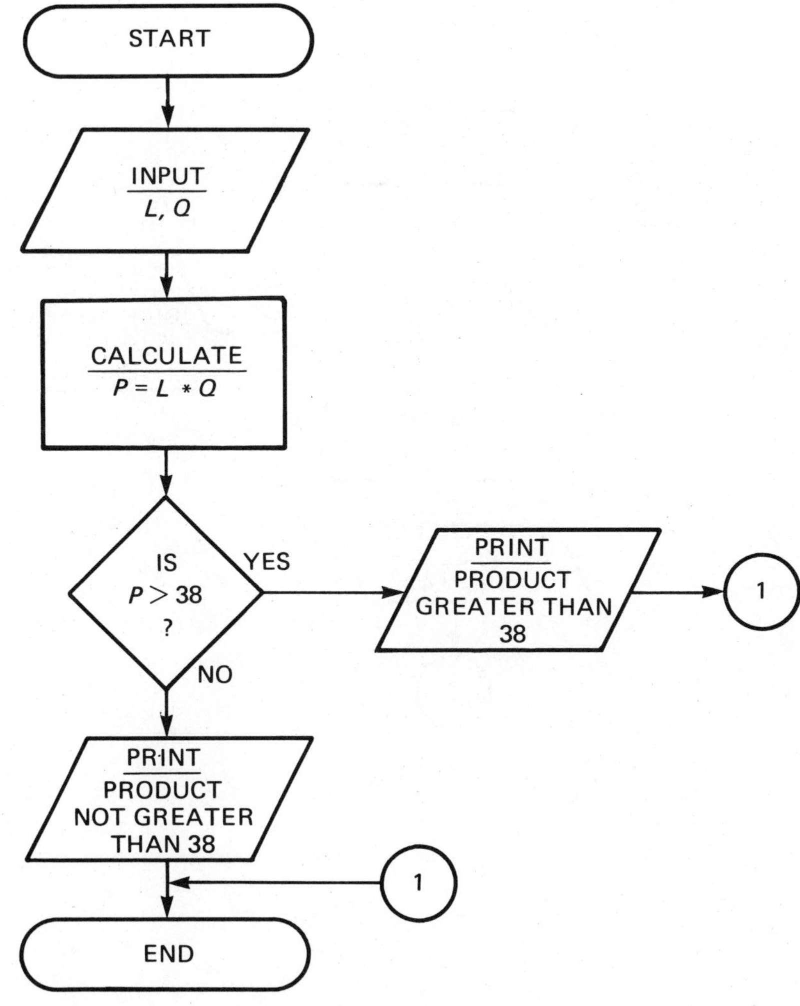

18. **Problem:** Enter two numbers. Determine if their difference is less than 47.

Test Data: a. 78, 36 b. 54, 5

Flowchart

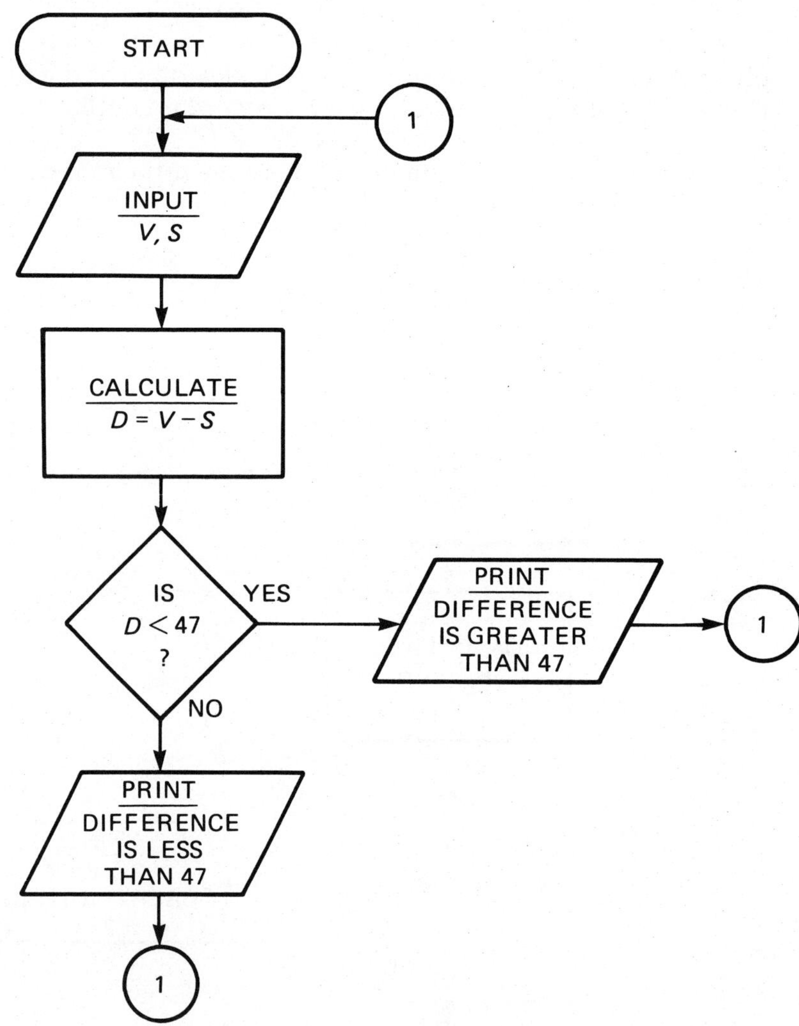

19. **Problem:** Enter three numbers. Determine if the sum of the second and third numbers is less than the sum of the first and second.

Test Data: a. 4, 6, 5 b. 10, 7, 4

Flowchart

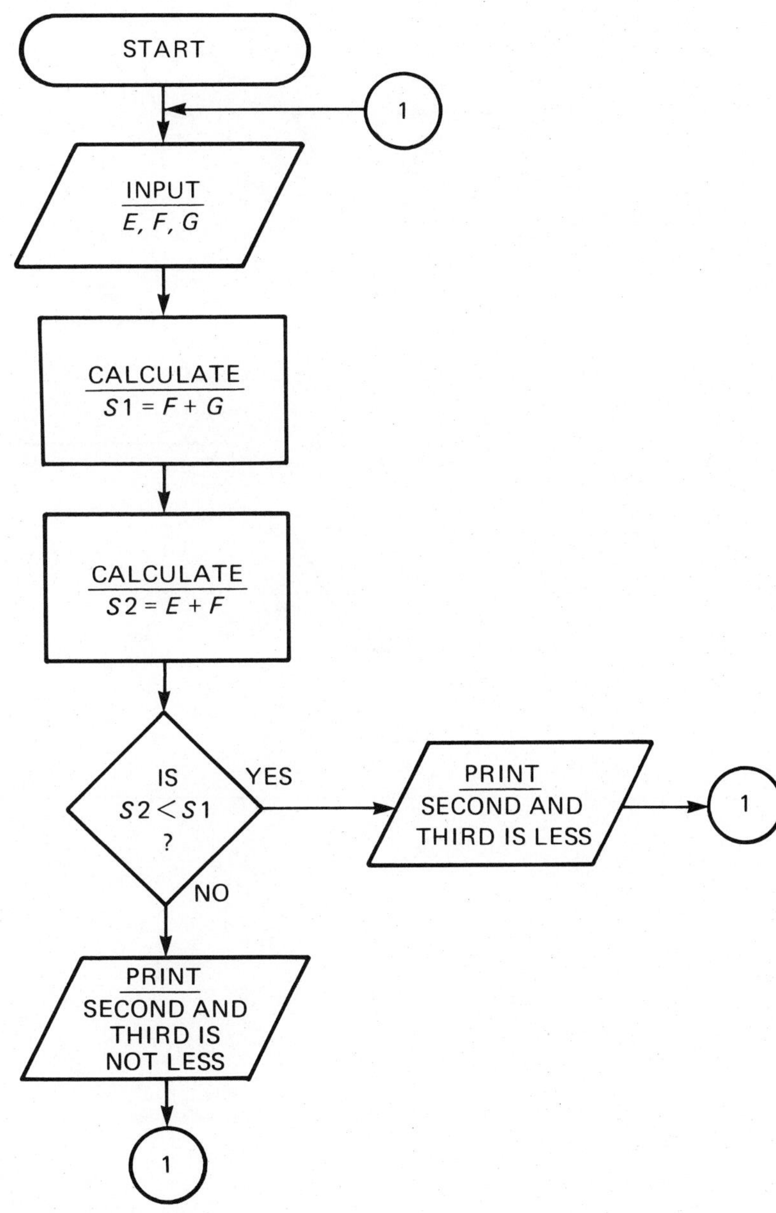

20. **Problem:** Enter two different numbers. Print the larger and then the smaller number.

Test Data: a. 6, 1 b. 8, 9

Flowchart

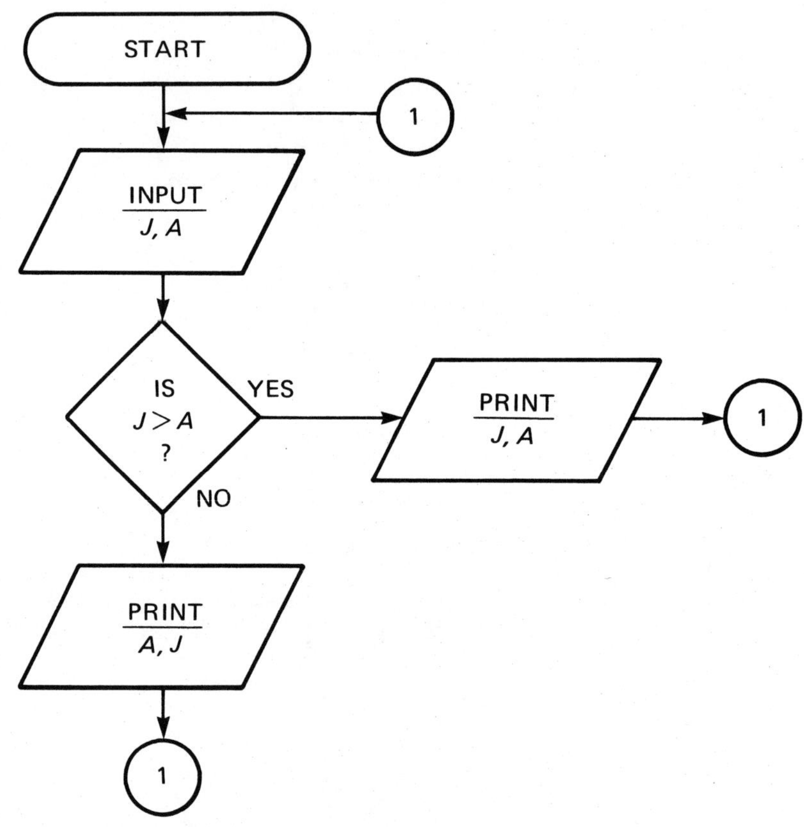

Section 3.2 DEVELOPING A FLOWCHART SOLUTION TO A PROBLEM

We will now learn how to write a flowchart solution to a problem.

Example 1

a. Develop a flowchart solution to the problem.
b. Trace the flowchart to see if it solves the problem.

Problem: Enter two numbers. Determine if the difference of the first number minus the second number is negative.

Solution: We write the flowchart after we analyze and plan a computer solution to the problem.

a. *Analysis:*
   1. A number is *negative* if it is less than 0. We get a negative number when we subtract a larger number from a smaller number.
   2. A *difference* is the answer when subtracting two numbers.
   3. We must see if the difference is less than 0.

   Algorithm to solve the problem:
   1. Using the numbers A, B, let the difference, D, equal $A-B$.
   2. Check to see if $D<0$. (D will be less than zero (negative) whenever B is the larger number. Why?)

b. *Plan:*
   1. Enter two numbers A, B.
   2. Find the difference D: $D=A-B$.
   3. Is $D<0$?
      a) If the answer is yes, print "the difference is negative." Go back to line 1.
      b) If the answer is no, print "the difference is not negative." Go back to line 1.

The plan is now translated into symbols.

c. *Flowchart*

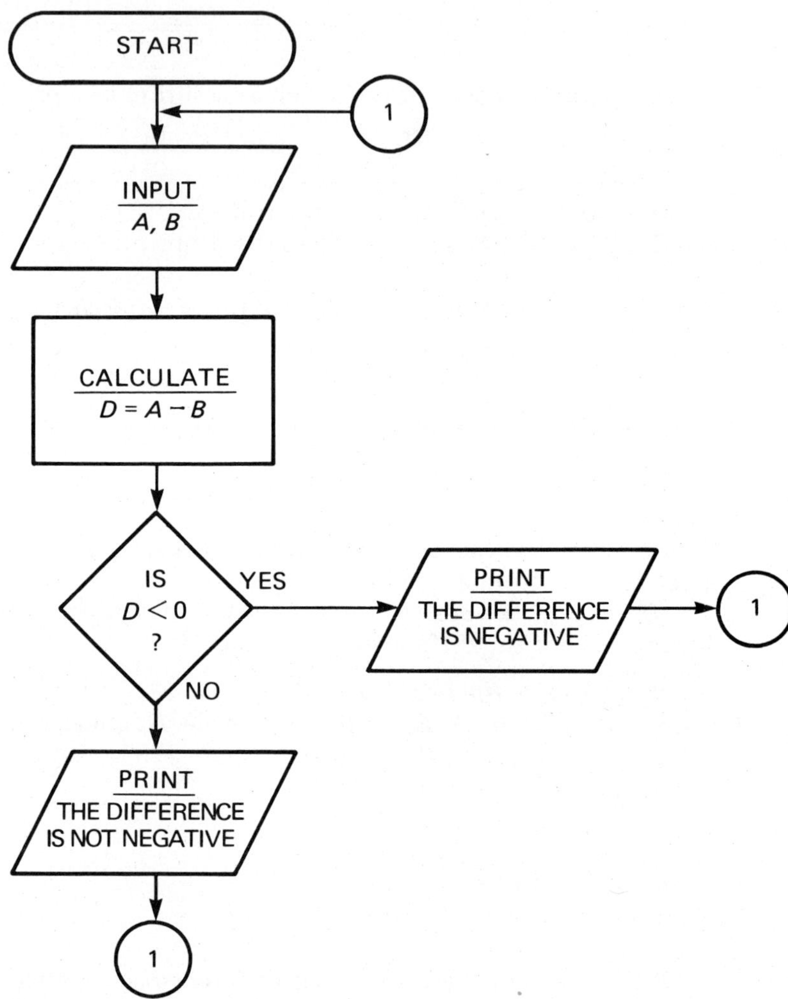

Trace the flowchart to see if it solves the problem. First choose the test data:

For $A = 8$ and $B = 3$ we should get the not negative message.

For $A = 3$ and $B = 8$ we should get the negative message.

Now we begin the trace.

1.

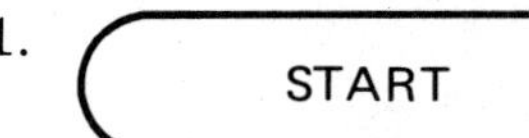

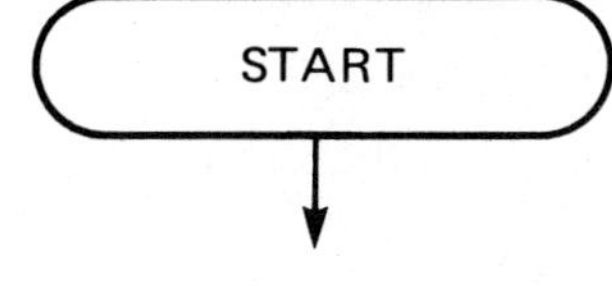

Test #1 (8, 3)

A	*B*	*D*

Test #2 (3, 8)

A	*B*	*D*

2.

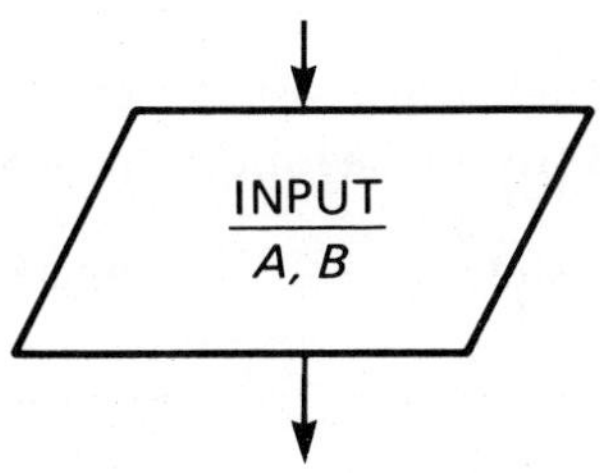

A	*B*	*D*
8	3	

A	*B*	*D*
3	8	

3.

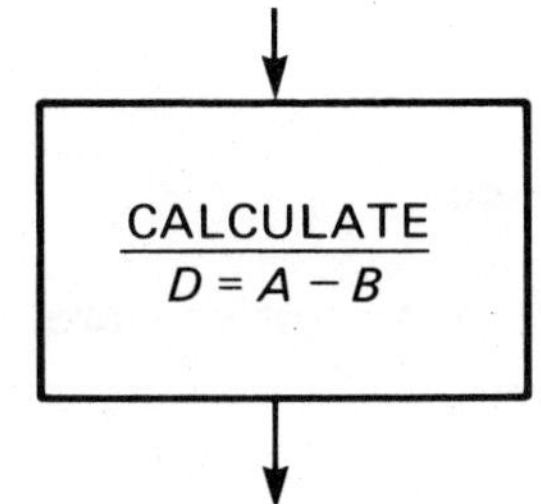

A	*B*	*D*
		$8 - 3 = 5$

A	*B*	*D*
		$3 - 8 = -5$

(*Note:* Since we are subtracting a larger number from a smaller number the difference is less than 0 or negative.)

4.

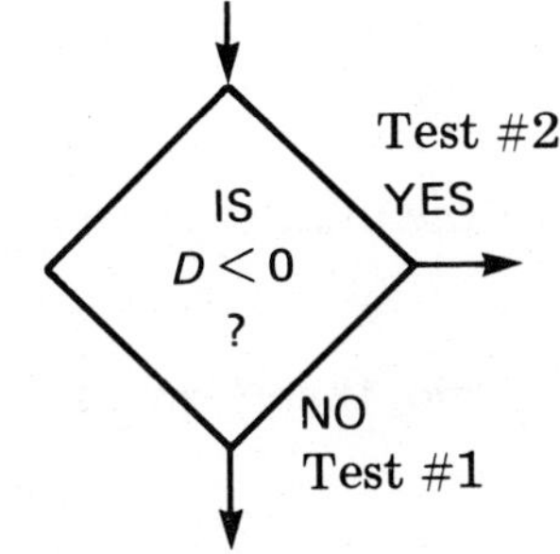

$D = 5$. Follow the NO route.

$D = -5$. Follow the YES route.

5. For Test #1 we print: THE DIFFERENCE IS NOT NEGATIVE

   For Test #2 we print: THE DIFFERENCE IS NEGATIVE

Since this is what we expected, the flowchart solves the problem.

Try the following problems.

Class Exercise 1 a. Develop a flowchart solution to the problem.
b. Trace the flowchart to see if it solves the problem.

1. **Problem:** Enter two numbers. Determine if their difference is positive. (Consider the difference to be the first number minus the second number.)

2. **Problem:** Enter two numbers. Determine if their product is greater than 71.

3. **Problem:** Enter two numbers. Determine if their quotient is less than or equal to one. (Consider the quotient the first number divided by the second number.)

4. **Problem:** Enter three numbers. Determine if the sum of the numbers is greater than or equal to the product of the first two numbers.

Sometimes problems involve more than one decision.

Example 2 a. Develop a flowchart solution to the problem.
b. Trace the flowchart to see if it solves the problem.

Problem: Enter a number. Determine if the number is positive, zero, or negative.

Solution:

a. *Analysis:*
1. A number is positive if it is greater than zero.
2. A number is negative if it is:
   a) less than zero or
   b) not positive and not zero.

Algorithm to solve the problem: Using N for the number, check:
1. Is $N > 0$? If yes, the number is positive.
2. Is $N = 0$? If yes, the number is zero.
3. If the answers to numbers 1 and 2 are both no, then N must be negative. (See Analysis 2b.)

b. *Plan:*
1. Enter a number, N.
2. Is $N > 0$?
   a) If yes, print "positive." Go to 1.
   b) If no, continue.
3. Is $N = 0$?
   a) If yes, print "zero." Go to 1.
   b) If no, continue.
4. Print "negative." Go to 1.

c. *Flowchart*

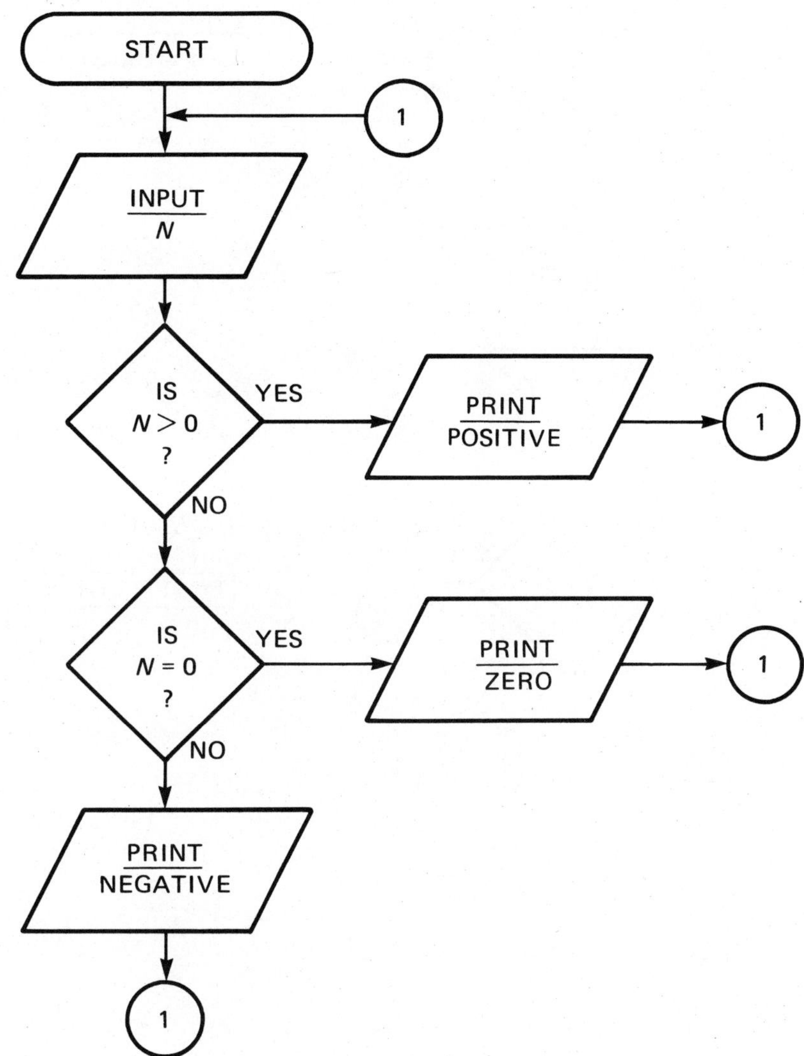

We choose the following test data:

12 should be "positive."
0 should be "zero."
−2 should be "negative."

Now we begin the trace.

		Test #1 (12)	*Test #2* (0)	*Test #3* (−2)
		N	N	N
1.	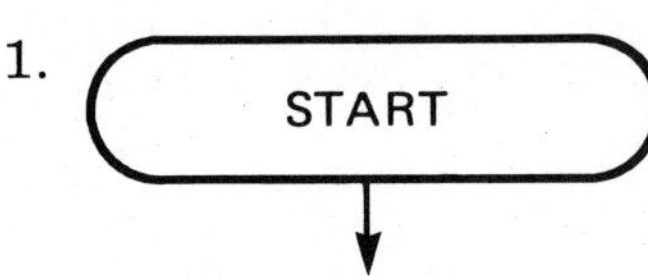			
2.	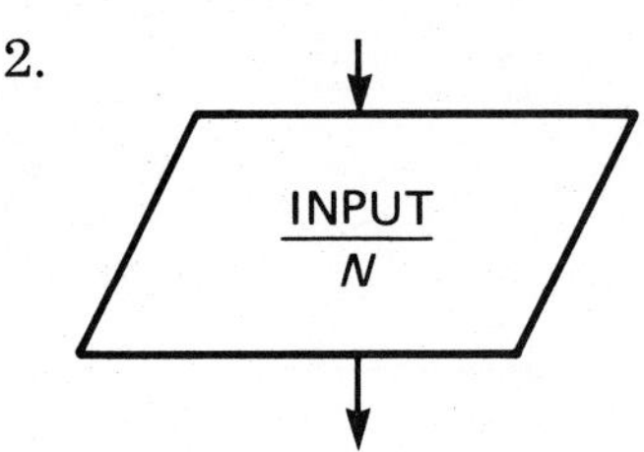	12	0	−2
3.	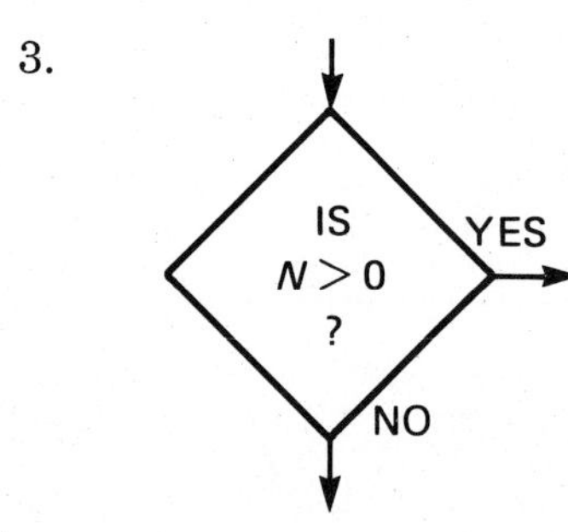	YES 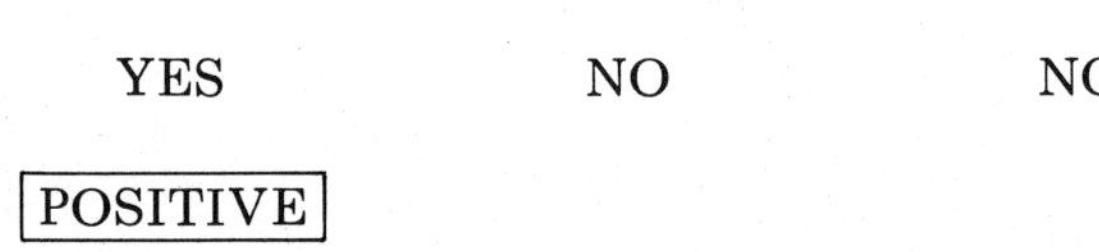 POSITIVE	NO	NO
4.	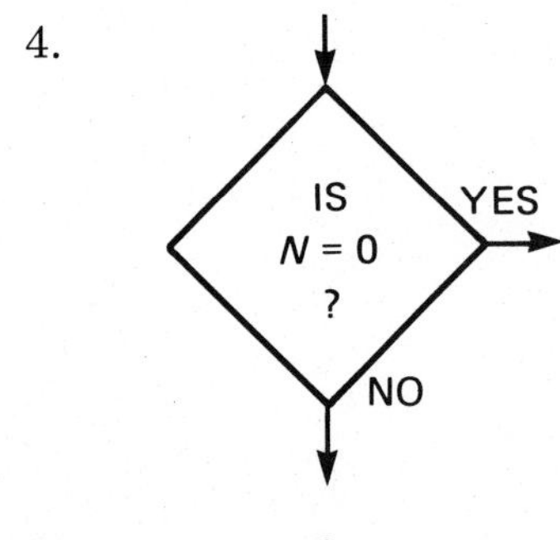		YES 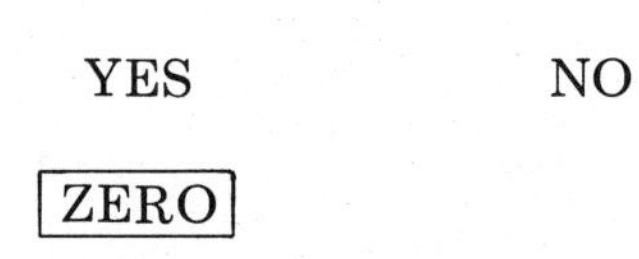 ZERO	NO
5.	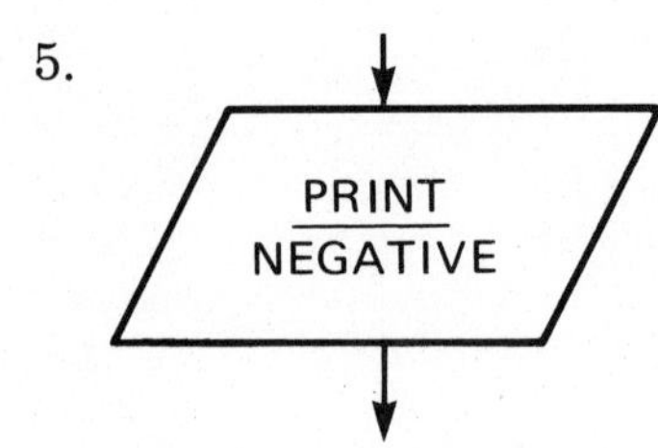			NEGATIVE

Now try the next exercise.

Class Exercise 2

a. Develop a flowchart solution to the problem.
b. Trace the flowchart to see if it solves the problem.

1. **Problem:** Enter a number. Determine if the number is greater than one, equal to one, or less than one.

2. **Problem:** Enter two numbers. Determine if their difference is positive, zero, or negative. (Consider the difference to be the first number minus the second number.)

3. **Problem:** Enter two numbers. Print the larger. If the numbers are equal, print "numbers are equal."

4. **Problem:** Enter a temperature (in Fahrenheit degrees). If the temperature is less than 32°, print "below freezing." If the temperature is from 32° to 48°, print "cold." If the temperature is greater than 48°, print "comfortable."

Exercise 3.2

a. Develop a flowchart solution to the problem.
b. Trace the flowchart to see if it solves the problem.

1. **Problem:** Enter two numbers. Determine if their difference is greater than 25. (Consider the difference to be the first number minus the second number.)

2. **Problem:** Enter two numbers. Determine if their product is less than or equal to 63.

3. **Problem:** Enter two numbers. Determine if their sum is equal to their product.

4. **Problem:** Enter two different numbers. Print the smaller number.

The following problems involve two decisions.

5. **Problem:** Enter a number. Determine if the number is negative, zero, or positive. (First check to see if it is negative.)

6. **Problem:** Enter a number. Determine if the number is zero, negative, or positive. (First check to see if it is zero.)

7. **Problem:** Enter the length of a side of a square. Determine if the perimeter is greater than, less than, or equal to 24.

8. **Problem:** Enter the length and width of a rectangle. Find the perimeter. (Be sure to include checks to determine if the sides have positive lengths.)

Section 3.3 PROGRAMMING THE COMPUTER TO MAKE DECISIONS (IF-THEN, GOTO)

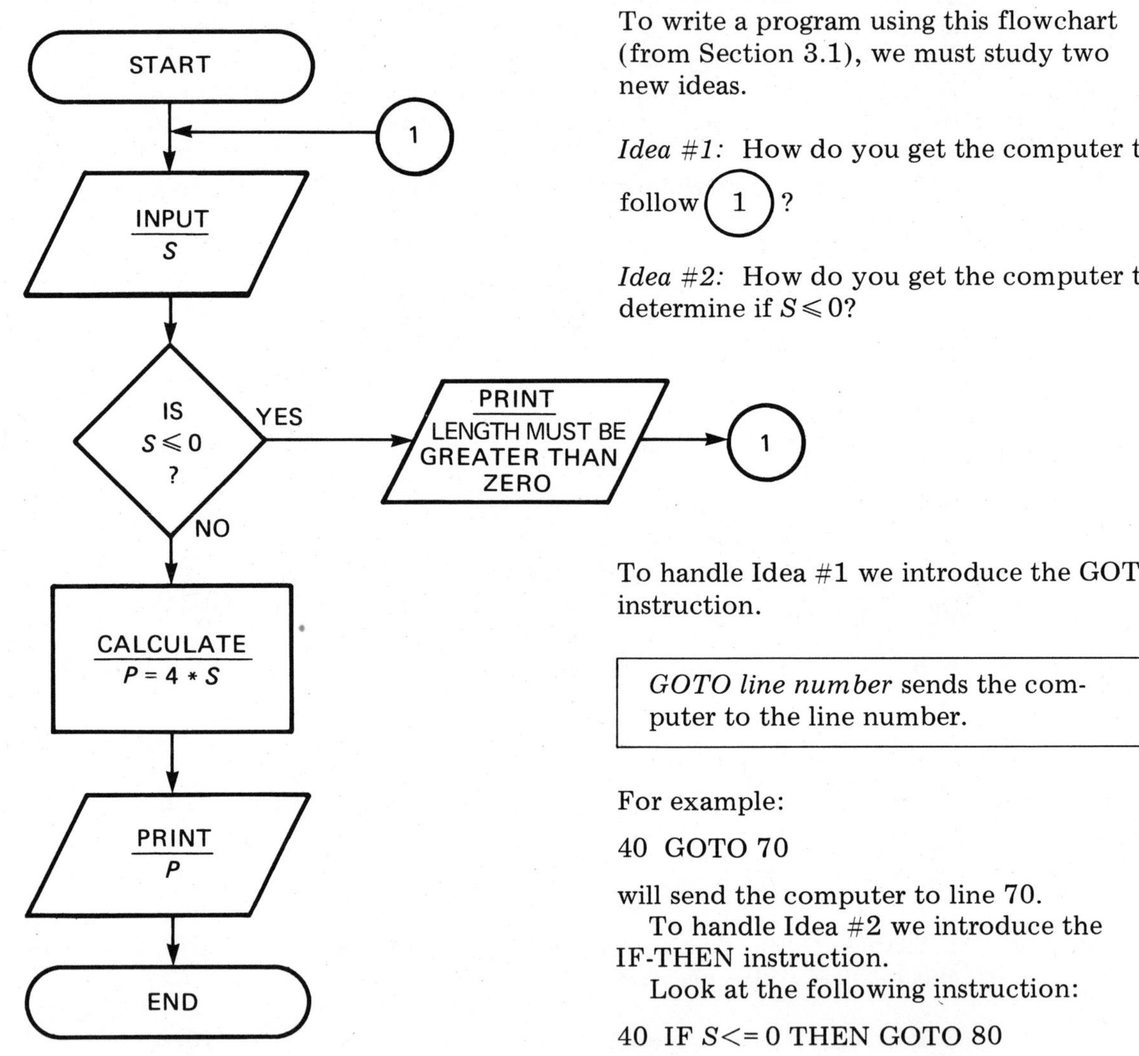

To write a program using this flowchart (from Section 3.1), we must study two new ideas.

Idea #1: How do you get the computer to follow (1)?

Idea #2: How do you get the computer to determine if $S \leqslant 0$?

To handle Idea #1 we introduce the GOTO instruction.

> *GOTO line number* sends the computer to the line number.

For example:

40 GOTO 70

will send the computer to line 70.

To handle Idea #2 we introduce the IF-THEN instruction.

Look at the following instruction:

40 IF *S*<=0 THEN GOTO 80

When *S* is less than or equal to zero, the computer goes to line 80.

Otherwise, the computer moves to the next line.

IF Condition THEN Instruction

1. When the condition is true the computer executes the instruction.
2. When the condition is false the computer executes the next line.

These are valid examples of this new statement:

a. 70 IF $A=0$ THEN GOTO 30

b. 60 IF $L<6*S$ THEN GOTO 140

c. 50 IF $C+B=R*L$ THEN 130

- (pointing to 130) The GOTO is *not* needed after THEN.
- (pointing to =) When an equal sign (=) is used in an IF-THEN statement, it compares the left and right sides. (It is *not* an assignment.)

See how the IF-THEN statement is used below in a program.

Example 1 Show the output when

1. $C=4$
2. $C=0$

Program

```
10  INPUT C
20  IF C=0 THEN GOTO 50
30  PRINT "C IS NOT ZERO"
40  GOTO 60
50  PRINT "C IS ZERO"
60  END
```

Solution:

1.

```
   RUN
■  ? 4
■■ C IS NOT ZERO
```

Discussion of the Planned Output:

■ 4 is placed in memory cell C.

■■ When the computer executes line 20, C is 4. Since $4 \neq 0$, it moves to line 30 and displays the message.

2.

```
   RUN
■  ? 0
■■ C IS ZERO
```

Discussion of the Planned Output:

■ 0 is placed in memory cell C.

■■ When the computer executes line 20, C is 0. Since $0=0$, it goes to line 50 and displays the message.

Class Exercise 1 Determine which are valid IF-THEN statements.

1. 60 IF $A > 7$ THEN GOTO 30
2. 30 IF A; THEN GOTO 80
3. 20 IF $E + F <> G$ THEN GOTO 70
4. 50 IF $R + 4 * C$; THEN 130

For each of the following programs show the output when $C = 2$, $B = 7$ and $D = 10$.

5.
```
10  INPUT C
20  IF C>1 THEN GOTO 50
30  PRINT "C IS NOT GREATER THAN 1"
40  GOTO 60
50  PRINT "C IS GREATER THAN 1"
60  END
```

6.
```
10  INPUT C, B
20  IF C*B=14 THEN 50
30  PRINT "NOT EQUAL"
40  GOTO 60
50  PRINT "PRODUCT IS 14"
60  END
```

7.
```
10  INPUT C, D
20  R=D/C
30  IF R>7 THEN GOTO 60
40  PRINT "VALUE IS"; R
50  GOTO 70
60  PRINT R; "IS GREATER THAN 7"
70  END
```

8.
```
10  INPUT B, C, D
20  IF B↑C=5*D THEN 50
30  PRINT "NOT EQUAL"
40  GOTO 60
50  PRINT "EQUAL"
60  END
```

Now we will code the program for the flowchart at the beginning of this section. We will also use the planned outputs below.

Planned Output: when $S = 6$

```
THE PERIMETER OF A SQUARE
ENTER THE LENGTH OF A SIDE
? 6
THE PERIMETER IS 24
```

Note: Program ends after the perimeter is displayed.

Planned Output: when $S = 0$

```
THE PERIMETER OF A SQUARE
ENTER THE LENGTH OF A SIDE
? 0
SIDE'S LENGTH MUST BE GREATER THAN 0
ENTER THE LENGTH OF A SIDE
?
```

Note: When S is 0, program has user put in a different value for S.

Program

```
   10   PRINT "THE PERIMETER OF A SQUARE"
   20   PRINT "ENTER THE LENGTH OF A SIDE"
   30   INPUT S
■  40   IF S<=0 THEN GOTO 80
   50   P=4*S
   60   PRINT "THE PERIMETER IS"; P
■■ 70   GOTO 100
   80   PRINT "SIDE'S LENGTH MUST BE GREATER THAN 0"
   90   GOTO 20
   100  END
```

Discussion of the Program:
■ The underline indicates that when we are writing line 40, we do not know where to send the computer. (When typing in the program you do not type the underline.)
■■ Again we leave a blank line. At this point we did not know the line number that will end the program.

This program will produce the planned output. Follow the suggestions below when writing your programs.

How to Write a Program from a Flowchart and a Planned Output

---

1. Program all the messages from the planned output that come before the INPUT statement.
2. *Program straight down the flowchart.*
3. Program the YES part of the IF-THEN statement *after* the NO part is programmed.
4. When you write a PRINT statement, look at the planned output to see the message that goes with the PRINT statement.

In our program after the perimeter is displayed, the program ends.

To get the computer to go to the beginning and ask us again to:

```
ENTER THE LENGTH OF A SIDE
?
```

we must edit (change) line 70 to read:

70 GOTO 20

Now after each execution of the program, the computer will automatically return to the beginning and wait for new data.

Try running the program with the new line 70.

If you wish to stop the program, type the following:

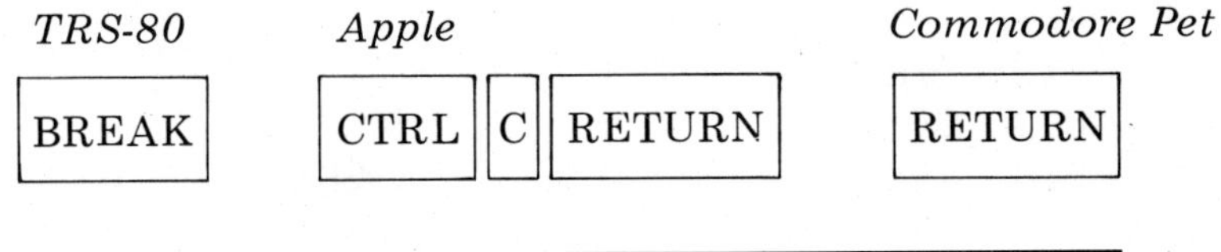

(For other computers, see your manual.)

The following exercise will give you practice in writing programs from flowcharts and planned outputs.

Class Exercise 2 For each problem statement and flowchart:

a. Write two programs, one for each planned output.
b. Run the programs using the test data from the planned outputs.

1. **Problem:** Find the perimeter of a square when we are given the length of a side. (The side's length must be positive.)

Flowchart

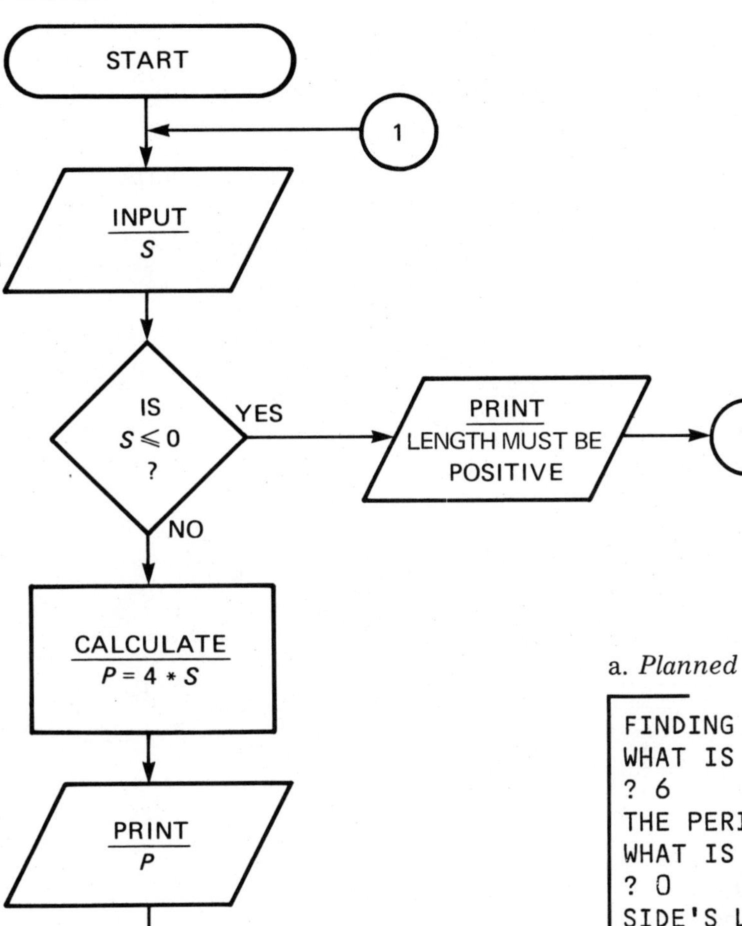

a. *Planned Output #1*

```
FINDING THE PERIMETER OF A SQUARE
WHAT IS THE LENGTH OF THE SIDE?
? 6
THE PERIMETER IS 24
WHAT IS THE LENGTH OF THE SIDE?
? 0
SIDE'S LENGTH MUST BE POSITIVE
WHAT IS THE LENGTH OF THE SIDE?
?
```

b. *Planned Output #2*

```
WHAT IS THE PERIMETER?
ENTER THE LENGTH OF THE SIDE
? 6
THE SIDE'S LENGTH IS 6
THE PERIMETER IS 24
ENTER THE LENGTH OF THE SIDE
? 0
ILLEGAL. ENTRY MUST BE POSITIVE
ENTER THE LENGTH OF THE SIDE
?
```

2. **Problem:** Enter two numbers. Determine if their product is less than 30.

Flowchart

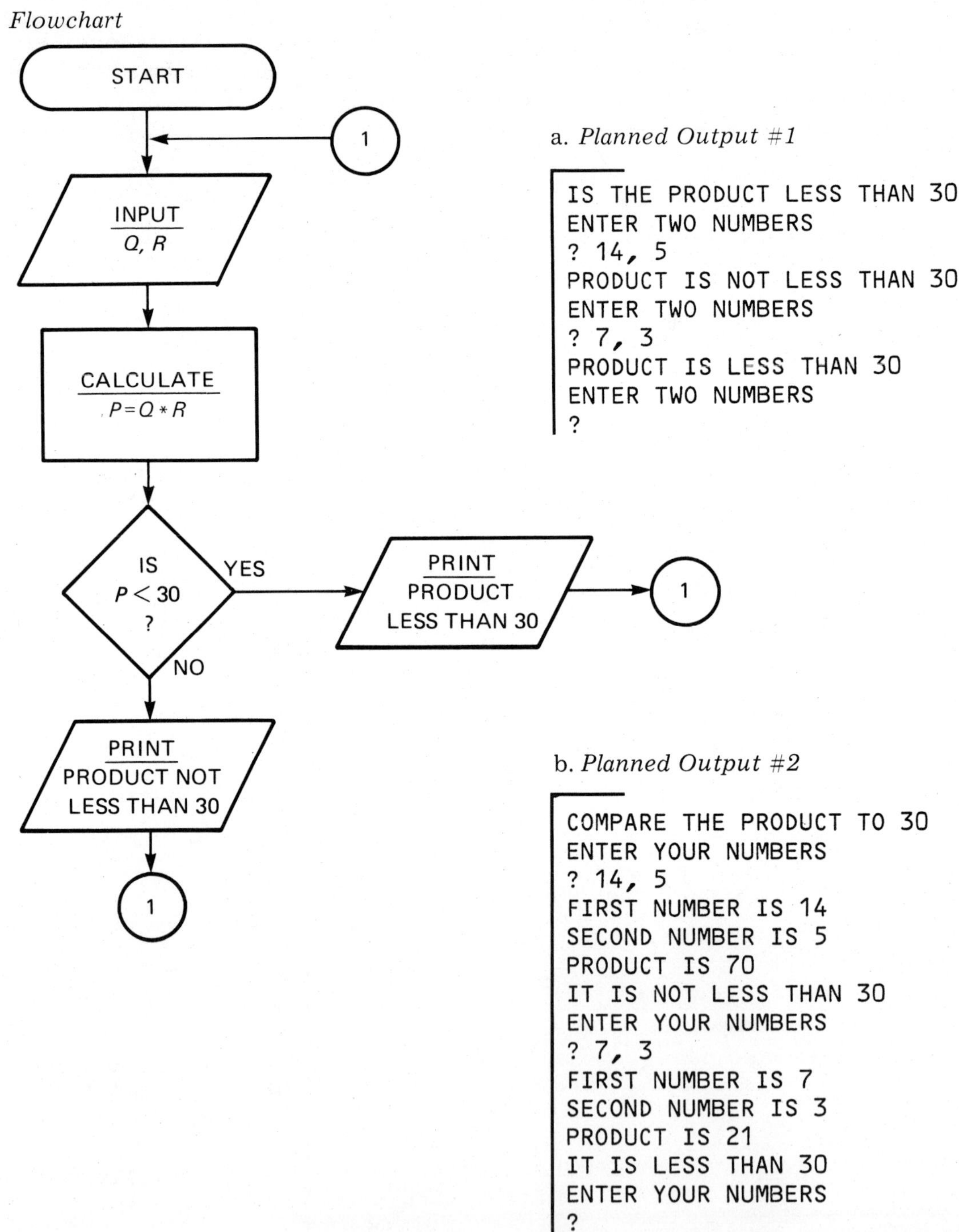

a. *Planned Output #1*

```
IS THE PRODUCT LESS THAN 30?
ENTER TWO NUMBERS
? 14, 5
PRODUCT IS NOT LESS THAN 30
ENTER TWO NUMBERS
? 7, 3
PRODUCT IS LESS THAN 30
ENTER TWO NUMBERS
?
```

b. *Planned Output #2*

```
COMPARE THE PRODUCT TO 30
ENTER YOUR NUMBERS
? 14, 5
FIRST NUMBER IS 14
SECOND NUMBER IS 5
PRODUCT IS 70
IT IS NOT LESS THAN 30
ENTER YOUR NUMBERS
? 7, 3
FIRST NUMBER IS 7
SECOND NUMBER IS 3
PRODUCT IS 21
IT IS LESS THAN 30
ENTER YOUR NUMBERS
?
```

A Little Gallery of Hardware
A Look at the Main Parts of a Computer System

Whether it is a huge supercomputer or a small desktop computer, the essential activities are the same. This gallery first takes a look at four types of computers—super, mainframe, minicomputer, and microcomputer—then presents examples of hardware demonstrating the four parts common to all computer systems: input, processing, output, and storage. One other essential component of a computer system is also shown: people.

Reprinted by permission from H.L. Capron and Brian K. Williams, COMPUTERS AND DATA PROCESSING, Menlo Park, California, The Benjamin/Cummings Publishing Company, Inc., 1982, Gallery 2.

1 "I've got the world on a screen . . .": Examples of both text and graphics appear on the terminal screen of this Ramtek 6214 Colorgraphic Computer. Not all terminals can provide text and graphics or even color. The terminal is one of the most common ways of communicating with the computer—and often for the computer to communicate back.

15 Most data processing organizations maintain a tape (or disk) library. This one belongs to the Information Services Division of TRW, Inc., which provides consumer credit reporting services on a nationwide basis to banks, retail stores, and other consumer lenders.

3. **Problem:** Enter two numbers. Determine if their sum is equal to 41.

Flowchart

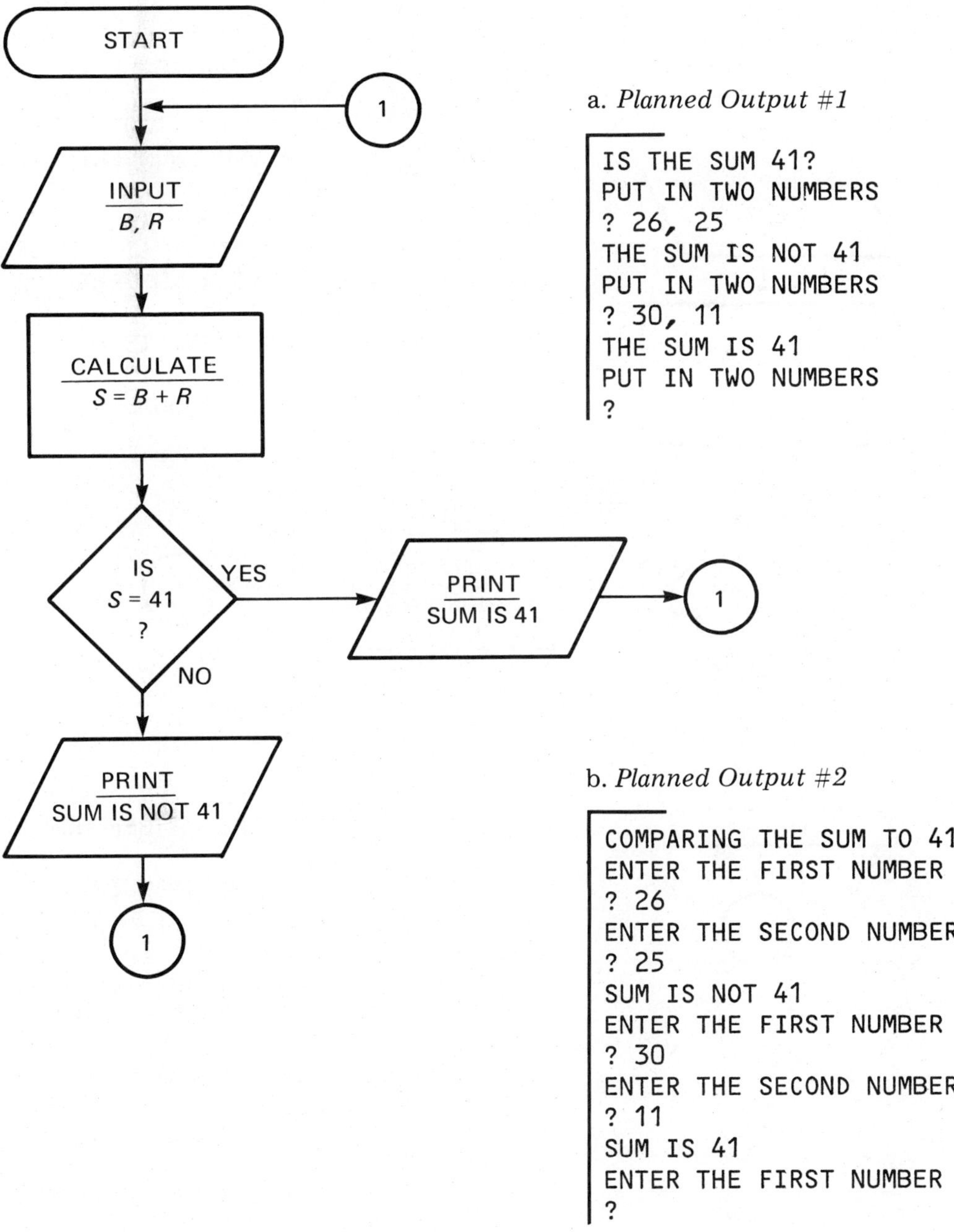

a. *Planned Output #1*

```
IS THE SUM 41?
PUT IN TWO NUMBERS
? 26, 25
THE SUM IS NOT 41
PUT IN TWO NUMBERS
? 30, 11
THE SUM IS 41
PUT IN TWO NUMBERS
?
```

b. *Planned Output #2*

```
COMPARING THE SUM TO 41
ENTER THE FIRST NUMBER
? 26
ENTER THE SECOND NUMBER
? 25
SUM IS NOT 41
ENTER THE FIRST NUMBER
? 30
ENTER THE SECOND NUMBER
? 11
SUM IS 41
ENTER THE FIRST NUMBER
?
```

4. **Problem:** Enter two numbers. Determine if the quotient is greater than 4.

Flowchart

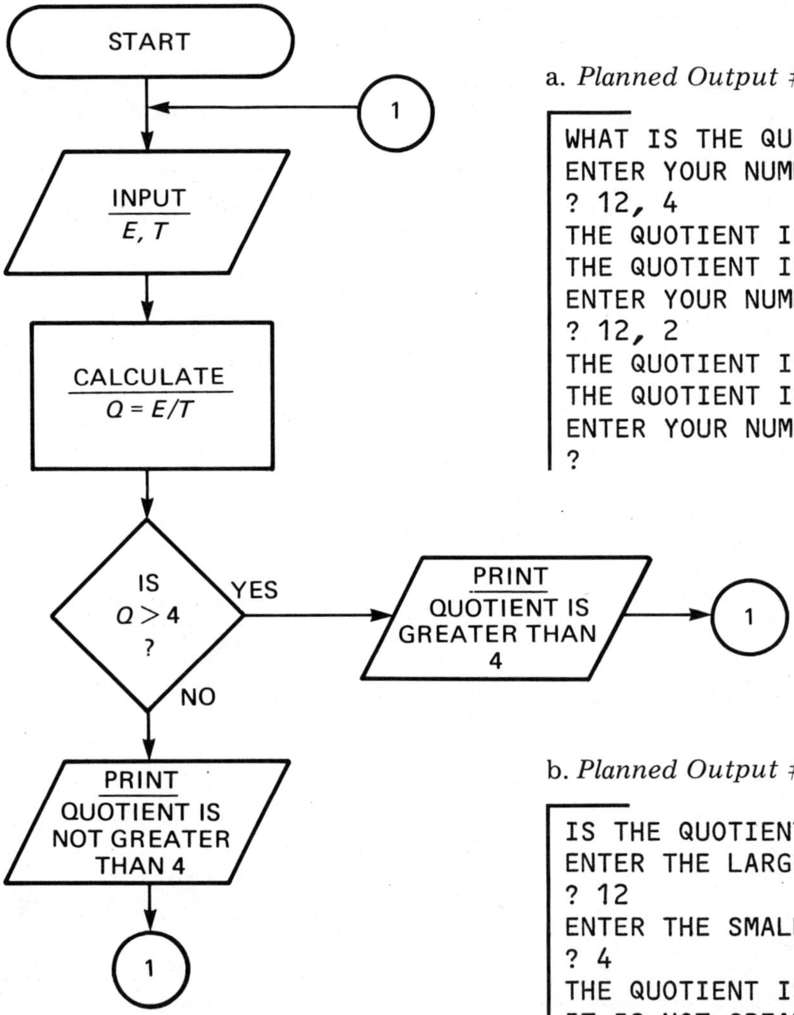

a. *Planned Output #1*

```
WHAT IS THE QUOTIENT?
ENTER YOUR NUMBERS
? 12, 4
THE QUOTIENT IS 3
THE QUOTIENT IS NOT GREATER THAN 4
ENTER YOUR NUMBERS
? 12, 2
THE QUOTIENT IS 6
THE QUOTIENT IS GREATER THAN 4
ENTER YOUR NUMBERS
?
```

b. *Planned Output #2*

```
IS THE QUOTIENT >4?
ENTER THE LARGER NUMBER
? 12
ENTER THE SMALLER NUMBER
? 4
THE QUOTIENT IS 3
IT IS NOT GREATER THAN 4
ENTER THE LARGER NUMBER
? 12
ENTER THE SMALLER NUMBER
? 2
THE QUOTIENT IS 6
IT IS GREATER THAN 4
ENTER THE LARGER NUMBER
?
```

Exercise 3.3

Determine which are valid IF-THEN statements:

1. 50 IF $X <> 6$ THEN GOTO 190
2. 70 IF Y: THEN 80
3. 30 $Y + D = L$ THEN GOTO 120
4. 50 IF $C/2$; THEN GOTO 20

For each of the following programs, show the output when $A = 3$, $B = 5$ and $C = 20$.

5.
```
10  INPUT A, B, C
20  IF A<C/B THEN GOTO 50
30  PRINT "HELLO"
40  GOTO 60
50  PRINT "GOOD-BYE"
60  END
```

6.
```
10  INPUT A, B, C
20  R=A+C/B
30  IF R=A+B THEN 60
40  PRINT "R IS NOT EQUAL TO A+B"
50  GOTO 70
60  PRINT "R IS EQUAL TO A+B"
70  END
```

7.
```
10  INPUT A
20  K=1
30  PRINT K
40  K=K+1
50  IF K>A THEN GOTO 70
60  GOTO 30
70  END
```

8.
```
  10  INPUT C
  20  N=14
■ 30  PRINT N;
  40  N=N+2
  50  IF N>C THEN GOTO 70
  60  GOTO 30
  70  END
```

■ What effect does the semi-colon have on the output? Try the program without the semi-colon and compare the outputs.

For each of the following problem statements and flowcharts:
a. Write two programs, one for each planned output.
b. Run the programs using the test data from the planned outputs.

9. **Problem:** Enter two numbers. Determine if the difference of the numbers is greater than the quotient.

Flowchart

START
1
INPUT
E, L
CALCULATE
D = E − L
CALCULATE
Q = E/L
IS
D > Q
?
YES
NO
PRINT
DIFFERENCE
> QUOTIENT
1
PRINT
DIFFERENCE IS
NOT > QUOTIENT
1

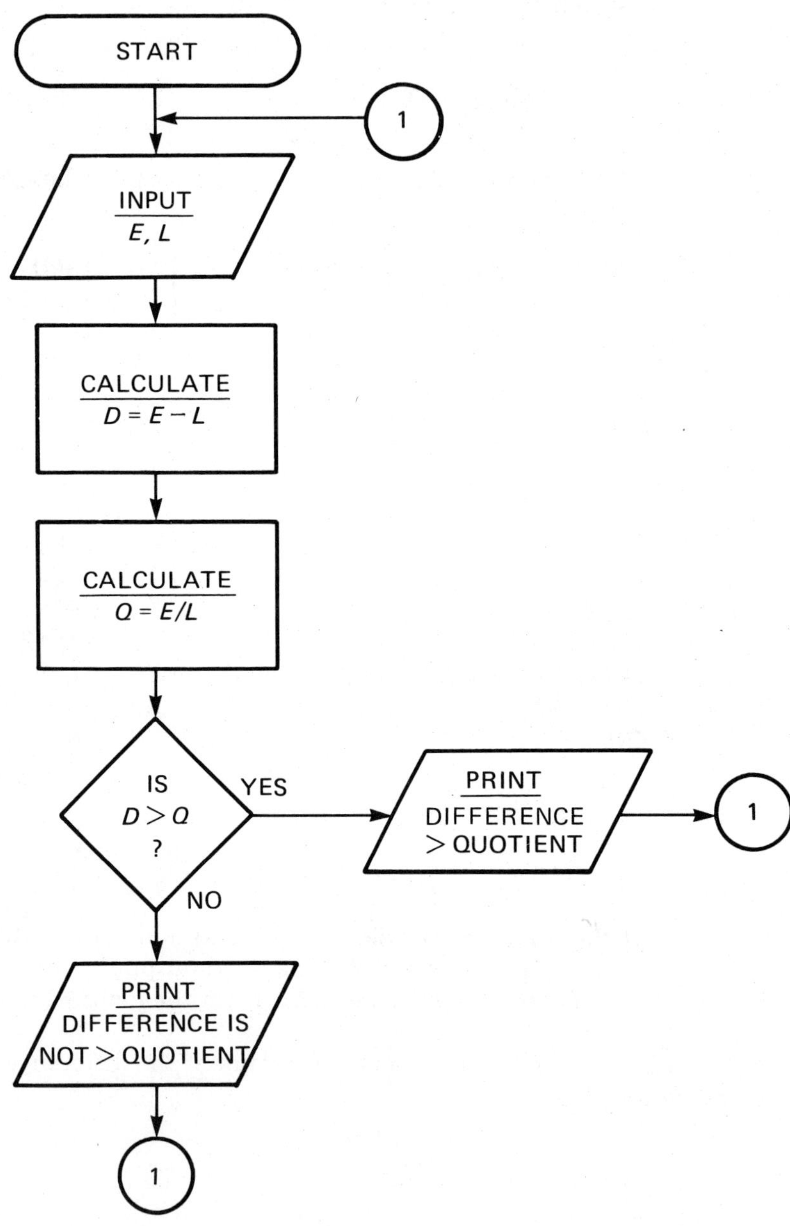

a. *Planned Output #1*

```
COMPARING THE DIFFERENCE AND QUOTIENT
ENTER TWO NUMBERS
? 12, 12
DIFFERENCE IS NOT GREATER THAN QUOTIENT
ENTER TWO NUMBERS
? 12, 3
DIFFERENCE IS GREATER THAN QUOTIENT
ENTER TWO NUMBERS
?
```

b. *Planned Output #2*

```
IS THE DIFFERENCE > QUOTIENT?
PUT IN TWO NUMBERS
? 12, 12
THE FIRST NUMBER IS 12
THE SECOND NUMBER IS 12
THE DIFFERENCE IS 0
THE QUOTIENT IS 1
DIFFERENCE IS NOT GREATER THAN QUOTIENT
PUT IN TWO NUMBERS
? 12, 3
THE FIRST NUMBER IS 12
THE SECOND NUMBER IS 3
THE DIFFERENCE IS 9
THE QUOTIENT IS 4
DIFFERENCE IS GREATER THAN QUOTIENT
PUT IN TWO NUMBERS
?
```

10. **Problem:** Enter three numbers. Determine if the sum of the first two numbers is less than the sum of the second and third.

Flowchart

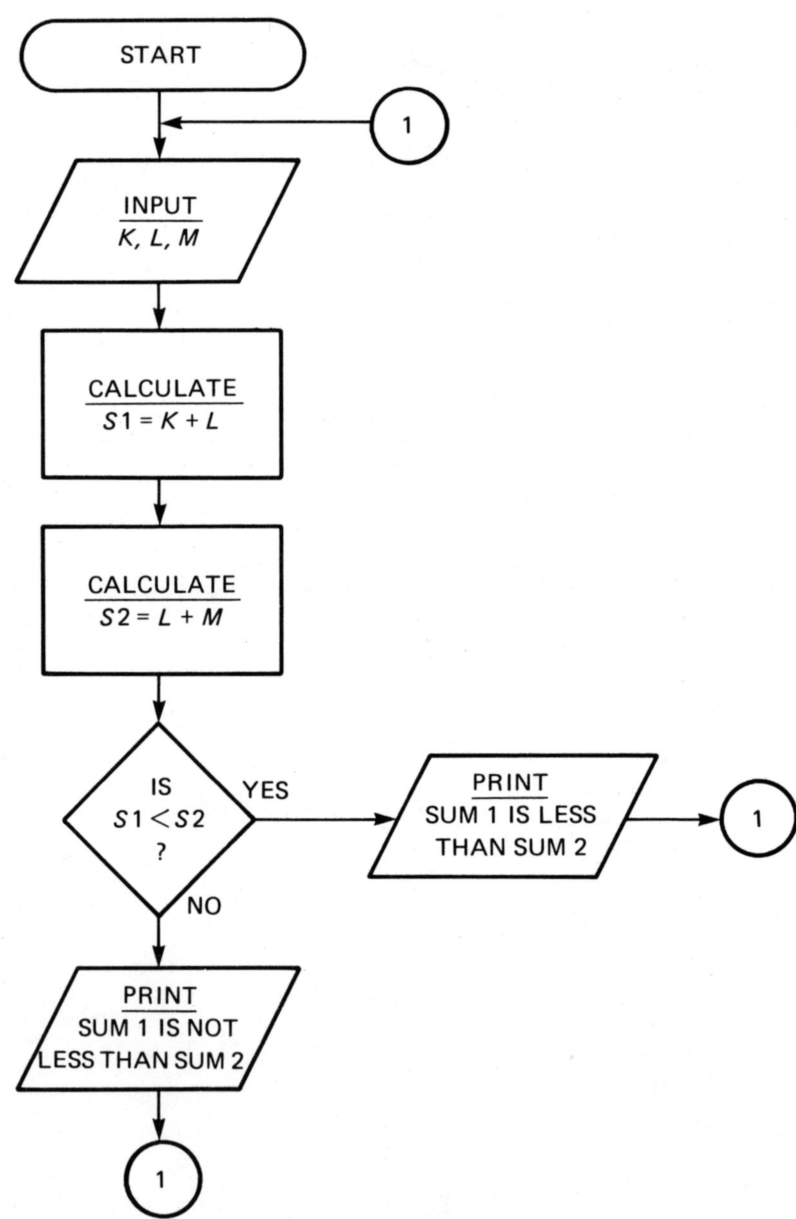

a. *Planned Output #1*

```
IS SUM 1 LESS THAN SUM 2?
PUT IN THREE NUMBERS
? 12, 11, 2
SUM 1 IS NOT LESS THAN SUM 2
PUT IN THREE NUMBERS
? 2, 11, 12
SUM 1 IS LESS THAN SUM 2
PUT IN THREE NUMBERS
?
```

b. *Planned Output #2*

```
COMPARING SUM 1 AND SUM 2
ENTER 3 NUMBERS
? 12, 11, 2
THE NUMBERS ARE 12, 11 AND 2
SUM 1=23
SUM 2=13
SUM 1 IS NOT LESS THAN SUM 2
ENTER 3 NUMBERS
? 2, 11, 12
THE NUMBERS ARE 2, 11 AND 12
SUM 1=13
SUM 2=23
SUM 1 IS LESS THAN SUM 2
ENTER 3 NUMBERS
?
```

11. **Problem:** Enter three numbers. Determine if the product of the first two is equal to the product of the first and third.

Flowchart

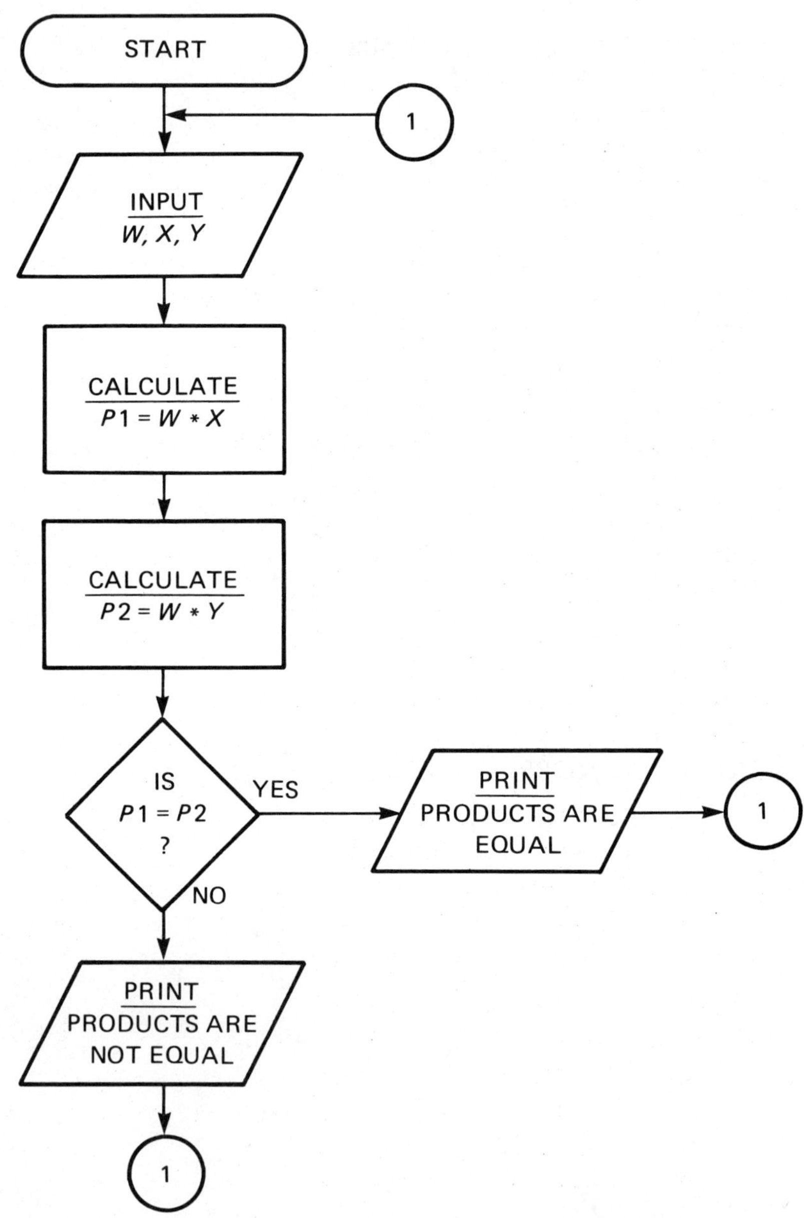

a. *Planned Output #1*

```
FINDING AND COMPARING PRODUCTS
WHAT ARE YOUR 3 NUMBERS?
? 8, 6, 8
FIRST PRODUCT IS 48
SECOND PRODUCT IS 64
PRODUCTS ARE NOT EQUAL
WHAT ARE YOUR 3 NUMBERS?
? 6, 8, 8
FIRST PRODUCT IS 48
SECOND PRODUCT IS 48
PRODUCTS ARE EQUAL
WHAT ARE YOUR 3 NUMBERS
?
```

b. *Planned Output #2*

```
COMPARING PRODUCTS FOR EQUALITY
ENTER 3 NUMBERS
? 8, 6, 8
8, 6 AND 8 ARE THE NUMBERS
PRODUCT 1 IS 48
64 IS PRODUCT 2
PRODUCTS ARE NOT EQUAL
ENTER 3 NUMBERS
? 6, 8, 8
6, 8 AND 8 ARE THE NUMBERS
PRODUCT 1 IS 48
48 IS PRODUCT 2
PRODUCTS ARE EQUAL
ENTER 3 NUMBERS
?
```

12. **Problem:** Enter three numbers. Determine if the first number times the sum of the second and third numbers is equal to the first number times the second number plus the first number times the third number.

Flowchart

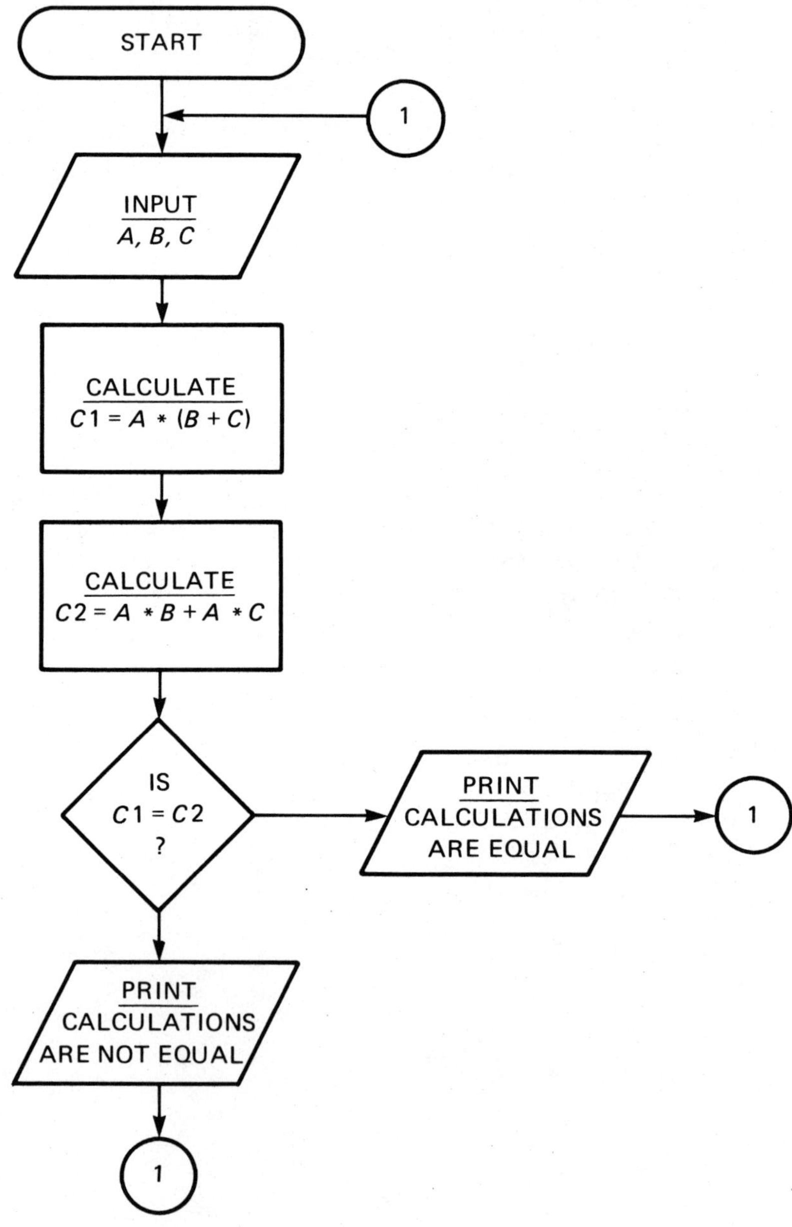

a. *Planned Output #1*

```
TESTING THE DISTRIBUTIVE LAW
ENTER THREE NUMBERS
? 3, 4, 5
CALCULATION 1 IS 27
CALCULATION 2 IS 27
CALCULATIONS ARE EQUAL
ENTER THREE NUMBERS
? 4, 3, 5
CALCULATION 1 IS 32
CALCULATION 2 IS 32
CALCULATIONS ARE EQUAL
ENTER THREE NUMBERS
?
```

(Why are the calculations always equal?)

b. *Planned Output #2*

```
THE DISTRIBUTIVE PROPERTY
ENTER 3 NUMBERS
? 3, 4, 5
C1=A*(B+C)=27
C2=A*B+A*C=27
IS A*(B+C)=A*B+A*C?
SINCE C1=C2, ANSWER IS YES
ENTER 3 NUMBERS
? 4, 3, 5
C1=A*(B+C)=32
C2+A*B+A*C=32
IS A*(B+C)=A*B+A*C?
SINCE C1=C2, ANSWER IS YES
ENTER 3 NUMBERS
?
```

The next set of problems involve *two* decisions.

13. **Problem:** Enter a student's grade. Determine if the grade is greater than, equal to, or less than 65.

Flowchart

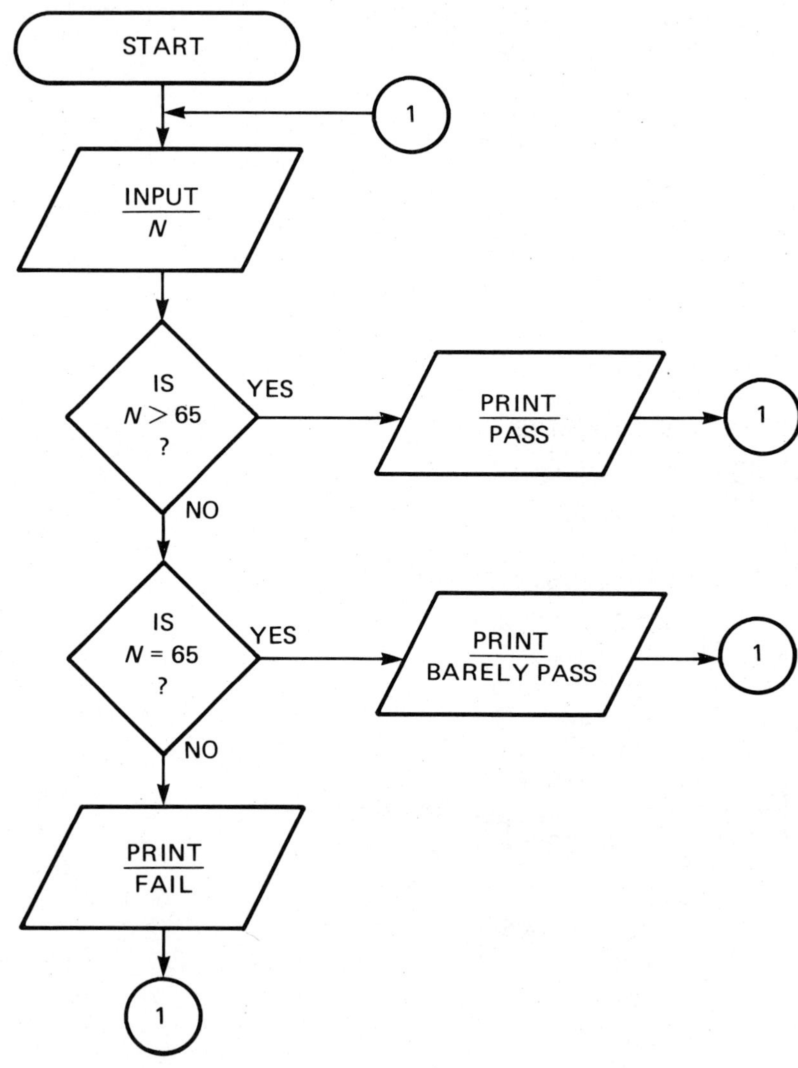

a. *Planned Output #1*

```
HOW DID YOU DO ON THE TEST?
ENTER YOUR TEST GRADE
? 60
FAILED. TOO BAD!
ENTER YOUR TEST GRADE
? 75
YOU PASSED
ENTER YOUR TEST GRADE
? 65
YOU BARELY PASSED
ENTER YOUR TEST GRADE
?
```

b. *Planned Output #2*

```
TELL ME ABOUT YOUR SCORE
PUT IN YOUR TEST GRADE
? 60
60. FAILED. TRY HARDER!
PUT IN YOUR TEST GRADE
? 75
75. YOU PASSED!
PUT IN YOUR TEST GRADE
? 65
65. JUST PASSED!
PUT IN YOUR TEST GRADE
?
```

14. **Problem:** Enter a number. Determine if the number is less than, greater than, or equal to zero. (Negative, positive, or zero.)

Flowchart

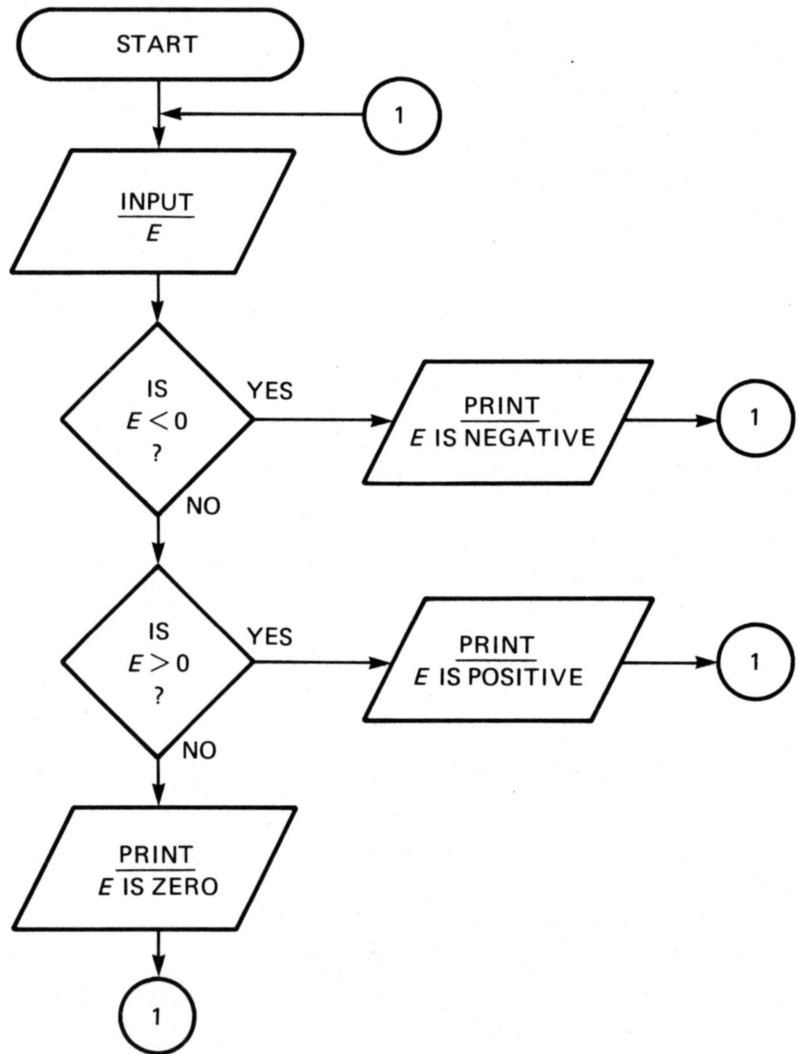

a. *Planned Output #1*

```
COMPARING THE NUMBER TO ZERO
ENTER THE NUMBER
? 0
THE NUMBER IS ZERO
ENTER THE NUMBER
? -4
THE NUMBER IS NEGATIVE
ENTER THE NUMBER
? 6
THE NUMBER IS POSITIVE
ENTER THE NUMBER
?
```

b. *Planned Output #2*

```
IS IT NEGATIVE, POSITIVE OR ZERO?
ENTER THE NUMBER
? 0
IT IS ZERO
ENTER THE NUMBER
? -4
-4 IS NEGATIVE
ENTER THE NUMBER
? 6
6 IS POSITIVE
ENTER THE NUMBER
?
```

15. **Problem:** Enter two numbers. Determine if the sum is greater than, less than, or equal to 15.

Flowchart

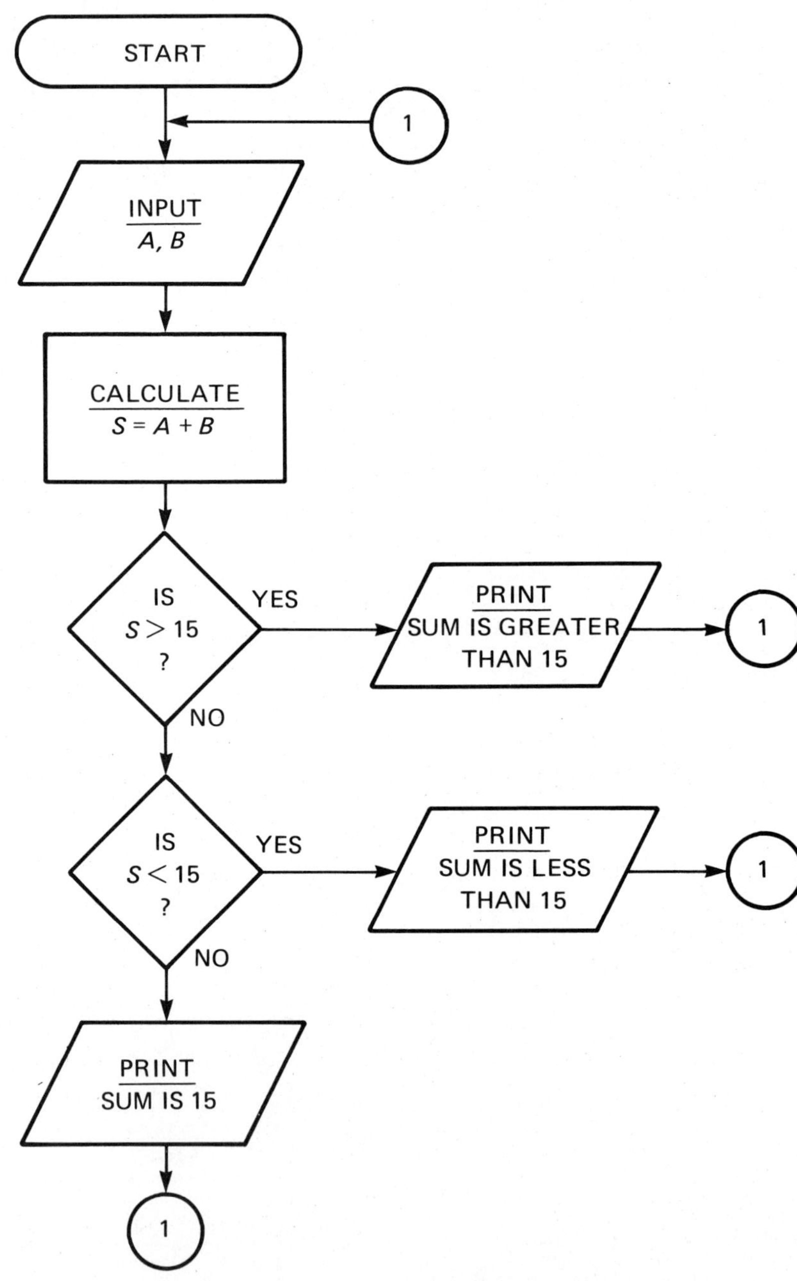

a. *Planned Output #1*

```
COMPARING THE SUM TO 15
ENTER TWO NUMBERS
? 9, 6
SUM IS 15
ENTER TWO NUMBERS
? 8, 9
SUM IS GREATER THAN 15
ENTER TWO NUMBERS
? 4, 6
SUM IS LESS THAN 15
ENTER TWO NUMBERS
?
```

b. *Planned Output #2*

```
IS THE SUM >, <, OR = 15?
ENTER THE FIRST NUMBER
? 9
ENTER THE SECOND NUMBER
? 6
YOUR SUM IS 15
ENTER THE FIRST NUMBER
? 8
ENTER THE SECOND NUMBER
? 9
YOUR SUM IS GREATER THAN 15
ENTER THE FIRST NUMBER
? 4
ENTER THE SECOND NUMBER
? 6
YOUR SUM IS LESS THAN 15
ENTER THE FIRST NUMBER
?
```

Flowchart

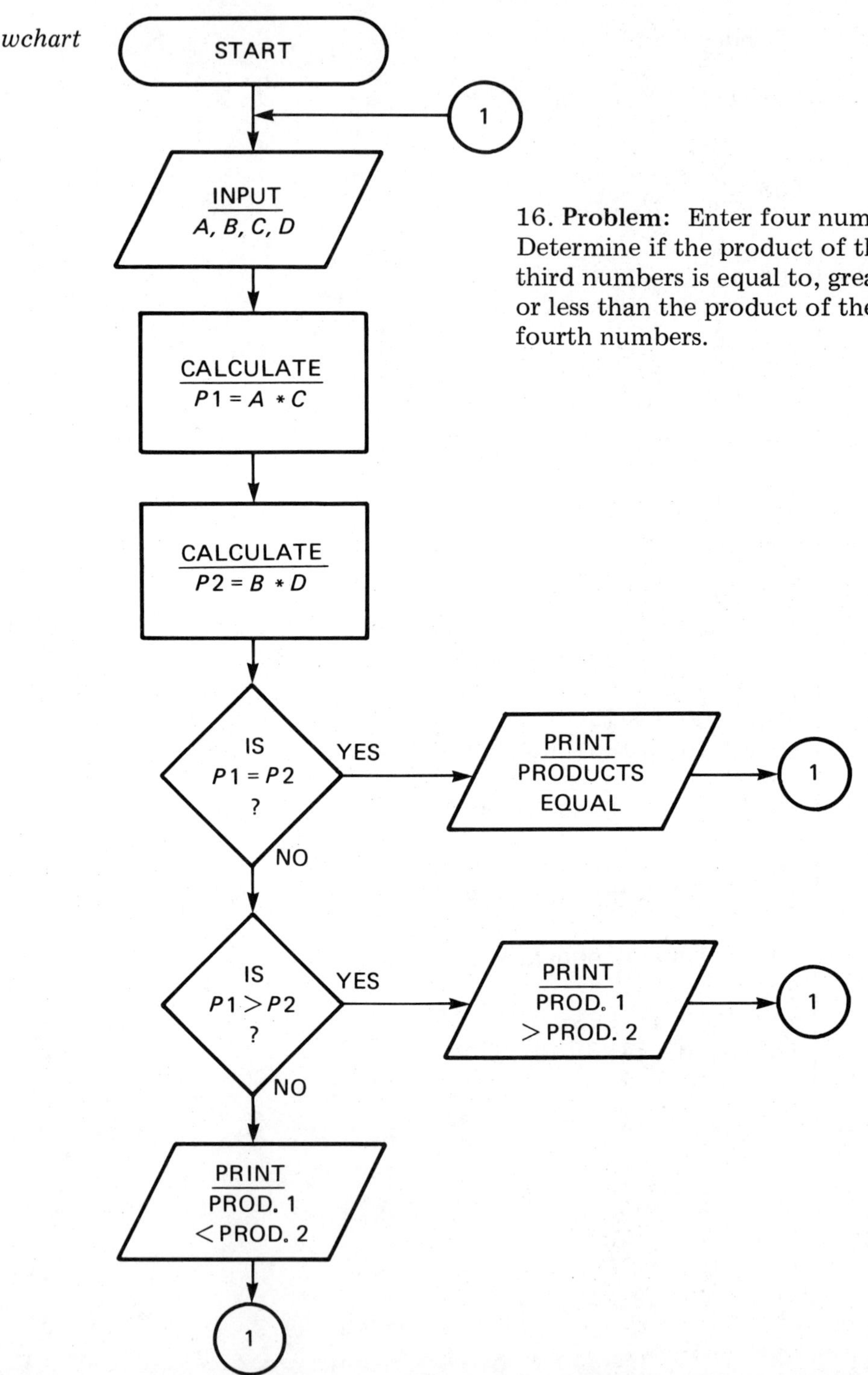

16. **Problem:** Enter four numbers. Determine if the product of the first and third numbers is equal to, greater than, or less than the product of the second and fourth numbers.

a. *Planned Output #1*

```
COMPARING PRODUCTS
ENTER FOUR NUMBERS
? 4, 5, 6, 7
PRODUCT 1 IS 24
PRODUCT 2 IS 35
PRODUCT 1 IS LESS THAN PRODUCT 2
ENTER FOUR NUMBERS
? 4, 8, 6, 3
PRODUCT 1 IS 24
PRODUCT 2 IS 24
PRODUCT 1 IS EQUAL TO PRODUCT 2
ENTER FOUR NUMBERS
? 5, 4, 6, 7
PRODUCT 1 IS 30
PRODUCT 2 IS 28
PRODUCT 1 IS GREATER THAN PRODUCT 2
ENTER FOUR NUMBERS
?
```

b. *Planned Output #2*

```
HOW DO THE PRODUCTS COMPARE?
PUT IN FOUR NUMBERS
? 4, 5, 6, 7
PRODUCT 1=24; PRODUCT 2=35
PROD. 1 < PROD. 2
PUT IN FOUR NUMBERS
? 4, 8, 6, 3
PRODUCT 1=24; PRODUCT 2=24
PROD. 1 = PROD. 2
PUT IN FOUR NUMBERS
? 5, 4, 6, 7
PRODUCT 1=30; PRODUCT 2=28
PROD. 1 > PROD. 2
PUT IN FOUR NUMBERS
?
```

Section 3.4 DEVELOPING COMPLETE COMPUTER SOLUTIONS TO PROBLEMS

We are now ready to develop a complete computer solution to a problem. That is, for the problem:

a. Write an analysis which includes the development of an algorithm.

b. Write a plan showing how you *use* the algorithm to solve the problem.

c. Develop a flowchart.

d. Design a planned output.

e. Code a program and RUN it on the computer.

Example 1 Develop a complete computer solution to the problem.

Problem Find the average temperature for the weekend (Saturday and Sunday). If the average temperature is greater than 65°, print "nice weather!" Otherwise, print "weather could be better." (Temperature is measured in Fahrenheit degrees.)

Solution:

a. *Analysis:* We must first find the average temperature. To find the average of temperature 1 (Saturday) and temperature 2 (Sunday), we calculate:

$AV = (T1 + T2)/2.$

$T1$, $T2$ — Temperatures for Saturday and Sunday.

AV — Average (mean).

b. *Plan:*

1. Put in values for $T1$ and $T2$.
2. Calculate: $AV = (T1 + T2) / 2$.
3. Check: Is $AV > 65$?
   a) If yes, print "nice weather!" Go to end.
   b) If no, continue.
4. Print "weather could be better."
5. End.

c. *Flowchart*

START
INPUT
T1, T2
AV = (T1 + T2)/2
IS
AV > 65
?
YES
NO
PRINT
NICE WEATHER!
1
PRINT
WEATHER COULD
BE BETTER
1
END

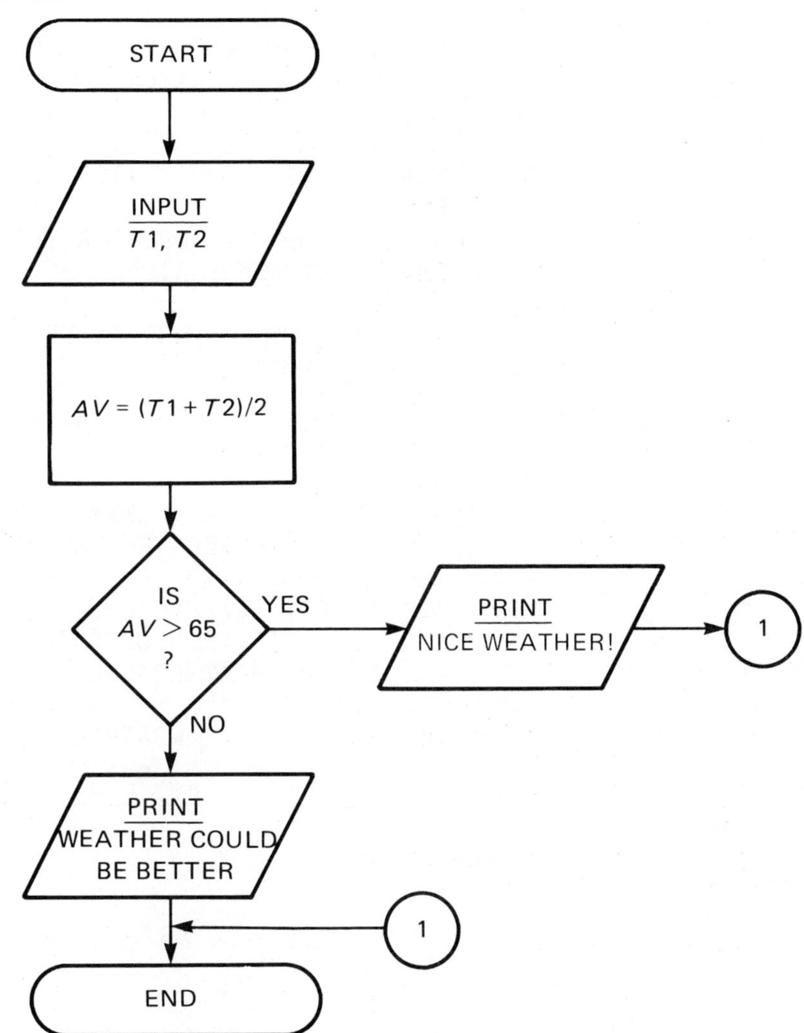

d. *Planned Output*

```
AVERAGE TEMPERATURE FOR THE WEEKEND
ENTER TWO TEMPERATURES
? 50, 60
AVERAGE TEMPERATURE IS 55
WEATHER COULD BE BETTER
RUN
AVERAGE TEMPERATURE FOR THE WEEKEND
ENTER TWO TEMPERATURES
? 70, 80
AVERAGE TEMPERATURE IS 75
NICE WEATHER!
```

e. *The Program*

```
10  PRINT "AVERAGE TEMPERATURE FOR THE WEEKEND"
20  PRINT "ENTER TWO TEMPERATURES"
30  INPUT T1, T2
40  AV=(T1+T2)/2
50  IF AV>65 THEN GOTO 80
60  PRINT "WEATHER COULD BE BETTER"
70  GOTO 90
80  PRINT "NICE WEATHER!"
90  END
```

Now try the next exercise.

Exercise 3.4 Revise the program that was written in this section to produce the planned outputs below.

1.

```
AVERAGING WEEKEND TEMPERATURES
PUT IN TWO TEMPERATURES
? 50, 60
SATURDAY'S TEMPERATURE WAS 50
SUNDAY'S TEMPERATURE WAS 60
55 WAS THE AVERAGE
WEATHER COULD BE BETTER
RUN
AVERAGING WEEKEND TEMPERATURES
PUT IN TWO TEMPERATURES
? 70, 80
SATURDAY'S TEMPERATURE WAS 70
SUNDAY'S TEMPERATURE WAS 80
75 WAS THE AVERAGE
NICE WEATHER!
```

2.

```
  AVERAGING TEMPERATURES
  WHAT WERE THE TEMPERATURES?
  ? 50, 60
  SATURDAY'S TEMP. = 50
  SUNDAY'S TEMP. = 60
  AVERAGE WAS 55
  WEATHER COULD BE BETTER
■ WHAT WERE THE TEMPERATURES?
  ? 70, 80
  SATURDAY'S TEMP. = 70
  SUNDAY'S TEMP. = 80
  AVERAGE WAS 75
  NICE WEATHER!
■ WHAT WERE THE TEMPERATURES?
  ?
```

Discussion of the Planned Output:
■ Note that RUN was not used to start the program again.

For each of the following problems:
a. Write an analysis which includes the development of an algorithm.
b. Write a plan showing how you *use* the algorithm to solve the problem.
c. Develop a flowchart.
d. Design a planned output.
e. Code a program and RUN it on the computer.

3. **Problem:** Find the average temperature for the weekend. If the average temperature is less than 75°, print "not hot." Otherwise, print "hot." (Temperature is measured in Fahrenheit degrees.)

4. **Problem:** Enter three test grades. Find the average. If the average is greater than or equal to 65, print "pass." Otherwise, print "fail."

5. **Problem:** Enter four numbers. Determine if the product of the first two numbers is less than the product of the third and fourth numbers.

6. **Problem:** Lara and Jeremy do babysitting on Friday and Saturday evenings.

a. Enter the amount that Lara earns each night ($L1$ and $L2$).
b. Enter the amount that Jeremy earns each night ($J1$ and $J2$).
c. Determine who earns more money. (*Hint:* They might earn the same amount.)

Section 3.5 DEVELOPING A PROGRAM FOR COMPUTER ASSISTED INSTRUCTION IN MULTIPLICATION

So far we have used the computer as a tool for solving problems. Computers are also used to assist teachers and students. This is called *C*omputer *A*ssisted *I*nstruction or CAI.

In this section we will develop a program that will give the user of the program (*user* for short) practice in multiplying two numbers.

Problem

Develop a program where the user:
a. Enters two numbers to be multiplied (A, B).
b. Enters the product, user's product (UP). The computer then tells the user if the answer is correct.

Solution: Read the planned output. It will help us to understand the problem better.

Planned Output

```
TO PRACTICE MULTIPLICATION
ENTER TWO NUMBERS
? 4, 3
ENTER YOUR PRODUCT
? 11
SORRY. THE ANSWER IS NOT CORRECT.
TRY AGAIN.
ENTER YOUR PRODUCT
? 12
CORRECT. VERY GOOD!
NOW TRY ANOTHER EXAMPLE.
ENTER TWO NUMBERS
?
```

We now analyze the problem.

a. *Analysis:* How can the computer determine if the user's product is correct?
  1. We must get the computer to calculate the actual product (AP):
     $AP = A * B$
  2. Then the computer must compare the user's product (UP) to the actual product (AP).

b. *Plan:*
  1. Enter two numbers, A, B.
  2. Calculate the product: $AP = A * B$.
  3. Enter the user's product, UP.
  4. Compare the actual product to the user's product.
     a) If UP equals AP, print "correct." Then, go to 1.
     b) If UP is not equal to AP, print "not correct." Then go to 3.
  5. End.

c. *Flowchart*

START
2
INPUT
A, B
CALCULATE
AP = A * B
1
INPUT
UP
IS
AP = UP
?
YES
PRINT
CORRECT
2
NO
PRINT
TRY AGAIN
1

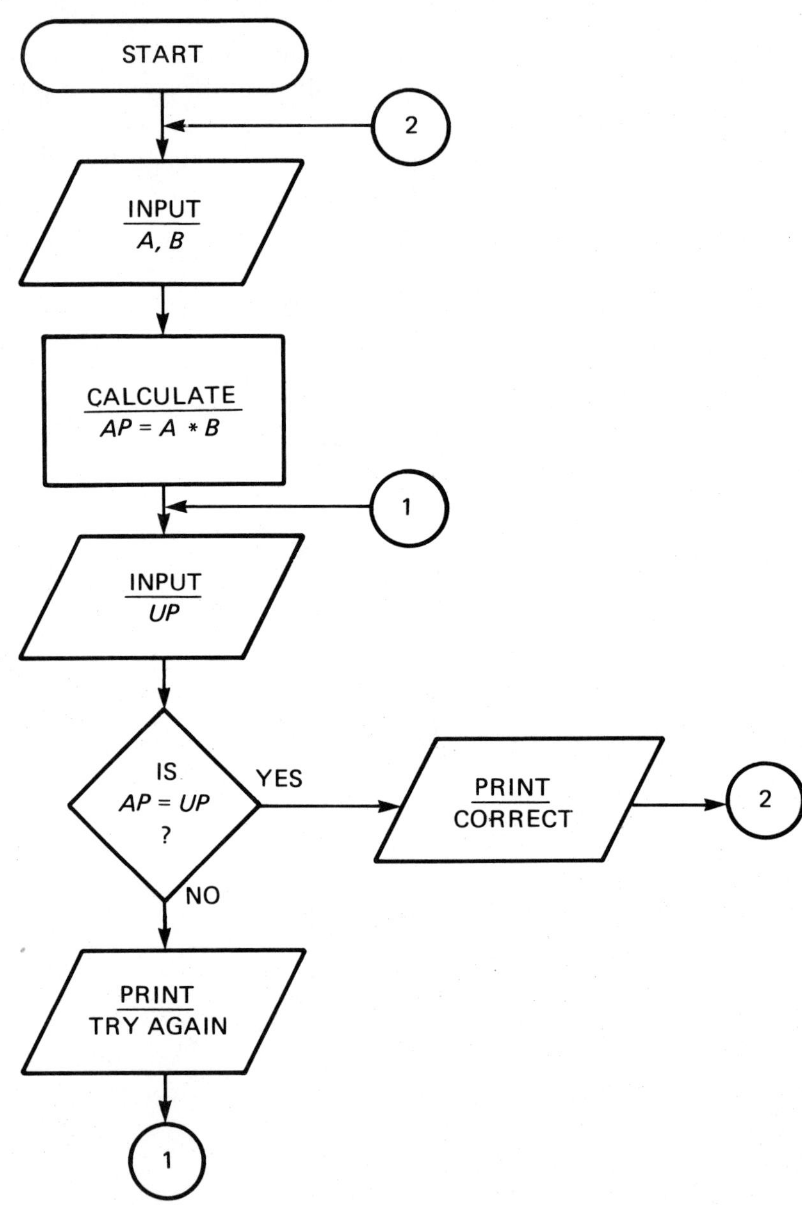

Using the flowchart and planned output shown previously, we code the program.

d. *The Program*

```
10   PRINT "TO PRACTICE MULTIPLICATION"
20   PRINT "ENTER TWO NUMBERS"
30   INPUT A, B
40   AP=A*B
50   PRINT "ENTER YOUR PRODUCT"
60   INPUT UP
70   IF AP=UP THEN GOTO 110
80   PRINT "SORRY. THE ANSWER IS NOT CORRECT."
90   PRINT "TRY AGAIN."
100  GOTO 50
110  PRINT "CORRECT. VERY GOOD!"
120  PRINT "NOW TRY ANOTHER EXAMPLE."
130  GOTO 20
140  END
```

Exercise 3.5 Revise the program that was written in this section to produce each of the planned outputs below.

1. *Planned Output*

```
MULTIPLICATION PRACTICE
PUT IN TWO NUMBERS
? 4, 3
WHAT IS YOUR PRODUCT?
? 11
TRY AGAIN.
WHAT IS YOUR PRODUCT?
? 12
THAT IS CORRECT.
TRY ANOTHER EXAMPLE.
PUT IN TWO NUMBERS
?
```

2. *Planned Output*

```
PRACTICING MULTIPLICATION
ENTER TWO NUMBERS
? 4, 3
ENTER YOUR PRODUCT FOR 4X 3
? 11
NOT CORRECT. TRY AGAIN.
ENTER YOUR PRODUCT FOR 4X 3
? 12
VERY GOOD. TRY ANOTHER ONE.
ENTER TWO NUMBERS
?
```

For the next problem, flowchart and planned output, code a program to solve the problem.

3. **Problem:** Enter two numbers *C*, *D*. Then enter a third number, *E*. Determine if the third number is less than the sum of the first two numbers.

Flowchart

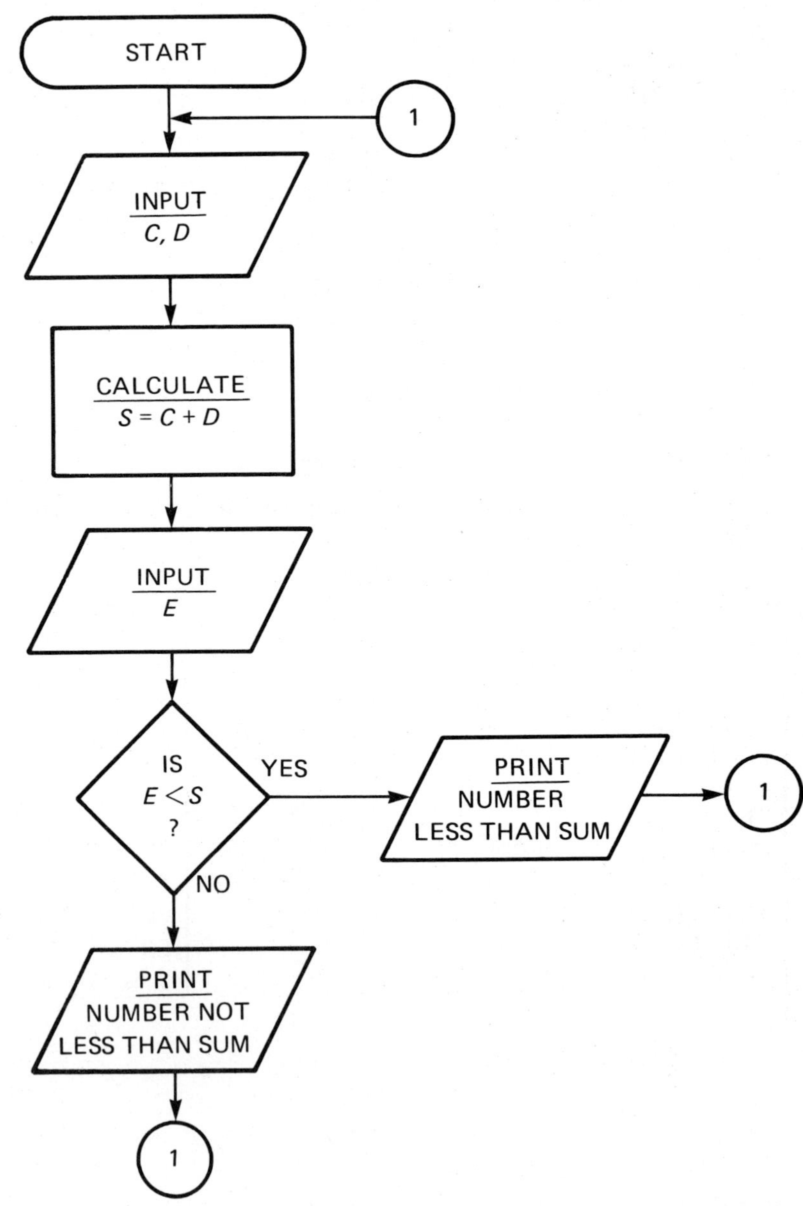

Planned Output

```
COMPARING A NUMBER TO A SUM
ENTER TWO NUMBERS
? 4, 5
ENTER ANOTHER NUMBER
? 12
THE NUMBER 12 IS NOT LESS THAN
THE SUM 9
ENTER TWO NUMBERS
? 8, 10
ENTER ANOTHER NUMBER
? 7
THE NUMBER 7 IS LESS THAN
THE SUM 18
ENTER TWO NUMBERS
?
```

For each of the following problems:
a. Write an analysis which includes the development of an algorithm.
b. Write a plan showing how you use the algorithm to solve the problem.
c. Develop a flowchart.
d. Design a planned output.
e. Code a program and RUN it on the computer.

4. **Problem:** Write a CAI program to give the user practice in adding two numbers. Have the user:

a. Enter two numbers (Q and T) to be added.
b. Enter the sum, user's sum (US). Then have the computer tell the user if the answer is correct.

5. **Problem:** Write a CAI program to give the user practice in subtracting two numbers. In this program consider the first number entered to be the larger number.

6. **Problem:** Write a CAI program to give the user practice in dividing the first number entered by the second number. In this program consider the first number entered to be the larger number. Also, remember that we cannot divide by zero!

7. **Problem:** Write a CAI program to give the user practice in multiplication and addition. Have the user first multiply the numbers. After getting the correct answer, the user is to add the numbers.

END OF CHAPTER EXERCISES

Section 3.1

Show whether the following statements are true or false. (The relation and arithmetic symbols are written in BASIC.)

1. 2/4<4/2
2. 4/4>4−4
3. 5 * 4−5 * 2=5 * (4−2)
4. 16−8 * 2<(16−8) * 2
5. 3+4↑2<>4↑2+3
6. 7 * 9+4>=(9+4) * 7

Write each using flowchart symbols:

7. $F = X/A$
8. INPUT Q, R
9. IS $Y <= 5$?
10. GOTO 20
11. PRINT J
12. $A = B \uparrow E$

In Problems 13 through 26 which follow, determine if the flowchart solves the problem. Use the sets of test data along with the procedure outlined for tracing a flowchart.

13. **Problem:** Enter two numbers. Determine if the first number minus the second number is less than or equal to 10.

Test Data: a. 25, 8 b. 30, 23

Flowchart

START
1
INPUT
B, C
CALCULATE
D = B − C
IS
D <= 10
?
YES
NO
PRINT
DIFFERENCE IS
LESS THAN OR
EQUAL TO 10
1
PRINT
DIFFERENCE IS
GREATER
THAN 10
1

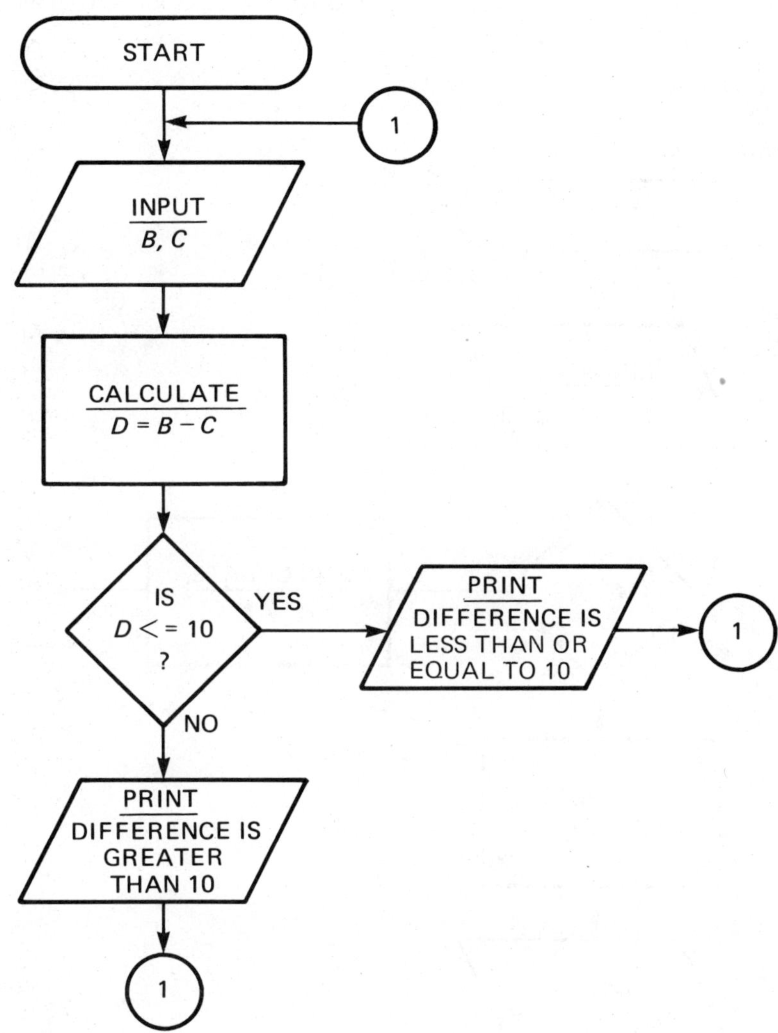

14. **Problem:** Enter two numbers. If the first number is greater than the second number, double the first number and print this new number. Otherwise, add five to the second number and print this new number.

Test Data: a. 4, 8 b. 8, 4

Flowchart

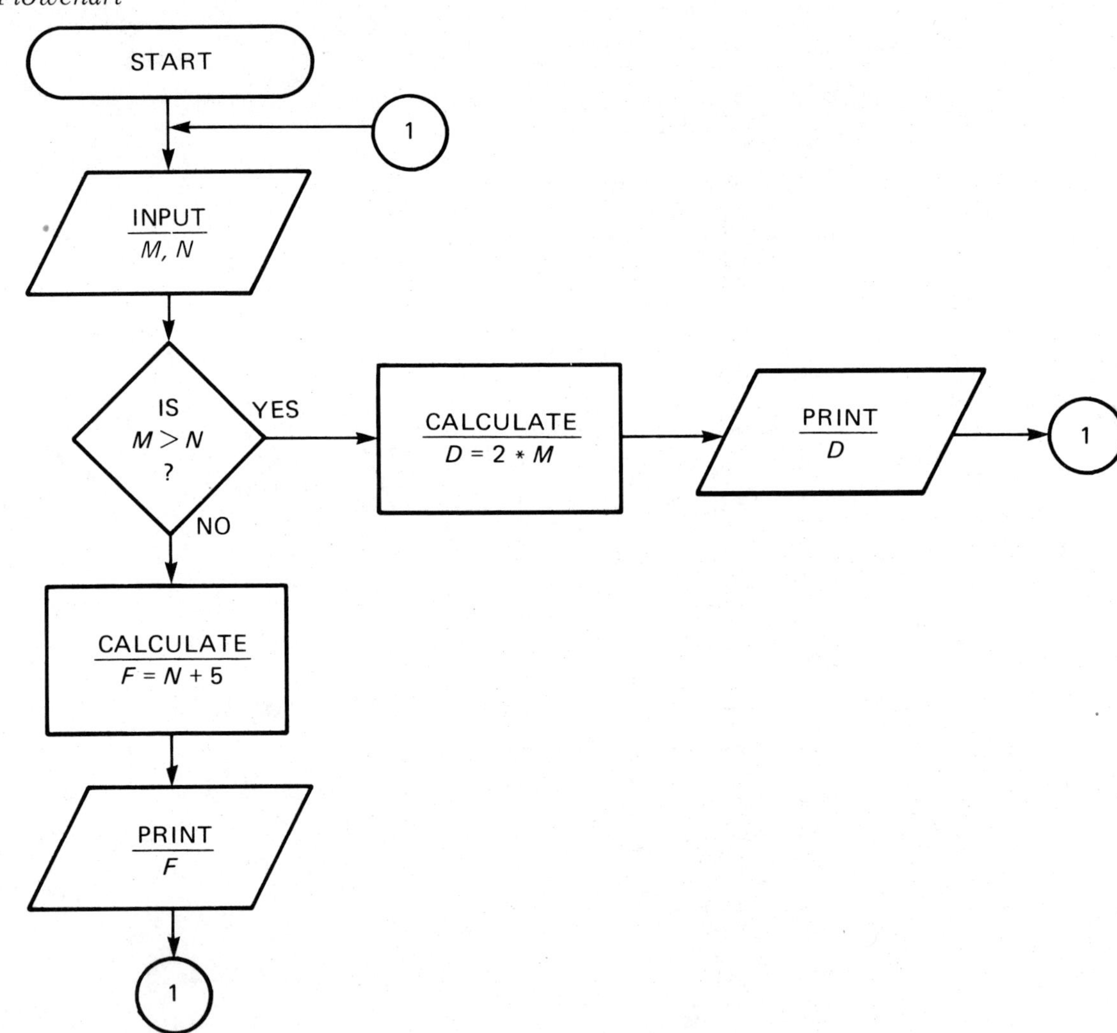

15. **Problem:** Enter three numbers. If the product of the first two is less than the third number, print the numbers in reverse order. Otherwise, print the sum of the numbers.

Test Data: a. 4, 2, 5 b. 3, 4, 13

Flowchart

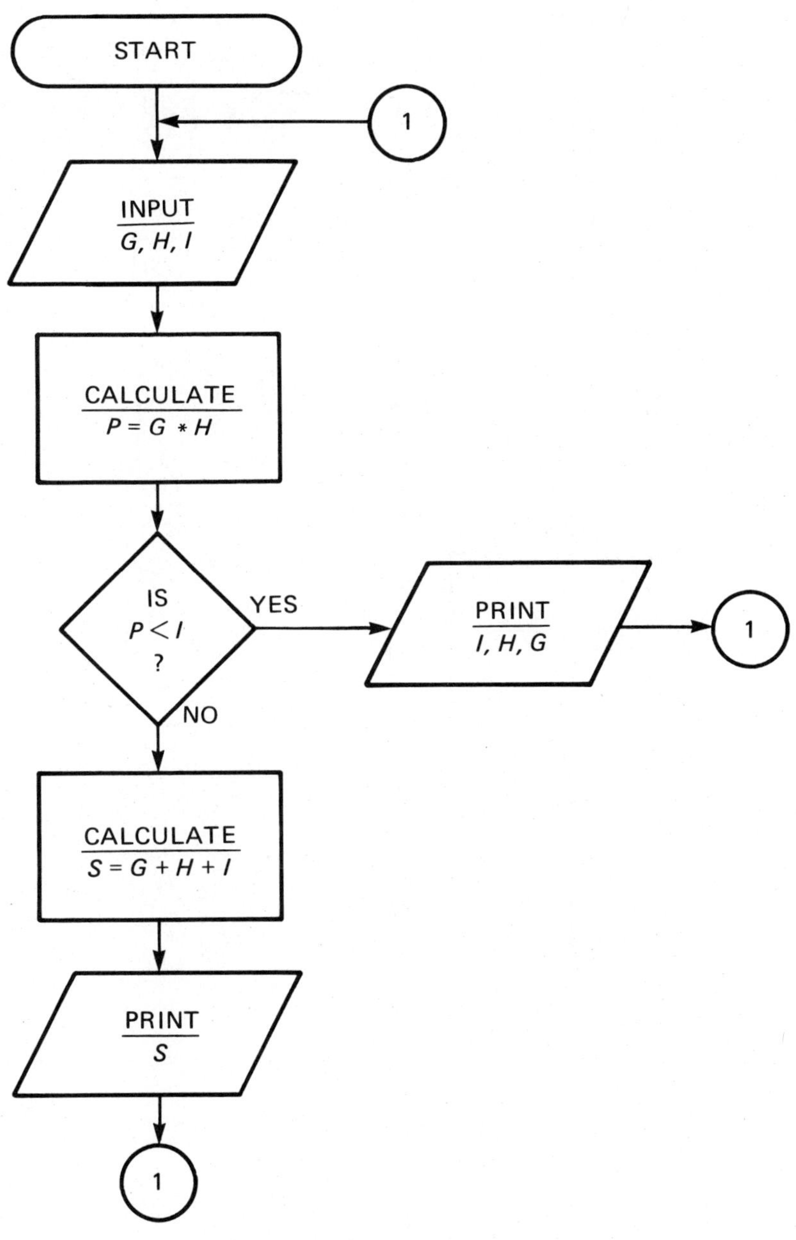

16. **Problem:** Enter two numbers. If the numbers are not equal, print the difference of the numbers. If the numbers are the same, print their sum.

Test Data: a. 17, 6 b. 5, 5

Flowchart

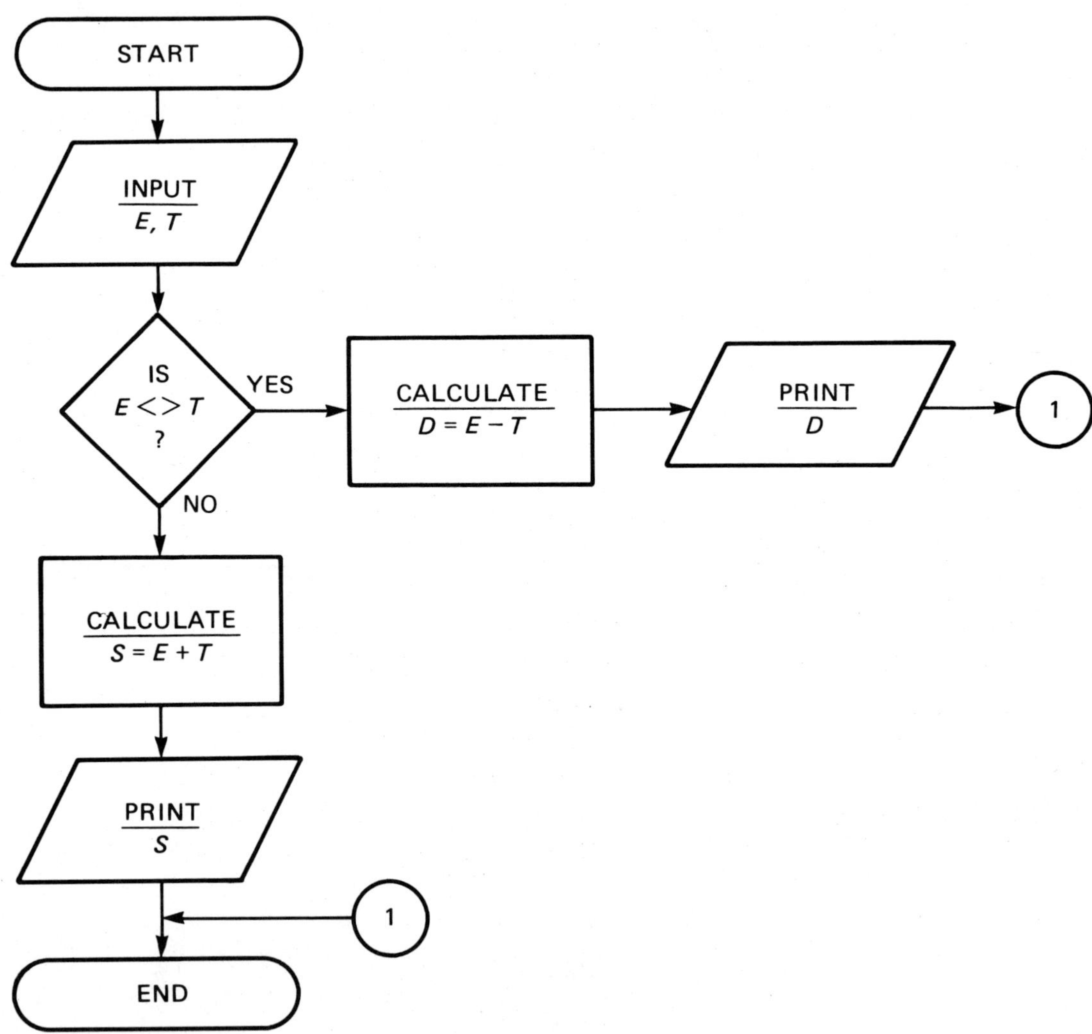

Section 3.2

For each of the following problems:
a. Develop a flowchart solution to the problem.
b. Trace the flowchart to see if it solves the problem.

17. **Problem:** Enter two numbers. Determine if their difference is zero. (Consider the difference to be the first number minus the second number.)

18. **Problem:** Enter two numbers. Determine if the numbers are equal. (Why could we use the same algorithm as in Problem 17?)

19. **Problem:** Enter three numbers. Determine if the sum of the numbers is less than the product of the first and third number.

20. **Problem:** Enter a number. If the number is greater than 16, triple the number and print the new number. Otherwise, add 4 to the number and print this new number.

These problems involve two decisions.

21. **Problem:** Enter two numbers. Print the larger and then the smaller number. If the numbers are the same, print "numbers are the same."

22. **Problem:** Enter a person's temperature, T (in Fahrenheit degrees). If the temperature is 98.6°, print "temperature is normal." If the temperature is greater than 98.6°, print "you have a fever." Otherwise, print "your temperature is below normal."

23. **Problem:** Enter the length of a side of a regular pentagon. Determine if the perimeter is equal to 25, is greater than 25, or is less than 25.

24. **Problem:** Enter a temperature (in Fahrenheit degrees). If the temperature is greater than 70°, print "hot." If the temperature is from 45° to 70°, print "nice weather." If the temperature is less than 45°, print "cold."

Section 3.3

Determine which are valid IF-THEN statements:

25. 70 IF L; GO 20
26. 40 IF $X=6$ THEN GOTO 130
27. 60 IF $A*B$; GOTO 150
28. 30 IF $C*D=A+B$ THEN 180

For each program that follows, show the output when $D = 2$, $E = 6$ and $F = 18$.

29. *Program*

```
10  INPUT D, E
20  T=E+3*D
30  IF T>D*E THEN 60
40  PRINT "E="; E
50  GOTO 70
60  PRINT "D="; D
70  END
```

30. *Program*

```
10  INPUT D, E, F
20  P1=D*E
30  P2=E*F
40  IF P1=P2 THEN GOTO 70
50  PRINT "D AND F ARE DIFFERENT"
60  GOTO 80
70  PRINT "D=F"
80  END
```

31. *Program*

```
10  INPUT E
20  C=1
30  PRINT C
40  C=C+1
50  IF C>E THEN GOTO 70
60  GOTO 30
70  END
```

32. *Program*

```
  10  INPUT E, F
  20  C=0
■ 30  PRINT C;
  40  C=C+E
  50  IF C>F THEN GOTO 70
  60  GOTO 30
  70  END
```

■ What effect does the semi-colon have on the output? Try the same program without the semi-colon.

For each of the following problem statements and flowcharts:
a. Write two programs, one for each planned output.
b. Run the programs using the test data from the planned outputs.

33. **Problem:** Enter two numbers. Determine if the product of the numbers is greater than the sum.

Flowchart

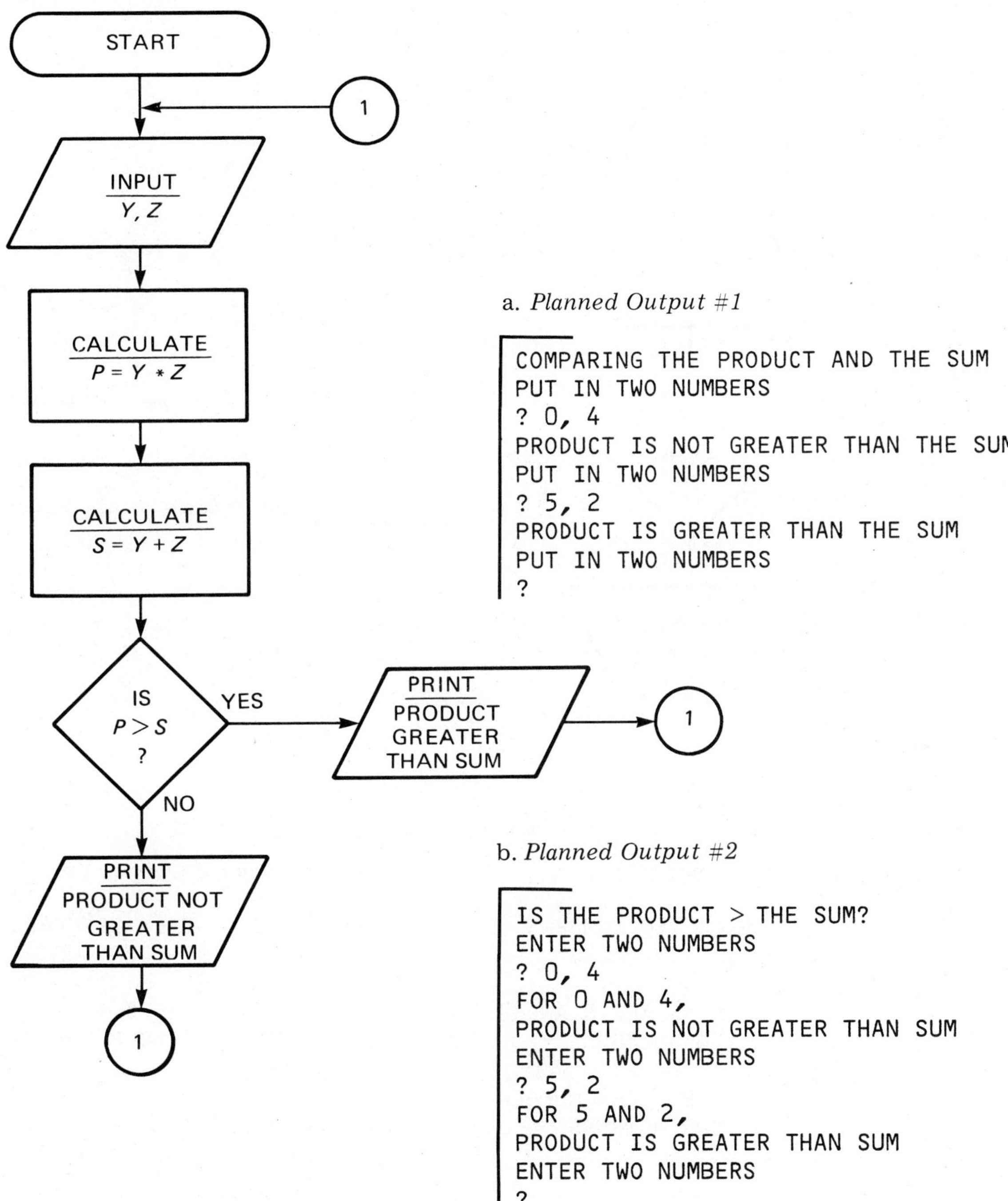

a. *Planned Output #1*

```
COMPARING THE PRODUCT AND THE SUM
PUT IN TWO NUMBERS
? 0, 4
PRODUCT IS NOT GREATER THAN THE SUM
PUT IN TWO NUMBERS
? 5, 2
PRODUCT IS GREATER THAN THE SUM
PUT IN TWO NUMBERS
?
```

b. *Planned Output #2*

```
IS THE PRODUCT > THE SUM?
ENTER TWO NUMBERS
? 0, 4
FOR 0 AND 4,
PRODUCT IS NOT GREATER THAN SUM
ENTER TWO NUMBERS
? 5, 2
FOR 5 AND 2,
PRODUCT IS GREATER THAN SUM
ENTER TWO NUMBERS
?
```

34. **Problem:** Enter two numbers. If the first number is less than the second number, double the first number and print this new result. If not, print both numbers.

Flowchart

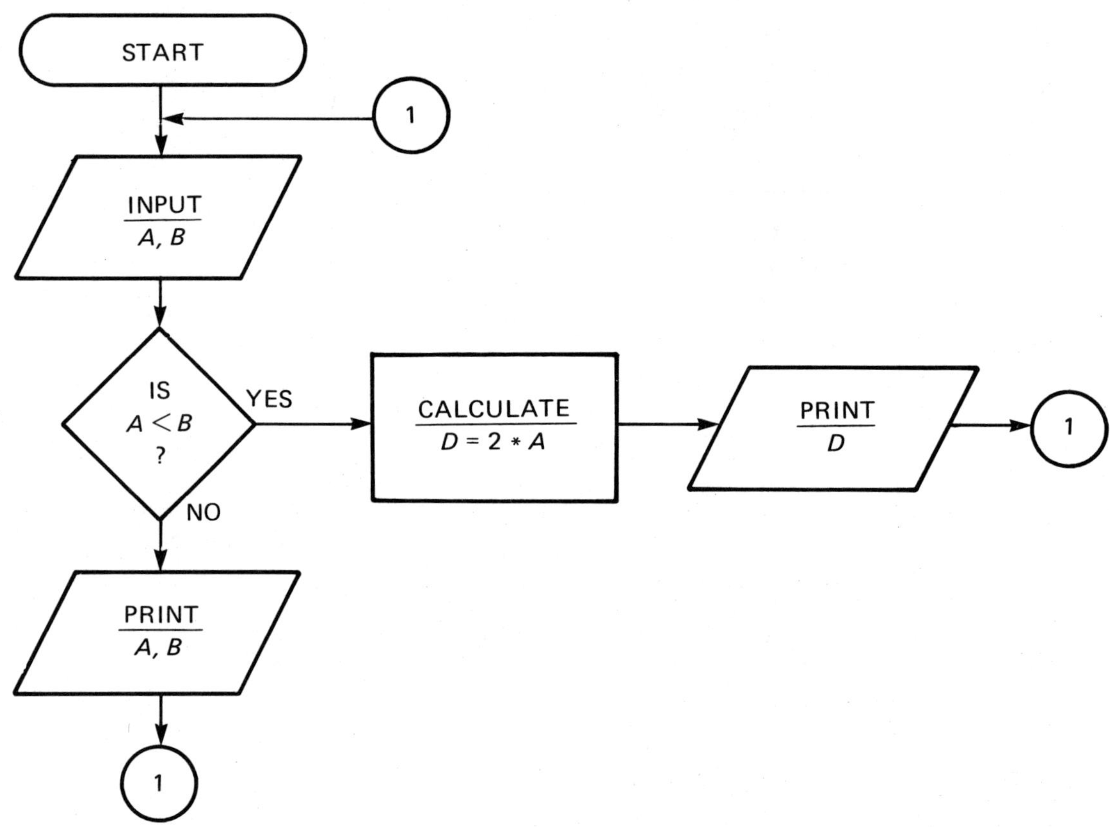

a. *Planned Output #1*

```
COMPARING TWO NUMBERS
ENTER YOUR NUMBERS
? 8, 3
THE NUMBERS ARE 8 AND 3
ENTER YOUR NUMBERS
? 7, 11
DOUBLE THE FIRST NUMBER IS 14
ENTER YOUR NUMBERS
?
```

b. *Planned Output #2*

```
HOW DO THEY COMPARE?
WHAT ARE YOUR NUMBERS?
? 8, 3
YOUR NUMBERS ARE 8 AND 3
WHAT ARE YOUR NUMBERS?
? 7, 11
7 IS LESS THAN 11
DOUBLE 7 IS 14
WHAT ARE YOUR NUMBERS?
?
```

35. **Problem:** Enter two numbers. If the first number is less than the second, add 12 to the first number and print the number and the new result. Otherwise, triple the second number and print that number and the new result.

Flowchart

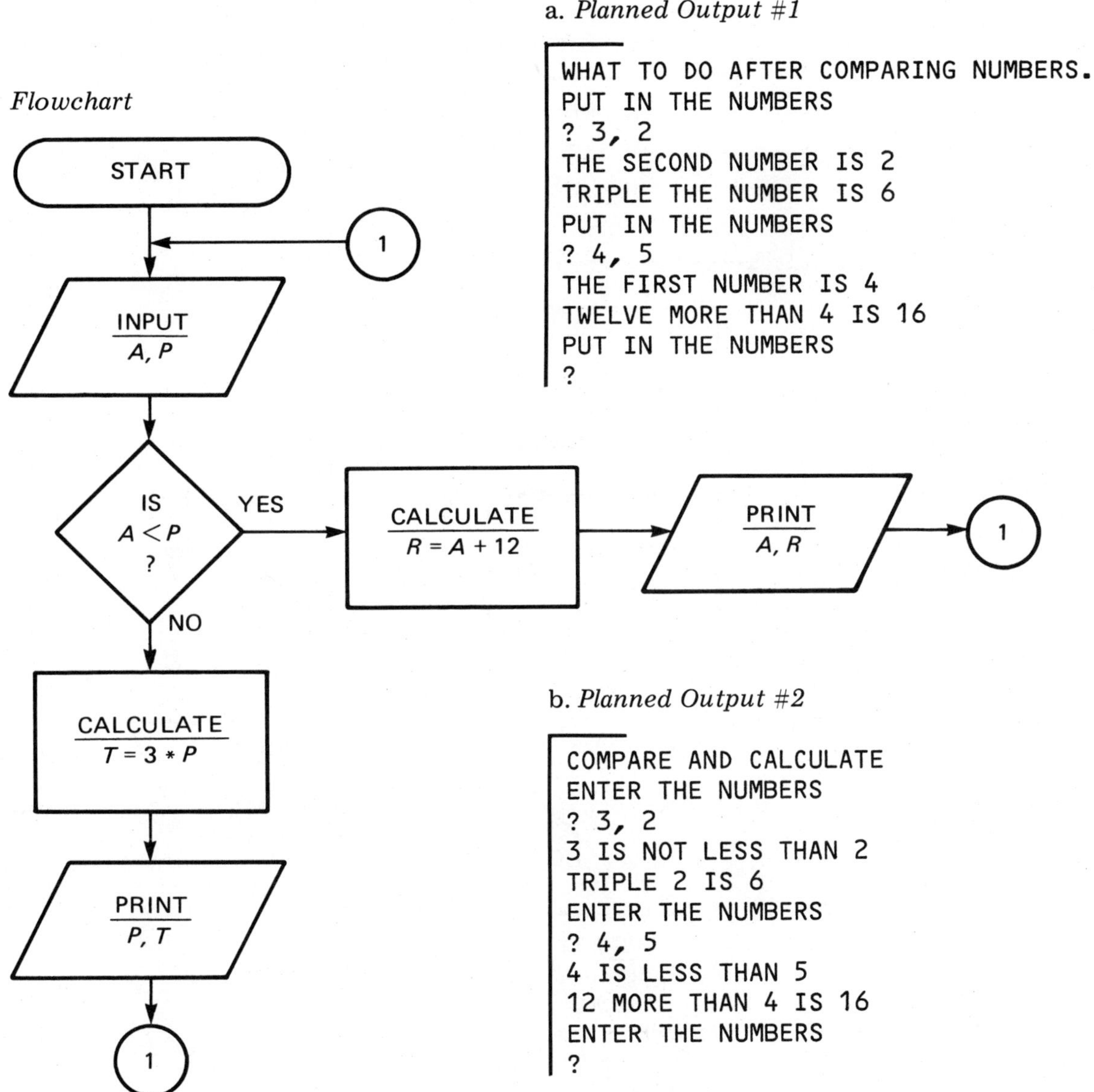

a. *Planned Output #1*

```
WHAT TO DO AFTER COMPARING NUMBERS.
PUT IN THE NUMBERS
? 3, 2
THE SECOND NUMBER IS 2
TRIPLE THE NUMBER IS 6
PUT IN THE NUMBERS
? 4, 5
THE FIRST NUMBER IS 4
TWELVE MORE THAN 4 IS 16
PUT IN THE NUMBERS
?
```

b. *Planned Output #2*

```
COMPARE AND CALCULATE
ENTER THE NUMBERS
? 3, 2
3 IS NOT LESS THAN 2
TRIPLE 2 IS 6
ENTER THE NUMBERS
? 4, 5
4 IS LESS THAN 5
12 MORE THAN 4 IS 16
ENTER THE NUMBERS
?
```

36. **Problem:** Enter four numbers. Determine if the sum of the first number and the third number is greater than or equal to the sum of the second and fourth numbers.

Flowchart

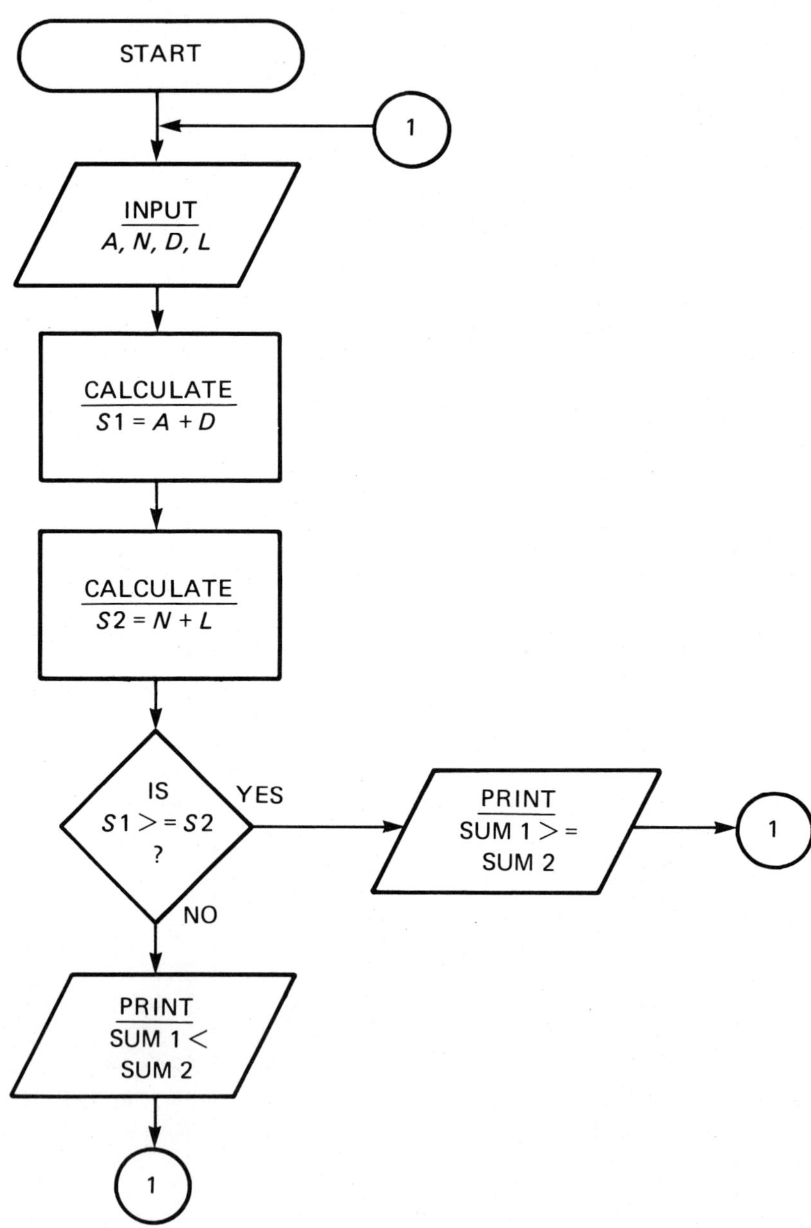

a. *Planned Output #1*

```
COMPARE THE SUMS
ENTER 4 NUMBERS
? 3, 2, 4, 6
THE NUMBERS ARE 3, 2, 4 AND 6.
SUM 1 IS LESS THAN SUM 2
ENTER 4 NUMBERS
? 6, 1, 2, 4
THE NUMBERS ARE 6, 1, 2 AND 4.
SUM 1 IS GREATER THAN OR = SUM 2
ENTER 4 NUMBERS
?
```

b. *Planned Output #2*

```
COMPARING ALTERNATE SUMS
PUT IN 4 NUMBERS
? 3, 2, 4, 6
3, 2, 4 AND 6 ARE THE NUMBERS.
SUM 1= 3+ 4= 7
SUM 2= 2+ 6= 8
SUM 1 < SUM 2
PUT IN 4 NUMBERS
? 6, 1, 2, 4
6, 1, 2 AND 4 ARE THE NUMBERS.
SUM 1= 6+ 2= 8
SUM 2= 1+ 4= 5
SUM 1 >= SUM 2
PUT IN 4 NUMBERS
?
```

37. **Problem:** Enter a temperature (in Fahrenheit degrees). Determine if the temperature is less than, equal to, or greater than zero.

Flowchart

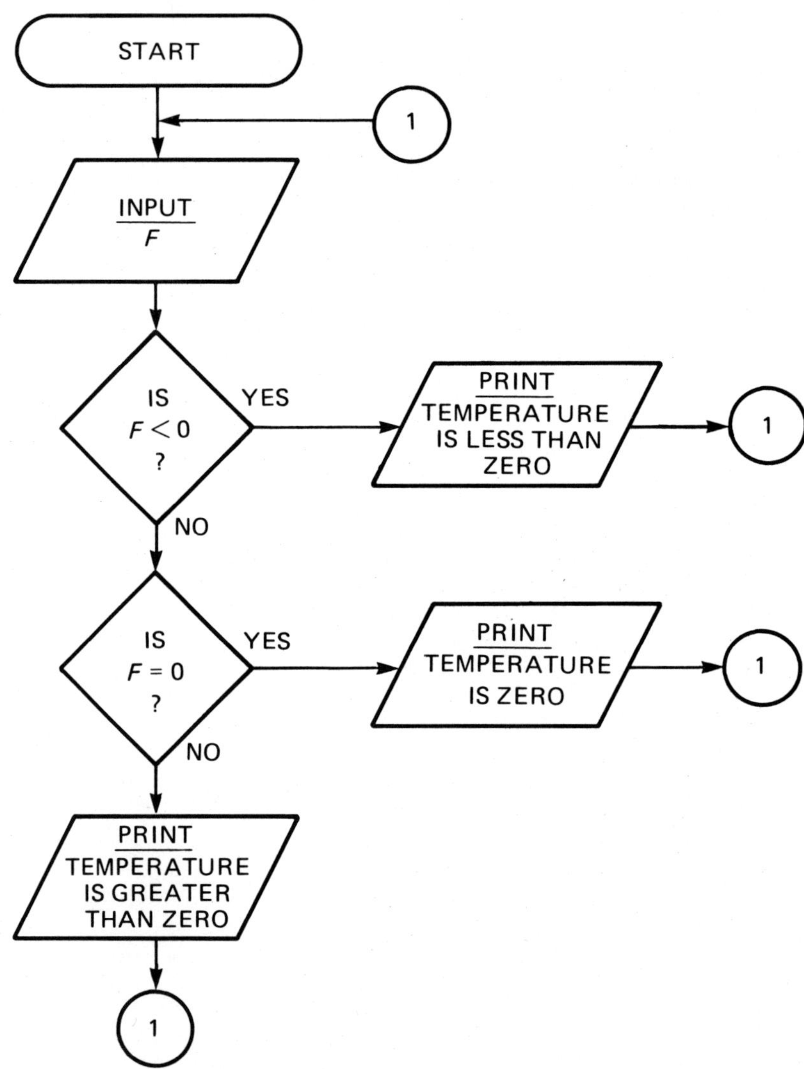

a. *Planned Output #1*

```
CHECKING THE TEMPERATURE
PUT IN THE TEMPERATURE
? 15
TEMPERATURE IS ABOVE ZERO
PUT IN THE TEMPERATURE
? -4
TEMPERATURE IS BELOW ZERO
PUT IN THE TEMPERATURE
? 0
TEMPERATURE IS ZERO
PUT IN THE TEMPERATURE
?
```

b. *Planned Output #2*

```
TELL ME ABOUT THE TEMPERATURE
ENTER THE TEMPERATURE
? 15
15 IS ABOVE ZERO
ENTER THE TEMPERATURE
? -4
-4 IS BELOW ZERO
ENTER THE TEMPERATURE
? 0
THE TEMPERATURE IS ZERO
ENTER THE TEMPERATURE
?
```

38. **Problem:** Newspaper deliverers receive extra money (a bonus) if they sell many papers. Below we can see how the extra money is given out.

If you sell:

a. more than 75 papers, you get a $20 bonus.
b. from 55 to 75 papers, you get a $10 bonus.
c. below 55 papers, you get no bonus.

Enter the number of papers sold. Print the bonus as outlined above.

Flowchart

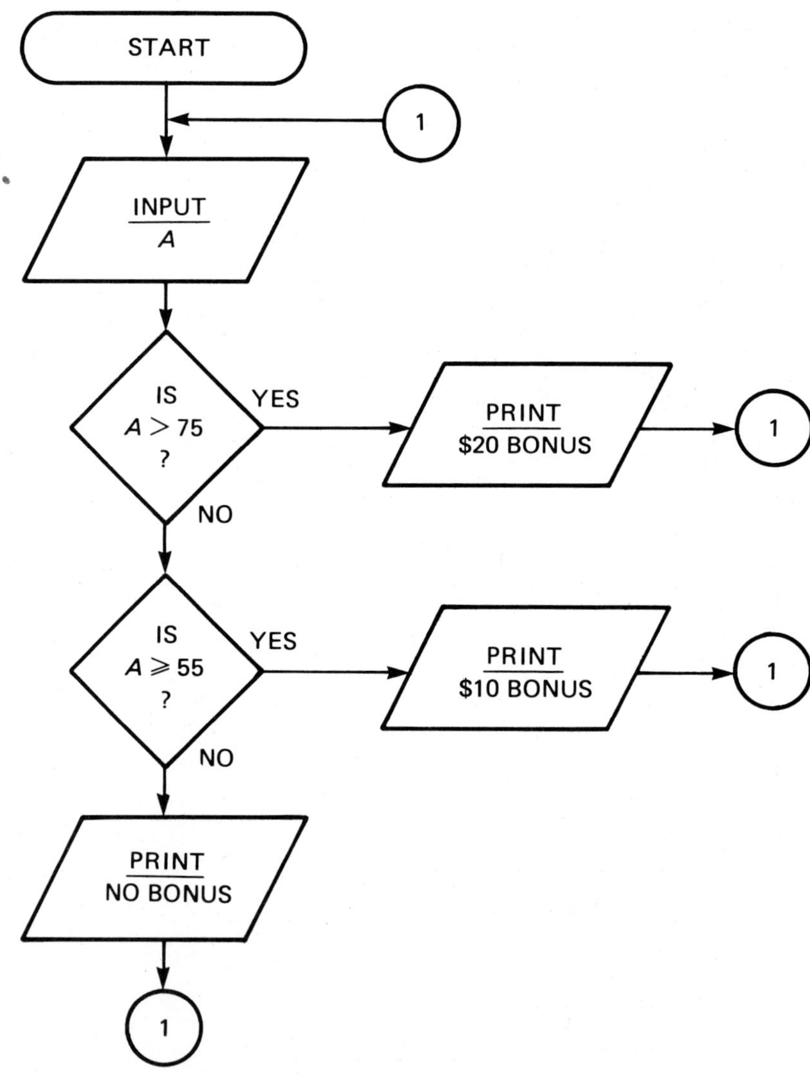

a. *Planned Output #1*

```
WHAT IS THE BONUS?
ENTER THE NUMBER SOLD
? 40
SORRY! NO BONUS
ENTER THE NUMBER SOLD
? 84
YOU GET A $20 BONUS
ENTER THE NUMBER SOLD
? 60
YOU GET A $10 BONUS
ENTER THE NUMBER SOLD
?
```

b. *Planned Output #2*

```
HOW MUCH IS THE BONUS?
HOW MANY PAPERS SOLD?
? 40
NOT ENOUGH FOR A BONUS.
HOW MANY PAPERS SOLD?
? 84
FOR 84 PAPERS, BONUS IS $20.
HOW MANY PAPERS SOLD?
? 60
FOR 60 PAPERS, BONUS IS $10.
HOW MANY PAPERS SOLD?
?
```

39. **Problem:** A bonus is given for selling boxes of candy as follows:
If the number of boxes sold is:

a. greater than 100, you receive a bonus of $15;
b. from 75 to 100, you receive a bonus of $8;
c. less than 75, there is no bonus.

Enter the number of boxes of candy sold. Print the bonus as outlined above.

Flowchart

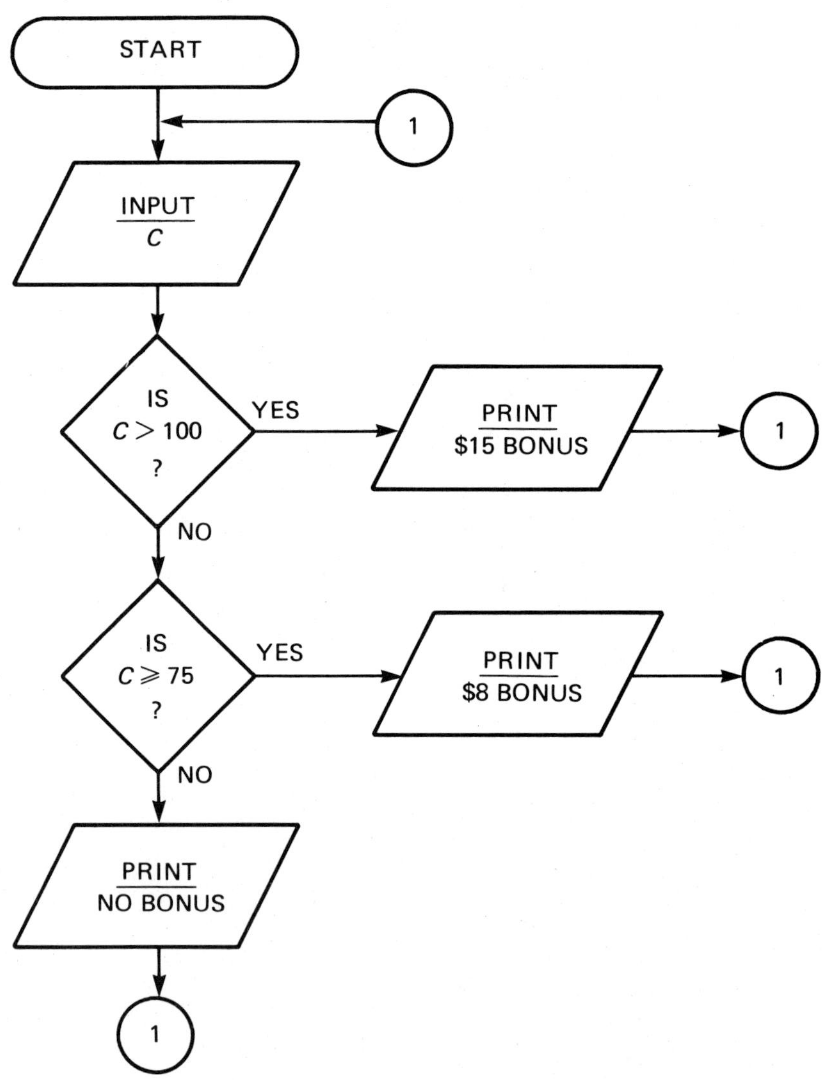

a. *Planned Output #1*

```
DETERMINE THE BONUS
HOW MANY BOXES SOLD?
? 41
NO BONUS
HOW MANY BOXES SOLD?
? 106
$15 BONUS
HOW MANY BOXES SOLD?
? 83
$8 BONUS
HOW MANY BOXES SOLD?
?
```

b. *Planned Output #2*

```
WHAT IS THE BONUS?
ENTER # OF BOXES SOLD
? 41
FOR 41, NO BONUS.
ENTER # OF BOXES SOLD
? 106
FOR 106, BONUS IS $15.
ENTER # OF BOXES SOLD
? 83
FOR 83, BONUS IS $8.
ENTER # OF BOXES SOLD
?
```

40. **Problem:** A bonus is given for selling boxes of cookies as follows:
If the number of boxes sold is:

a. greater than 80, you receive a bonus of $12;
b. from 62 to 80, you receive a bonus of $7;
c. less than 62, there is no bonus.

Enter the number of boxes of cookies sold. Print the bonus as outlined above. If someone does not get a bonus, print the number of additional boxes he must sell in order to receive a $7 bonus.

Flowchart

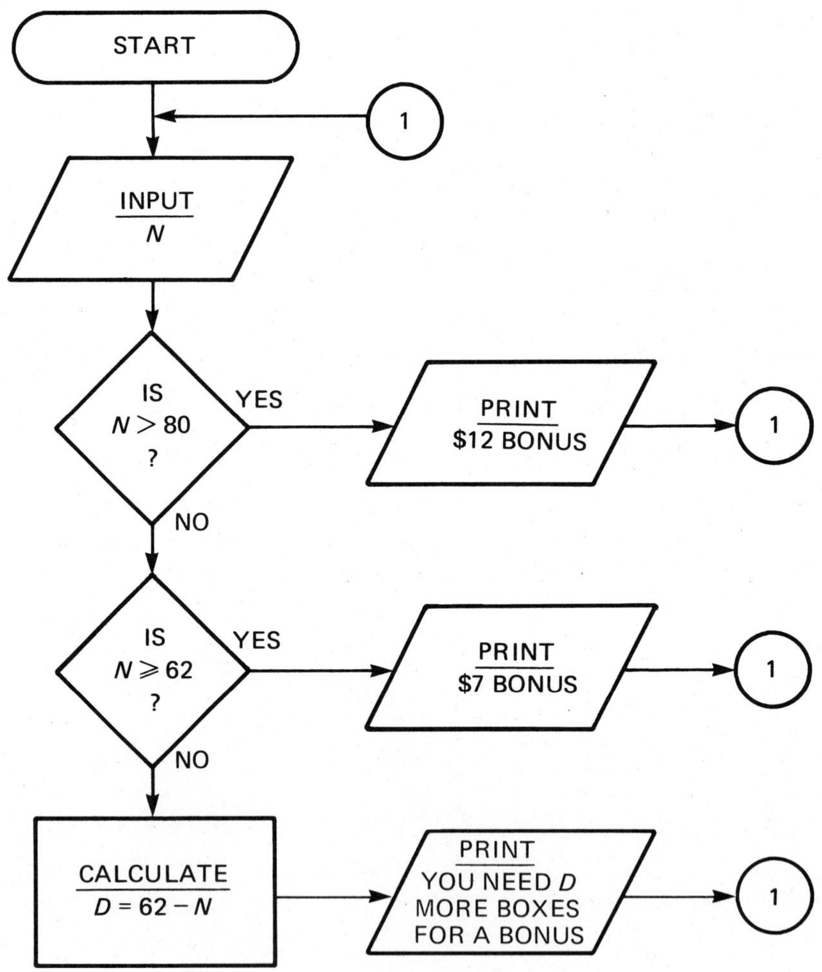

a. *Planned Output #1*

```
DID YOU GET A BONUS?
ENTER NUMBER SOLD
? 40
FOR A BONUS SELL AN ADDITIONAL 22
ENTER NUMBER SOLD
? 108
$12 BONUS
ENTER NUMBER SOLD
? 65
$7 BONUS
ENTER NUMBER SOLD
?
```

b. *Planned Output #2*

```
WHAT IS THE BONUS?
PUT IN NUMBER SOLD
? 40
40 IS NOT ENOUGH FOR A BONUS
SELL 22 MORE BOXES
PUT IN NUMBER SOLD
? 108
FOR 108 BONUS IS $12
PUT IN NUMBER SOLD
? 65
FOR 65 BONUS IS $7
PUT IN NUMBER SOLD
?
```

Section 3.4

For each of the following problems:
a. Write an analysis which includes the development of an algorithm.
b. Write a plan showing how you *use* the algorithm to solve the problem.
c. Develop a flowchart.
d. Design a planned output.
e. Code a program and RUN it on the computer.

41. **Problem:** David and Lisa each had three tests. Determine if Lisa's grades were greater than or equal to David's grades.

42. **Problem:** Enter three numbers. Determine if the quotient of the first number divided by the second number is less than the third number.

43. **Problem:** Enter a test grade. If the grade is:

a. greater than 85, print "very good;"
b. from 65 to 85, print "passing;"
c. less than 65, print "failing."

44. **Problem:** Enter three test grades. If the average is:

a. greater than 90, print "*A*;"
b. from 75 to 90, print "*B*;"
c. less than 75, print "work harder."

Section 3.5 For the problem, flowchart, and planned output, code a program to solve the problem.

45. **Problem:** Enter two numbers A, B. Then enter a third number C. Determine if the third number is equal to the product of the first two numbers.

Flowchart

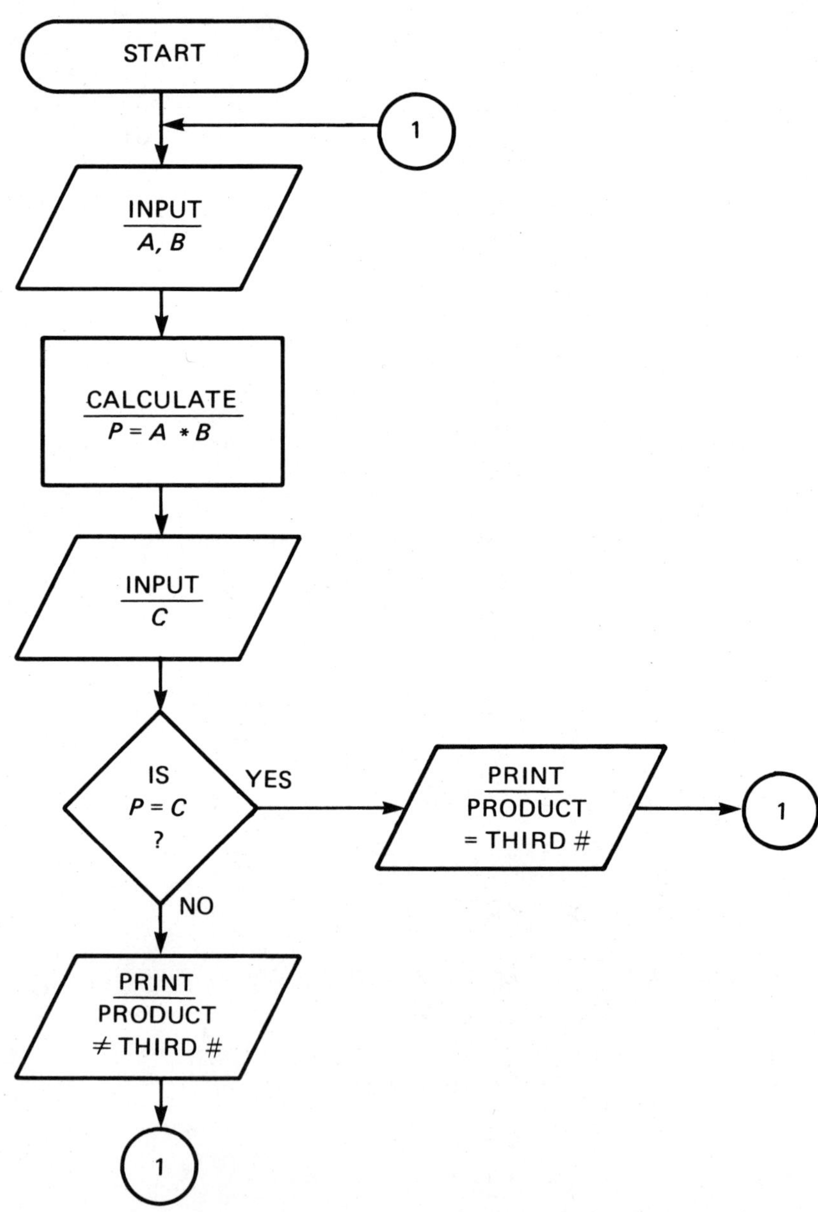

Planned Output

```
COMPARING A NUMBER TO A PRODUCT
PUT IN TWO NUMBERS
? 4, 5
ENTER ANOTHER NUMBER
? 21
THE PRODUCT 20 IS NOT EQUAL TO 21
PUT IN TWO NUMBERS
? 7, 5
ENTER ANOTHER NUMBER
? 35
THE PRODUCT 35 IS EQUAL TO 35
PUT IN TWO NUMBERS
?
```

For each of the following problems:

a. Write an analysis which includes the development of an algorithm.
b. Write a plan showing how you use the algorithm to solve the problem.
c. Develop a flowchart.
d. Design a planned output.
e. Code a program and RUN it on the computer.

46. **Problem:** Write a CAI program to give the user practice in subtracting two numbers. All differences must be positive. Do not assume that the first number is the largest.

47. **Problem:** Write a CAI program to give the user practice in dividing two numbers. All quotients must be greater than one. Therefore, do not assume that the first number is the largest. (Also, remember we cannot divide by zero!)

48. **Problem:** Write a CAI program to give the user practice in subtraction and division. Have the user first subtract the numbers. After getting the correct answer, the user is to divide the two numbers.

49. **Problem:** Write a CAI program to give the user practice in *either* addition or multiplication. (*Hint:* Set up a code where 1 means addition and 2 means multiplication.)

Chapter 4

Applications of Intermediate Programming Concepts

Section 4.1 SOLVING EQUATIONS USING DIVISION AND ADDITION OR SUBTRACTION (SKIPPING LINES)

Solving certain problems can be made easier if we translate the words into symbols.

Look at the chart. It shows common words used in mathematics and their translation into symbols.

Translating Words into Symbols

Words	*Symbols*
1. *increased by*	1. $+$
a. six increased by two	a. $6 + 2$
b. x increased by four	b. $x + 4$
2. *decreased by*	2. $-$
a. four decreased by one	a. $4 - 1$
b. n decreased by six	b. $n - 6$
3. *times*	3. $\times$, $\cdot$, ( ) ( )
a. five times seven	a. 5×7, $5 \cdot 7$, $(5)(7)$ or $5(7)$
b. two times C	b. $2 \times C$, $2 \cdot C$, $(2)(C)$ or $2C$

Class Exercise 1 Translate into symbols:

1. eleven increased by six
2. nine decreased by five
3. four times eight
4. two times six increased by four
5. x decreased by three
6. C increased by seven
7. six times n increased by four
8. nine times m decreased by five

Example 1 Translate into symbols. Use n to represent the number:

1. a number increased by six
2. a number decreased by two
3. five times a number
4. two times a number increased by one
5. Three times a number decreased by seven is equal to eight.

Solution: We replace the word "number" with the letter n.

1. $n + 6$
2. $n - 2$
3. $5 \cdot n$ or $5n$
4. $2n + 1$
5. $3n - 7 = 8$

See how well you can translate words into symbols. Try the next exercise.

Class Exercise 2 Translate into symbols. Use n to represent the number.

1. a number increased by thirteen
2. a number decreased by six
3. four times a number
4. six times a number increased by twelve
5. Nine times a number decreased by one is equal to seventeen.
6. Seven times a number increased by two is equal to twenty-six.

Now that we can translate words into symbols, we can use this skill to solve the next problem.

Problem Two times a number increased by fourteen is thirty-two. Find the number.

Solution: We let n (or any other variable) represent the number. Translate the sentence into mathematical symbols.

$2n + 14 = 32$

Solving the problem now becomes solving this new type of equation.

$$2n + 14 = 32$$

Subtract 14 from both sides:

$$\underline{\quad - 14 = -14}$$

$$2n = 18$$

Divide both sides by 2:

$$\frac{2n}{2} = \frac{18}{2}$$

$$n = 9$$

Going back to the problem, we replace the words "a number" with our answer, 9, to see if the statement is true:

$$2 \cdot 9 + 14 = 32$$

$$18 + 14 = 32$$

$$32 = 32$$

So, the solution to this problem is 9.

Before solving more problems, let us practice solving equations similar to the equation above.

Example 2 Solve the equation. Check the result.

$3a - 21 = 18$

Solution:

$$3a - 21 = 18$$

Add 21 to both sides:

$$\underline{\quad + 21 = 21}$$

$$3a = 39$$

Divide both sides by 3: $\frac{3a}{3} = \frac{39}{3}$

$a = 13$

Check:

1. We write the equation again: $3a - 21 = 18$
2. Replace "a" with the answer: $3(13) - 21 = 18$
3. Use the order of operations: $39 - 21 = 18$
4. Since both sides are equal, the answer 13 is correct. $18 = 18$

Class Exercise 3 Solve the equations. Check each result.

1. $3b - 15 = 15$
2. $6n + 17 = 65$
2. $7n + 81 = 123$
4. $8d - 416 = 384$

Class Exercise 4 Solve the following problems. Check your results.

1. Nine times a number increased by two is equal to 74. Find the number.
2. Four times a number decreased by eleven is equal to 45. Find the number.
3. Seven increased by three times a number is equal to 31. Find the number.
4. Two increased by triple a number is 123. Find the number.

Before using the computer we must be able to represent in one form all the equations we wish to solve. For example, we want to solve equations like $6n + 17 = 65$ and $7n + 81 = 123$. Therefore, we can generalize this type of equation as:

$an + b = c$

If $a = 6$, $b = 17$ and $c = 65$, we get $\overset{a}{6}n + \overset{b}{17} = \overset{c}{65}$

If $a = 7$, $b = 81$ and $c = 123$, we get $\overset{a}{7}n + \overset{b}{81} = \overset{c}{123}$

Now we are ready to turn to the computer.

Problem

Solve equations of the form $an+b=c$ for n.

Solution:
a. *Analysis:* How do you solve $an + b = c$? We will follow the same procedure that we used when we solved $2n + 14 = 32$.

	Original Equation	*General Form*
	$2n + 14 = 32$	$an + b = c$
1. Subtract on both sides:	$- 14 = -14$	$- b = -b$
2. Divide both sides:	$\frac{2n}{2} = \frac{18}{2}$	$\frac{an}{a} = \frac{c-b}{a}$
	$n = 9$	$n = \frac{c-b}{a}$

Our algorithm to solve the problem is $n = \frac{c-b}{a}$.

Let us see how the algorithm solves the problem.

Example 3 Use the algorithm $n = \frac{c-b}{a}$ to solve the equation:

$2n + 14 = 32$

Solution: For the equation $2n + 14 = 32$, $a = 2$, $b = 14$ and $c = 32$. We substitute these values in the algorithm as follows:

$$n = \frac{c-b}{a}$$

$$n = \frac{(32-14)}{2}$$

$$n = \frac{18}{2} = 9$$

This answer is correct. It is the same answer that we got when we solved the equation before.

Class Exercise 5 Use the algorithm $n = \frac{c - b}{a}$ to solve the equations. Check each result.

1. $5n + 1 = 41$
2. $8n + 78 = 158$
3. $11n + 7 = 84$
4. $3n + 89 = 113$

Now that you know how to use the algorithm to solve equations, we move to the plan.

b. *Plan:*

1. Put in values for A, B, and C.

   (What if $A=0$? When we calculate $N=\frac{C-B}{A}$, we would be dividing by zero which cannot be done. We must therefore check to be sure a zero is not entered for "A.")
2. Check to see if $A=0$.
   a) If $A=0$ then go to 1 and enter new values.
   b) If A is not 0 then continue.
3. $N=\frac{C-B}{A}$.
4. Print N.
5. End.

c. *Flowchart*

START
1
INPUT
A, B, C
IS
A=0
?
YES
PRINT
A CANNOT
BE ZERO
1
NO
CALCULATE
N=(C−B)/A
PRINT
N
1

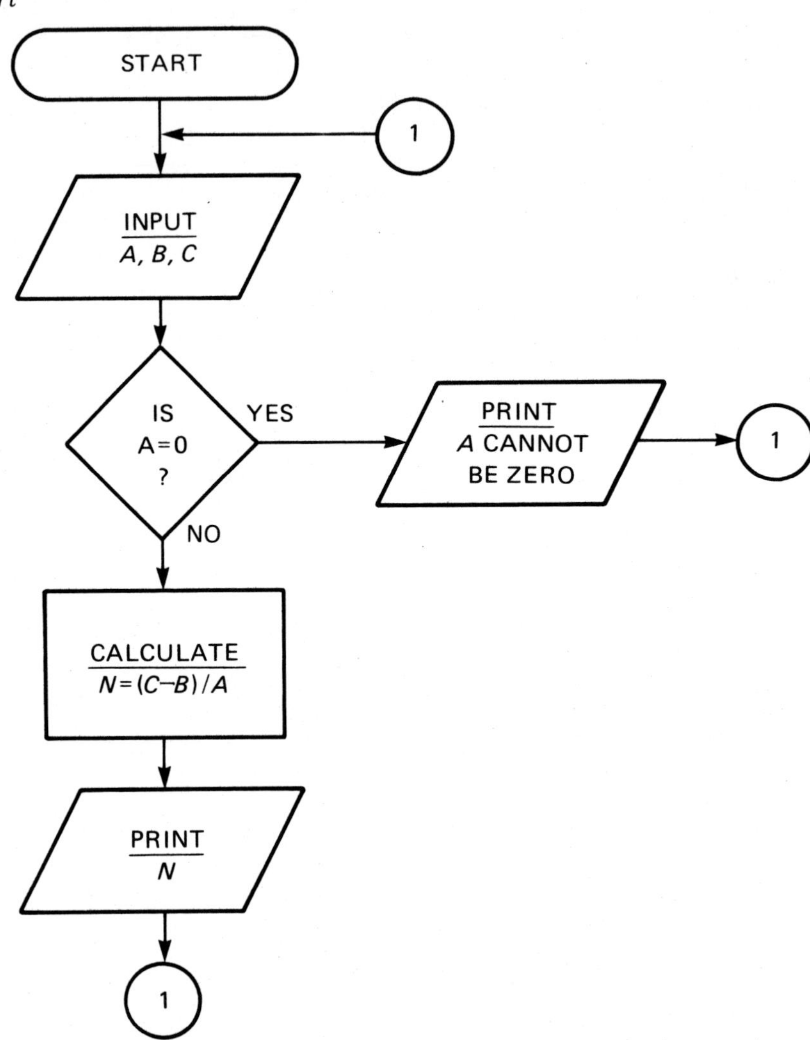

d. *Planned Output*

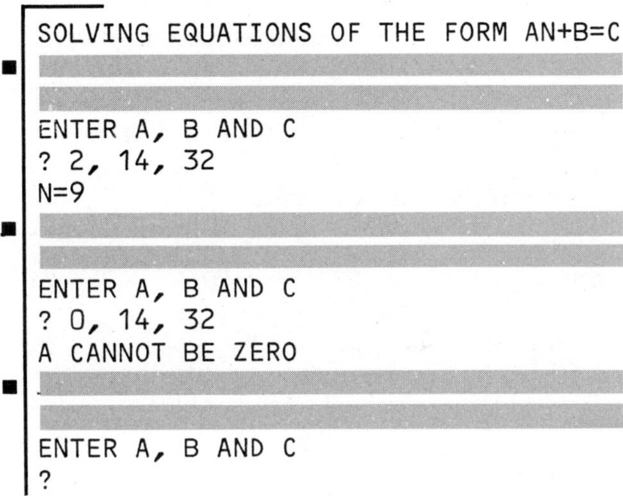

```
   SOLVING EQUATIONS OF THE FORM AN+B=C
■


   ENTER A, B AND C
   ? 2, 14, 32
   N=9
■


   ENTER A, B AND C
   ? 0, 14, 32
   A CANNOT BE ZERO
■


   ENTER A, B AND C
   ?
```

Discussion of the Planned Output:

■ Note that we wish to skip two lines. When the computer sees a PRINT statement, the cursor moves to the next line. *Placing a PRINT statement in a program with nothing after it will get the computer to skip one line.*

e. *The Program*

```
     10   PRINT "SOLVING EQUATIONS OF THE FORM AN+B=C"
■    20   PRINT
■    30   PRINT
     40   PRINT "ENTER A, B AND C"
     50   INPUT A, B, C
     60   IF A=0 THEN GOTO 100
     70   N=(C-B)/A
     80   PRINT "N="; N
■■   90   GOTO 20
     100  PRINT "A CANNOT BE ZERO"
■■   110  GOTO 20
     120  END
```

Discussion of the Program:

■ Each PRINT skips a line.

■■ Sends the computer back to skip two lines.

Exercise 4.1

Translate into symbols:

1. two increased by nine
2. ten decreased by eight
3. seven times four
4. three times one increased by two
5. m decreased by three
6. d increased by six
7. four times e increased by ten
8. six times x decreased by one

Translate into symbols. Use x to represent the number.

9. a number increased by eleven
10. a number decreased by one
11. seven times a number
12. four times the number decreased by six
13. Three times a number increased by two is equal to twenty-three.
14. Nine times a number decreased by six is equal to sixty-six.

Solve the equations. Check each result.

15. $8x + 1 = 57$
16. $3a - 21 = 0$
17. $12b + 5 = 149$
18. $6c - 53 = 7$

Solve the following problems. Check your results.

19. Three times a number decreased by ten is 11. Find the number.
20. Six times a number increased by forty-seven is 77. Find the number.
21. One increased by double a number is 161. Find the number.
22. Double a number decreased by seventeen is 17. Find the number.

Use the algorithm $n = \frac{c - b}{a}$ to solve the following equations. Check each result.

23. $7n + 18 = 123$
24. $3n + 63 = 543$
25. $12n + 75 = 219$
26. $5n + 86 = 186$

Revise the program that was written in this section to produce the following outputs.

27.

```
SOLVING AN+B=C



PUT IN VALUES FOR A, B AND C
? 2, 14, 32
N IS 9



PUT IN VALUES FOR A, B AND C
? 0, 14, 32



PUT IN VALUES FOR A, B AND C
?
```

28.

```
FINDING N FOR AN+B=C


ENTER VALUES FOR A, B AND C
? 2, 14, 32
A=2     B=14     C=32
THE SOLUTION IS 9


ENTER VALUES FOR A, B AND C
? 0, 14, 32
A=0     B=14     C=32
SORRY! A CANNOT BE ZERO!


ENTER VALUES FOR A, B AND C
?
```

For the given problem, flowchart and planned output, code a program to solve the problem.

29. **Problem:** Find the area of a square when we are given the length of a side. (Remember the length of a side must be positive.)

Flowchart

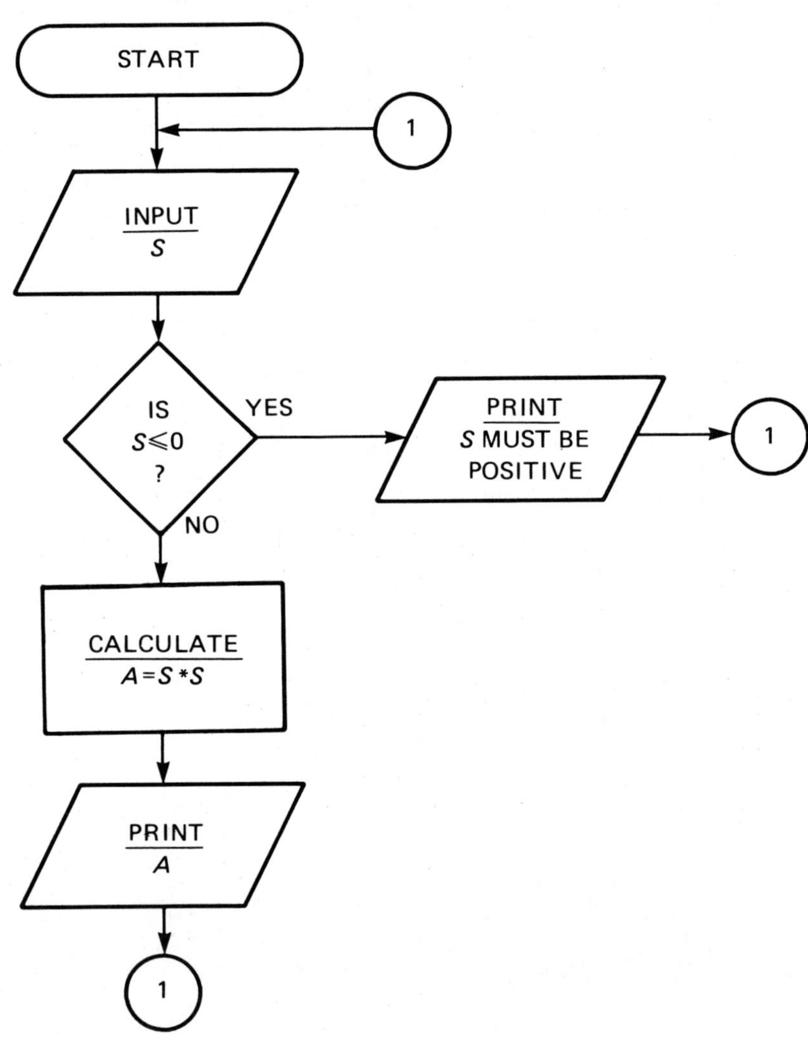

Planned Output

```
FINDING THE AREA OF A SQUARE


PUT IN THE LENGTH OF THE SIDE
? 10
THE AREA IS 100


PUT IN THE LENGTH OF THE SIDE
? 0
SIDE MUST HAVE POSITIVE LENGTH


PUT IN THE LENGTH OF THE SIDE
?
```

For each of the following problems:
a. Write an analysis which includes the development of an algorithm.
b. Write a plan for a computer solution to the problem.
c. Develop a flowchart.
d. Design a planned output.
e. Code a program to solve the problem and RUN it on the computer.

30. **Problem:** Enter the length of a side of a square. Print the perimeter and area. (Make sure that only positive numbers are entered for the side.)

31. **Problem:** Solve equations of the form $ax=b$ for x. (Be sure that a value of 0 will not lead to an error.)

32. **Problem:** Solve equations of the form $ax-b=c$ for x. (Be sure to handle the special case when $a=0$.)

Section 4.2 CLASSIFYING TRIANGLES BY ANGLES (MULTIPLE STATEMENTS ON A LINE)

We have already learned that triangles can be grouped by the lengths of their sides (Section 1.5). Sometimes it is convenient to group (classify) triangles by the measures of their angles.

An *acute* angle is one of measure less than 90° (but greater than 0°). Here are some acute angles:

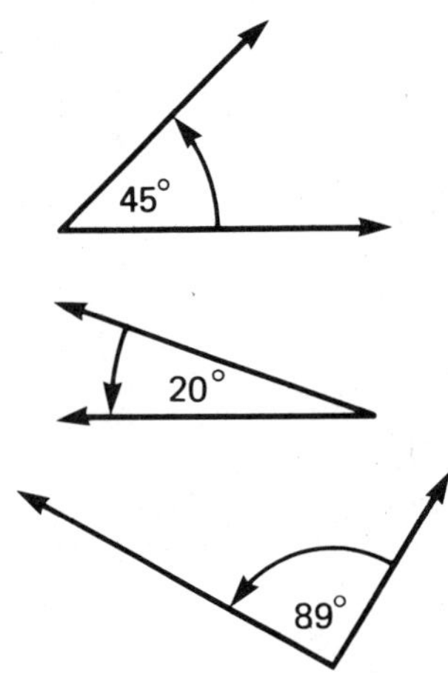

We know that a right angle is an angle that contains 90°.

A *right triangle* is a triangle that has a right angle.

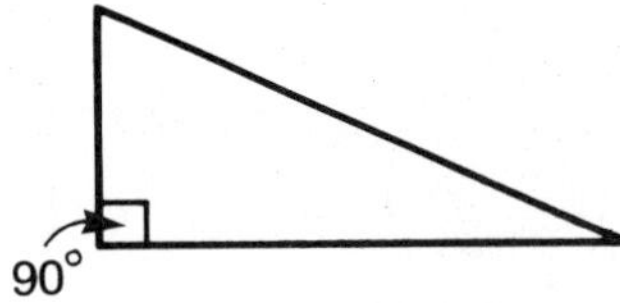

A triangle with all acute angles is called an *acute triangle.*

An angle of measure greater than 90° but less than 180° is called an *obtuse* angle. Here are some examples of obtuse angles:

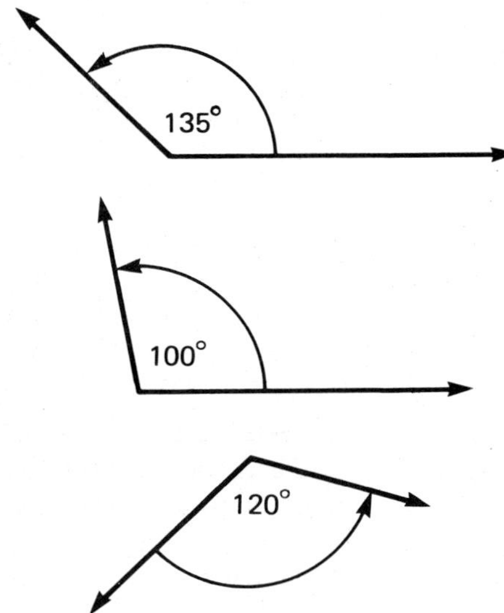

A triangle with an obtuse angle is called an *obtuse triangle.*

Class Exercise 1 Classify the following triangles as acute, right or obtuse.

1.

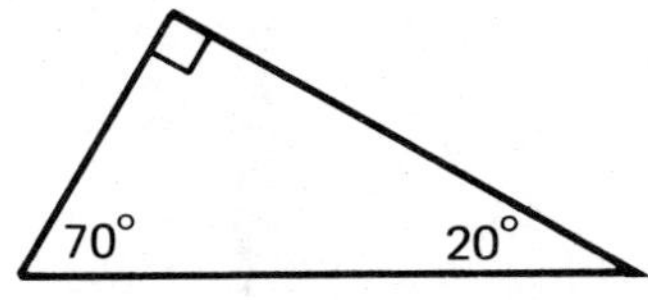

2.

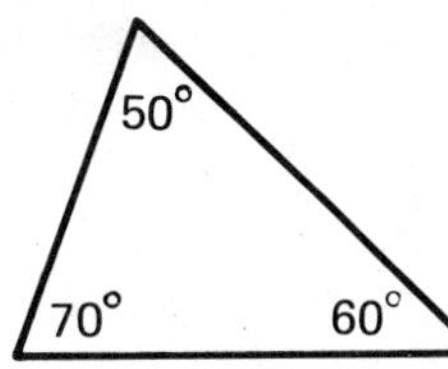

3.

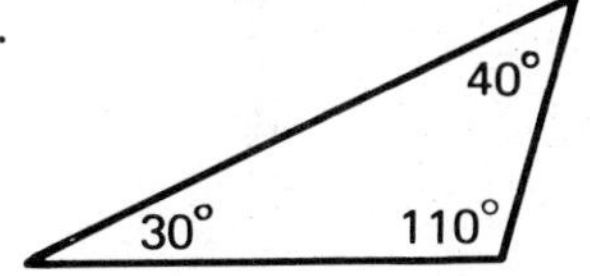

4.

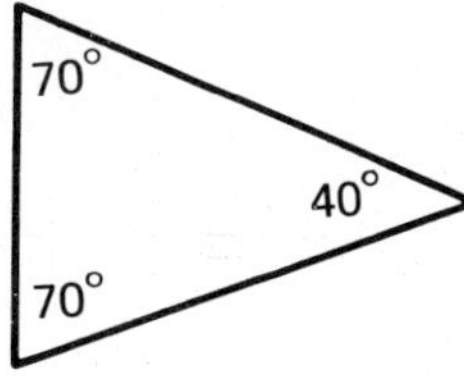

Example 1 The following are the measures of two angles of a triangle: 30° and 40°. State whether the triangle is acute, right or obtuse.

Solution: We must find the measure of the third angle of the triangle. Since the sum of the measures of the angles of a triangle is 180°, the third angle measures:

$$180-(30+40)=180-70=110^\circ$$

Since one of the angles of the triangle is obtuse (the 110° angle), the triangle is an obtuse triangle.

Class Exercise 2 Each of the following are the measures of two angles of a triangle. State whether the triangle is acute, right or obtuse.

1. 30°, 60°   2. 50°, 50°   3. 35°, 45°   4. 89°, 15°

Let us see how we can get the computer to classify angles and triangles.

Problem

Enter the measure of an angle (A). Determine if the angle is acute. (Assume the angle measure is positive.)

Solution:

a. *Analysis:* An acute angle has a measure that is less than 90°. We must, therefore, check to see if $A < 90$.

b. *Plan:*
1. Put in a value for A.
2. Is $A < 90$?
   a) If yes, print "acute." Go to 1.
   b) If no, print "not acute." Go to 1.
3. End.

c. *Flowchart*

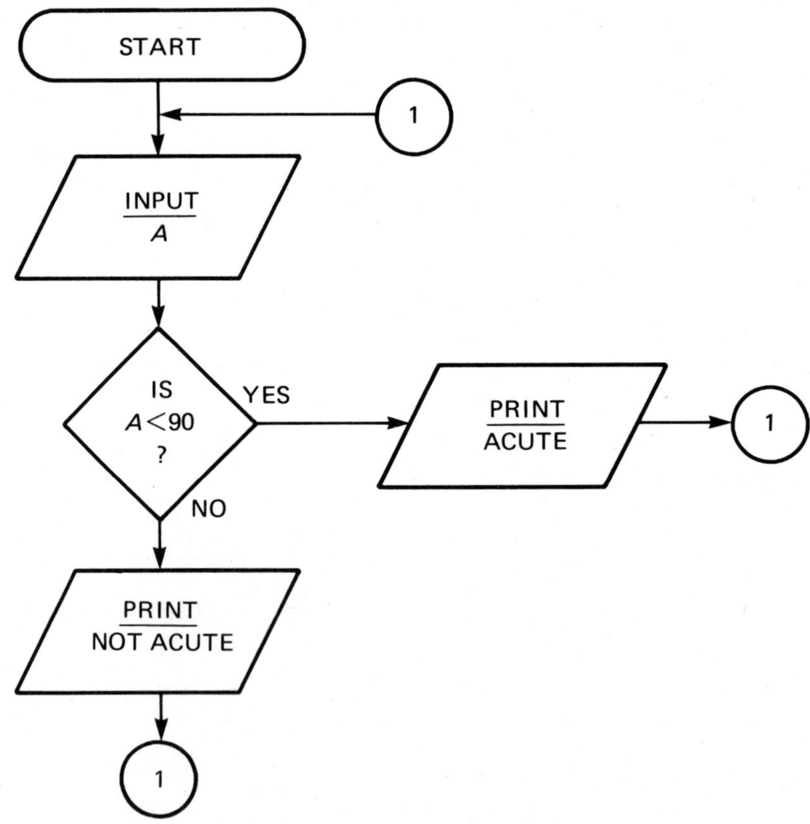

d. *Planned Output*

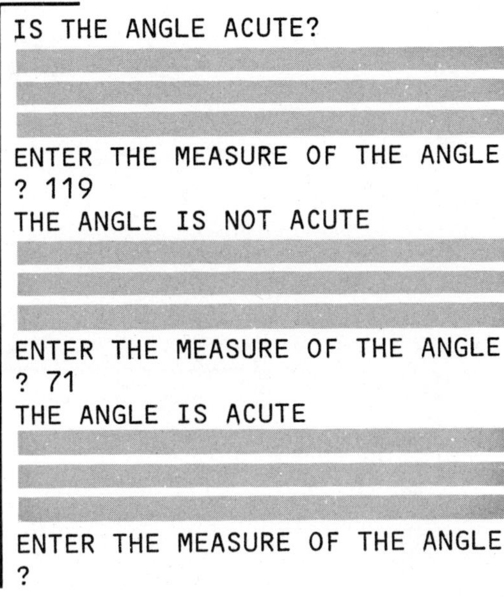

```
IS THE ANGLE ACUTE?



ENTER THE MEASURE OF THE ANGLE
? 119
THE ANGLE IS NOT ACUTE



ENTER THE MEASURE OF THE ANGLE
? 71
THE ANGLE IS ACUTE



ENTER THE MEASURE OF THE ANGLE
?
```

e. *The Program*

```
  10   PRINT "IS THE ANGLE ACUTE?"
■ 20   PRINT: PRINT: PRINT
  30   PRINT "ENTER THE MEASURE OF THE ANGLE"
  40   INPUT A
  50   IF A<90 THEN GOTO 80
  60   PRINT "THE ANGLE IS NOT ACUTE"
  70   GOTO 20
  80   PRINT "THE ANGLE IS ACUTE"
  90   GOTO 20
  100  END
```

Discussion of the Program:

■ In line 20 we used the *colon* for the first time to separate three PRINT statements. The colon allows us to place several statements on the same line, instead of assigning each a line number.

Exercise 4.2 Classify the following triangles as acute, right or obtuse.

1. 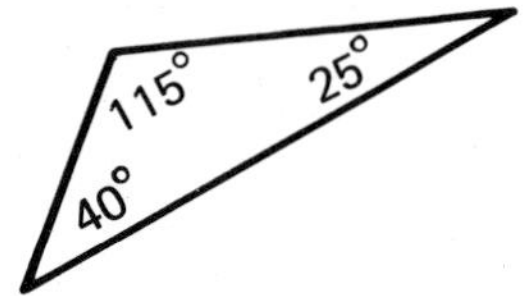

2. 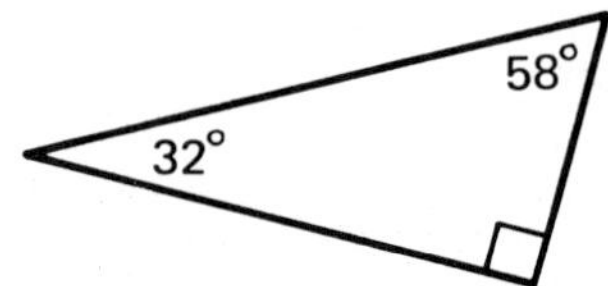

3. 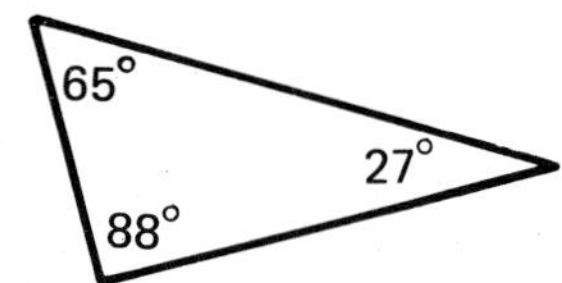

4. 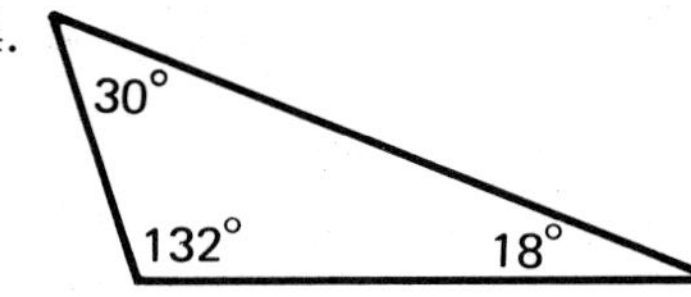

Each of the following is the measures of two angles of a triangle. State whether the triangle is acute, right or obtuse.

5. 45°, 45°
6. 20°, 60°
7. 72°, 45°
8. 126°, 35°

Revise the program that was written in this section to produce the following outputs.

9.
```
IS IT AN ACUTE ANGLE?


PUT IN THE ANGLE'S MEASURE
? 119
THE ANGLE MEASURES 119 DEGREES.
IT IS NOT ACUTE.

PUT IN THE ANGLE'S MEASURE
? 71
THE ANGLE MEASURES 71 DEGREES.
IT IS ACUTE.

PUT IN THE ANGLE'S MEASURE
?
```

10.

```
CLASSIFYING ACUTE ANGLES



WHAT IS THE MEASURE OF THE ANGLE?
? 119
119 IS NOT ACUTE


WHAT IS THE MEASURE OF THE ANGLE?
? 71
71 IS ACUTE


WHAT IS THE MEASURE OF THE ANGLE?
?
```

For the following problem, flowchart and planned output, code a program to solve the problem.

11. **Problem:** Enter the measure of an angle (A). Determine if the angle is obtuse. (Assume the measure of the angle to be positive.)

Flowchart

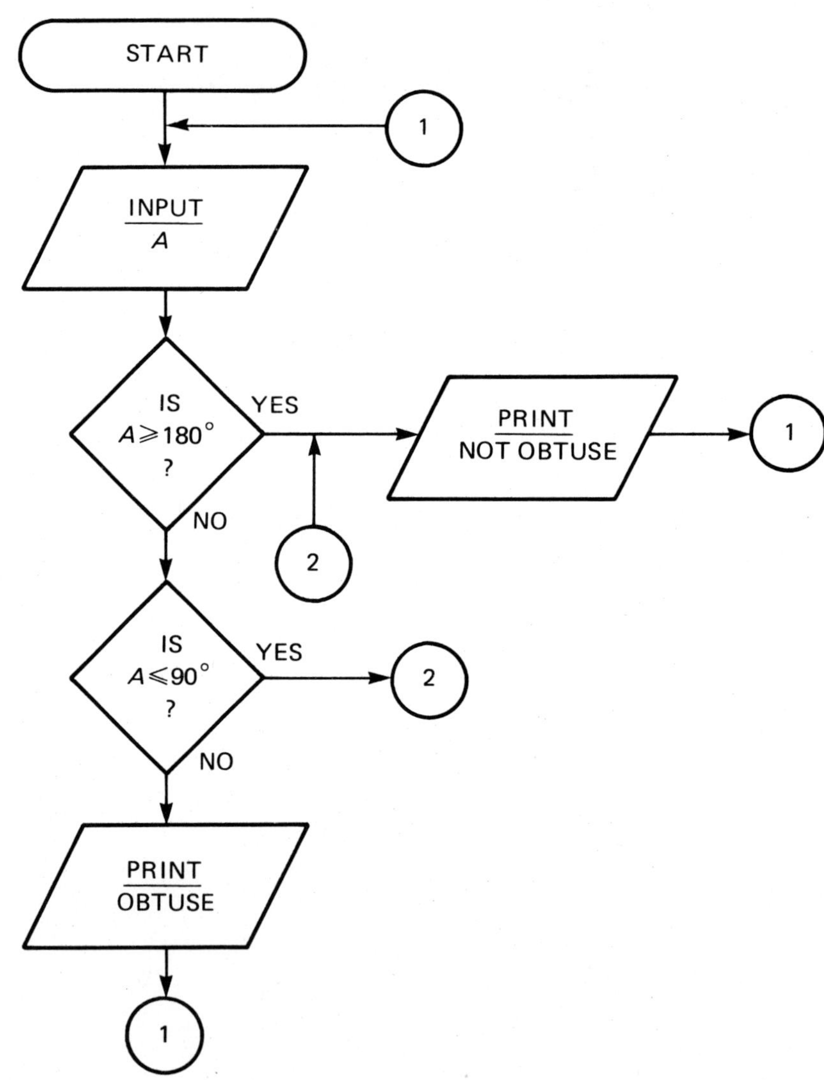

Planned Output

```
IS THE ANGLE OBTUSE?


ENTER THE MEASURE OF THE ANGLE.
? 120
120 DEGREES IS AN OBTUSE ANGLE MEASURE.


ENTER THE MEASURE OF THE ANGLE.
? 212
212 DEGREES IS NOT AN OBTUSE ANGLE MEASURE.


ENTER THE MEASURE OF THE ANGLE.
? 20
20 DEGREES IS NOT AN OBTUSE ANGLE MEASURE.


ENTER THE MEASURE OF THE ANGLE.
?
```

For each of the following problems:

a. Write an analysis which includes the development of an algorithm.
b. Write a plan for a computer solution to the problem.
c. Develop a flowchart.
d. Design a planned output.
e. Code a program to solve the problem and RUN it on the computer.

12. **Problem:** Enter the measure of an angle (A). Determine if the angle is a right angle.

13. **Problem:** Enter three numbers. Determine if the numbers may represent the measures of three angles of a triangle.

14. **Problem:** Enter the measure of an angle (A). Determine if the angle is obtuse. Solve this problem by programming the definition of an obtuse angle. That is, determine if the angle measures greater than 90° but less than 180°. (Do not use the same flowchart as was given in Problem 11.)

Section 4.3 CONVERTING MEASURES (COMBINING PRINT AND INPUT STATEMENTS)

When converting from one unit of measure to another, we must first be told which measure is to be converted. We then need to know the number of units to be converted. This process requires two separate items of information.

Let us see how the computer can be used to solve a problem of this kind. Consider the following:

Problem

If a temperature is entered in Fahrenheit degrees, convert it to Celsius. If the temperature is entered in Celsius degrees, convert it to Fahrenheit.

Solution:

a. *Analysis:* Before accepting a temperature, we must ask the user if the temperature being entered is Fahrenheit or Celsius.

We must set up a code so that a response of 1 is Fahrenheit and 2 is Celsius. A check of the number entered would tell which conversion to use.

b. *Plan:*

1. Print a message that tells the user to enter a 1 for Fahrenheit or 2 if a Celsius temperature is being entered.
2. Enter N.
3. Check:
   a) If $N=1$ then go to 8.
   b) If not, continue.
4. Enter a Celsius temperature (C).
5. $F = \frac{9}{5}C + 32$.
6. Print F.
7. Go to 1.
8. Enter a Fahrenheit temperature (F).
9. $C = \frac{5}{9}(F - 32)$.
10. Print C.
11. Go to 1.

c. *Flowchart*

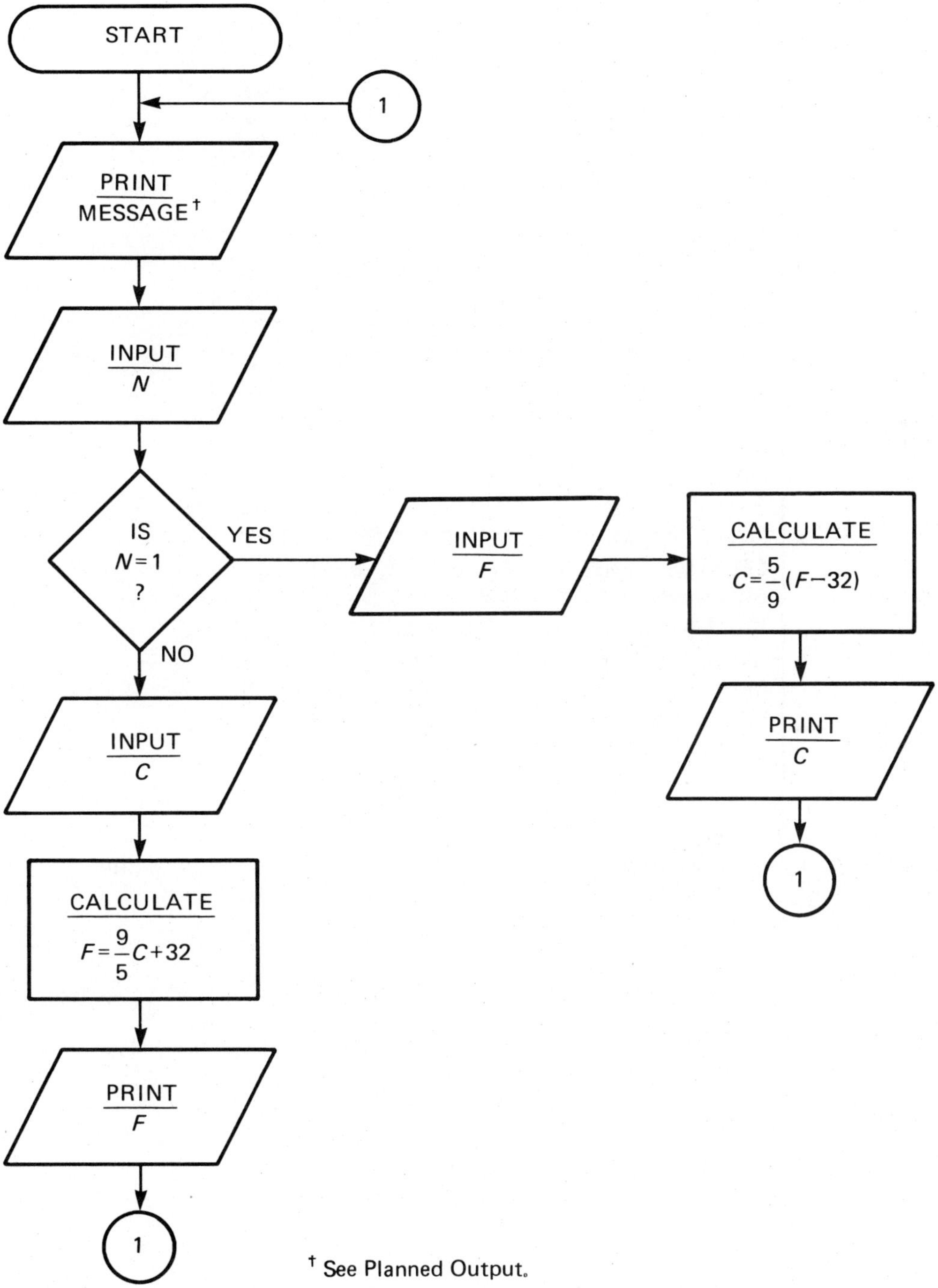

† See Planned Output.

d. *Planned Output*

```
CONVERTING FAHRENHEIT & CELSIUS TEMPS.



ARE YOU ENTERING FAHRENHEIT OR CELSIUS?
ENTER 1 FOR FAHRENHEIT.
ENTER 2 FOR CELSIUS.
? 2
ENTER CELSIUS TEMPERATURE ? 100
FAHRENHEIT TEMPERATURE IS 212 DEGREES.



ARE YOU ENTERING FAHRENHEIT OR CELSIUS?
ENTER 1 FOR FAHRENHEIT.
ENTER 2 FOR CELSIUS.
? 1
ENTER FAHRENHEIT TEMPERATURE ? 212
CELSIUS TEMPERATURE IS 100 DEGREES.



ARE YOU ENTERING FAHRENHEIT OR CELSIUS?
ENTER 1 FOR FAHRENHEIT.
ENTER 2 FOR CELSIUS.
?
```

Discussion of the Planned Output:

- Note that the temperatures are being entered on the same line as the message. See how this is done in the following program.

e. *Program*

```
   10   PRINT "CONVERTING FAHRENHEIT & CELSIUS TEMPS."
   20   PRINT: PRINT: PRINT
   30   PRINT "ARE YOU ENTERING FAHRENHEIT OR CELSIUS?"
   40   PRINT "ENTER 1 FOR FAHRENHEIT."
   50   PRINT "ENTER 2 FOR CELSIUS."
   60   INPUT N
   70   IF N=1 THEN GOTO 120
 ■ 80   INPUT "ENTER CELSIUS TEMPERATURE"; C
   90   F=9/5*C+32
■■ 100  PRINT "FAHRENHEIT TEMPERATURE IS"; F; "ƀDEGREES."
   110  GOTO 20
 ■ 120  INPUT "ENTER FAHRENHEIT TEMPERATURE"; F
   130  C=5/9*(F-32)
■■ 140  PRINT "CELSIUS TEMPERATURE IS"; C; "ƀDEGREES."
   150  GOTO 20
   160  END
```

Discussion of the Program:

■ This INPUT statement signals the computer to print the message inside the quotes and then wait for an INPUT. Note that the message in quotes must be followed by a semi-colon. (On some computers a question mark appears on the screen after the message is printed, as in our planned output. On other computers you will see a blinking cursor. In this book we will follow the message with a question mark.)

■■ ƀ indicates a blank space. Press the space bar to get the blank. (Some computers automatically leave a space.)

Exercise 4.3 Revise the program that was written in this section to produce the following outputs.

1.

```
FAHRENHEIT AND CELSIUS



FAHRENHEIT OR CELSIUS?
ENTER 1 FOR FAHRENHEIT
ENTER 2 FOR CELSIUS ? 2


WHAT IS THE CELSIUS TEMPERATURE? 100
212 DEGREES IS THE FAHRENHEIT TEMP.



FAHRENHEIT OR CELSIUS?
ENTER 1 FOR FAHRENHEIT
ENTER 2 FOR CELSIUS ? 1


WHAT IS THE FAHRENHEIT TEMPERATURE? 212
100 DEGREES IS THE CELSIUS TEMP.



FAHRENHEIT OR CELSIUS?
ENTER 1 FOR FAHRENHEIT
ENTER 2 FOR CELSIUS ?
```

2.

```
CONVERTING F AND C TEMPERATURES


ENTER 1 FOR FAHRENHEIT OR
2 FOR CELSIUS ? 2

ENTER CELSIUS TEMP. ? 100
100 CELSIUS IS 212 FAHRENHEIT


ENTER 1 FOR FAHRENHEIT OR
2 FOR CELSIUS ? 1

ENTER FAHRENHEIT TEMP. ? 212
212 FAHRENHEIT IS 100 CELSIUS


ENTER 1 FOR FAHRENHEIT OR
2 FOR CELSIUS ?
```

For the following problem, flowchart and planned output, code a program to solve the problem.

3. **Problem:** If a number is entered in centimeters, convert it to millimeters. If a number is entered in millimeters, convert it to centimeters.

Discussion of the Problem:
a. To convert centimeters (larger units) to millimeters (smaller units), we multiply. Our algorithm becomes:

$$MM = 10 \cdot CM$$

millimeters (MM) — centimeters (CM)

b. To convert millimeters (smaller units) to centimeters (larger units), we divide. Our algorithm becomes:

$$CM = MM/10$$

centimeters (CM) — millimeters (MM)

Flowchart

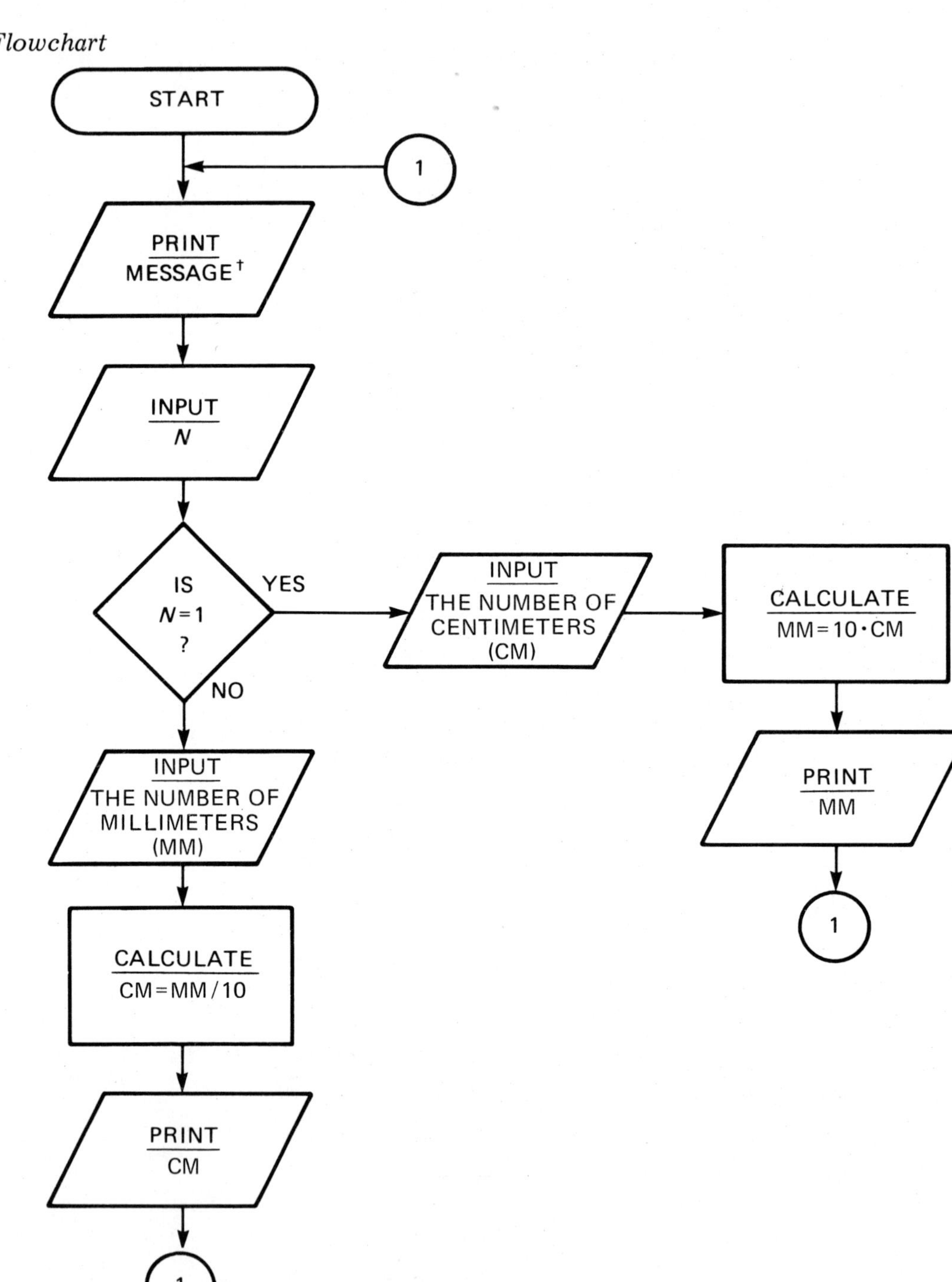

† See Planned Output.

Planned Output

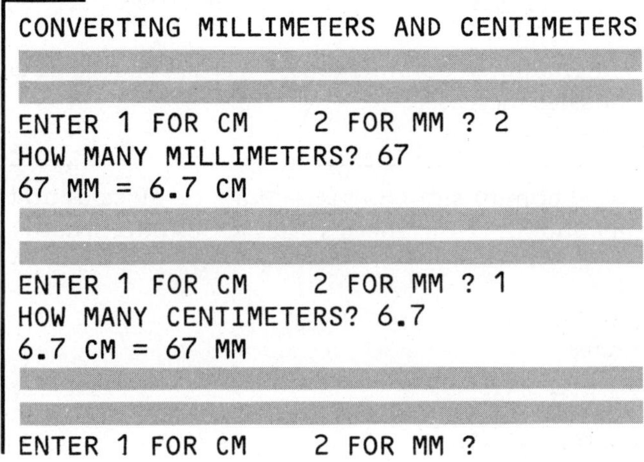

```
CONVERTING MILLIMETERS AND CENTIMETERS


ENTER 1 FOR CM    2 FOR MM ? 2
HOW MANY MILLIMETERS? 67
67 MM = 6.7 CM


ENTER 1 FOR CM    2 FOR MM ? 1
HOW MANY CENTIMETERS? 6.7
6.7 CM = 67 MM


ENTER 1 FOR CM    2 FOR MM ?
```

For each of the following problems:
a. Write an analysis which includes the development of an algorithm.
b. Write a plan for a computer solution to the problem.
c. Develop a flowchart.
d. Design a planned output.
e. Code a program to solve the problem and RUN it on the computer.

4. **Problem:** If a number is entered in feet, convert it to inches. If a number is entered in inches, convert it to feet.

5. **Problem:** If a number is entered in yards, convert it to feet. If a number is entered in feet, convert it to yards.

6. **Problem:** For a square, if the length of a side is entered, find the perimeter. If the perimeter is entered, find the length of a side. (Be sure that the numbers entered are valid sides and perimeters.)

Section 4.4 GRADUATED COMMISSIONS (REM STATEMENTS)

Some salespersons are paid graduated commissions. That is, the rate of commission increases as the amount of sales increases.

Example 1 Elizabeth is paid a graduated commission. She is paid 3% on all sales up to $5000 and 5% on sales in excess of (over) $5000. If Elizabeth's sales are $7000, what is her commission?

Solution:

a. *Determine the excess sales:*

excess sales = total sales − first sales
= 7000 − 5000
excess sales = $2000

b. *Calculate straight commission on the first sales:*

commission on first sales = rate of commission on first sales × first sales
= .03 × 5000
commission on first sales = $150

c. *Calculate straight commission on the excess sales:*

commission on excess sales = rate of commission on excess sales × excess sales
= .05 × 2000
commission on excess sales = $100

d. *Calculate the total commission:*

total commission = commission on first sales + commission on excess sales
= 150 + 100
total commission = $250

Use this example as a model and try the next exercise.

Class Exercise 1 Solve the following:

1. Ronald is paid a graduated commission. He is paid 3% on all sales up to $6000 and 5% of all sales in excess of $6000. If Ronald's sales are $9000, what is his commission?

2. Margaret is paid 4% commission on the first $7000 of her monthly sales and 7% on all sales in excess of $7000. Last month her sales were $9500. What was her commission?

Find the total commission for each of the following:

Sales	*Commission*
3. $8000	6% on the first $5000; 8% on the excess over $5000
4. $12,000	5% on the first $7500; 7% on the excess over $7500

These problems take a long time to solve. Let us put the computer to work to solve the problems more quickly.

Problem

Enter:
a. Total sales (*TS*).
b. First sales (*FS*).
c. Rate of commission on first sales (*FR*).
d. Rate of commission on excess sales (*ER*).

Find the total commission (*CT*).

Solution:

a. *Analysis:* To find the total commission we must:
1. Determine the excess sales.
2. Calculate the commission on the first sales.
3. Calculate the commission on the excess sales.
4. Add the answers from 2 and 3.

(If the total sales are not greater than the first sales there will be no excess sales. Therefore, #3 would not be needed.)

b. *Plan:*

1. Put in values for:
   a) the total sales (*TS*);
   b) the first sales (*FS*);
   c) the rate of commission on first sales (*FR*);
   d) the rate of commission on excess sales (*ER*).
2. Is total sales > first sales?
   a) If yes, go to 6.
   b) If no, continue.
3. Total commission = rate of commission on first sales × total sales.
4. Print total commission.
5. Go to 1.
6. Excess sales = total sales − first sales.
7. Commission on first sales = rate of commission on first sales × first sales.
8. Commission on excess sales = rate of commission on excess sales × excess sales.
9. Total commission = commission on first sales + commission on excess sales.
10. Print total commission.
11. Go to 1.

Before writing the flowchart, let us prepare a legend (a dictionary of terms).

Legend

Variable	*Meaning*
TS	Total Sales
FS	First Sales
FR	Rate of Commission on First Sales
ER	Rate of Commission on Excess Sales
ES	Excess Sales
CF	Commission on First Sales
CE	Commission on Excess Sales
CT	Total Commission

c. *Flowchart*

START
1
INPUT
TS, FS, FR, ER
IS
TS>FS
?
YES
NO
CALCULATE
ES=TS−FS
CALCULATE
CF=FR*FS
CE=ER*ES
CALCULATE
CT=CF+CE
2
CALCULATE
CT=FR*TS
2
PRINT
CT
1

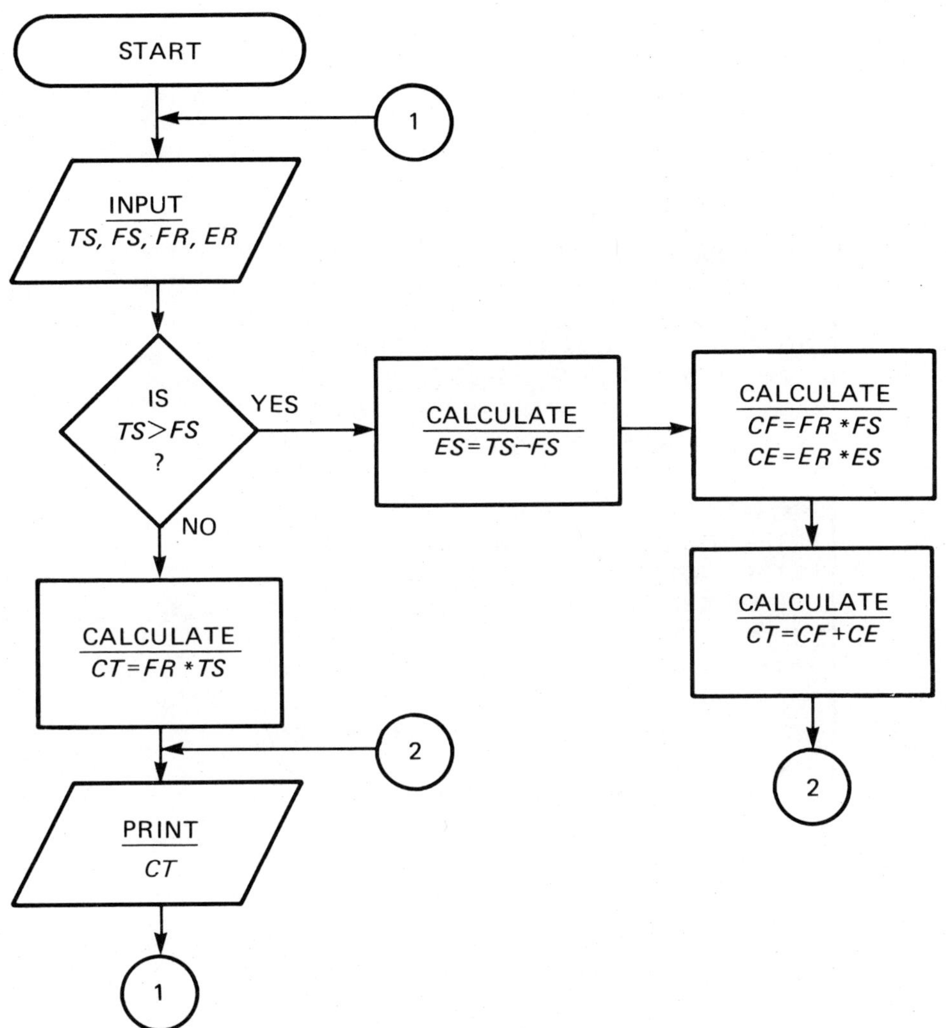

d. *Planned Output*

```
CALCULATING GRADUATED COMMISSIONS



ENTER TOTAL AND FIRST SALES ? 7000, 5000

ENTER RATE OF FIRST COMM. ? .03
ENTER RATE OF EXCESS COMM. ? .05


TOTAL COMMISSION IS 250 DOLLARS



ENTER TOTAL AND FIRST SALES ? 4000, 5000

ENTER RATE OF FIRST COMM. ? .03
ENTER RATE OF EXCESS COMM. ? .05


TOTAL COMMISSION IS 120 DOLLARS



ENTER TOTAL AND FIRST SALES ?
```

e. *The Program*

```
  10   PRINT "CALCULATING GRADUATED COMMISSIONS"
  20   PRINT: PRINT: PRINT
■ 30   INPUT "ENTER TOTAL AND FIRST SALES"; TS, FS
  40   PRINT
■ 50   INPUT "ENTER RATE OF FIRST COMM."; FR
■ 60   INPUT "ENTER RATE OF EXCESS COMM."; ER
  70   PRINT: PRINT
  80   IF TS>FS THEN GOTO 120
  90   CT=FR*TS
  100  PRINT "TOTAL COMMISSION IS"; CT; "ƀDOLLARS"
  110  GOTO 20
  120  ES=TS-FS
  130  CF=FR*FS
  140  CE=ER*ES
  150  CT=CF+CE
  160  GOTO 100
  170  END
```

Discussion of the Program:

■ In lines 30, 50 and 60 we used the form of the INPUT statement that will first print the message in quotes and then await the INPUT.

A programmer may wish to put remarks in a program but not have them executed. For example, in our program above we might wish to place our legend into the program. We can place *REM*ARKS into a program with a REM statement.

> REM is used to insert remarks into a program. When a line begins with REM it is not executed.

For example, we might want to insert the following lines into our program:

1 REM *TS* = TOTAL SALES
2 REM *FS* = FIRST SALES
3 REM *FR* = RATE OF COMM. ON FIRST SALES
4 REM *ES* = EXCESS SALES
5 REM *ER* = RATE OF COMM. ON EXCESS SALES
6 REM *CF* = COMM. ON FIRST SALES
7 REM *CE* = COMM. ON EXCESS SALES
8 REM *CT* = TOTAL COMMISSION

These lines will not show up when the program is RUN. However, we will always have a record (documentation) of what the variables in the program mean.

Exercise 4.4 Find the total commission for each.

	Sales	*Commission*
1.	$4600	4% on first $3200; 6% on excess
2.	$5000	12% on first $5500; 14% on excess
3.	$7200	8% on first $5000; 13% on excess
4.	$14,000	7% on first $10,000; 9% on excess

Revise the program that was written in this section to produce the following outputs.

5.
```
GETTING GRADUATED COMMISSIONS



ENTER TOTAL AND FIRST SALES
? 7000, 5000
ENTER RATE OF FIRST AND EXCESS COMM.
? .03, .05
THE FIRST COMMISSION IS $ 150
THE EXCESS COMMISSION IS $ 100
THE TOTAL COMMISSION IS $ 250



ENTER TOTAL AND FIRST SALES
? 4000, 5000
ENTER RATE OF FIRST AND EXCESS COMM.
? .03, .05
THE TOTAL COMMISSION IS $ 120



ENTER TOTAL AND FIRST SALES
?
```

6.

```
FINDING GRADUATED COMMISSIONS



ENTER TOTAL SALES ? 7000
ENTER FIRST SALES ? 5000

ENTER RATE OF FIRST COMM. ? .03
ENTER RATE OF EXCESS COMM. ? .05


5000 AT .03 IS 150
2000 AT .05 IS 100


TOTAL COMMISSION IS 150+100=250



ENTER TOTAL SALES ? 4000
ENTER FIRST SALES ? 5000

ENTER RATE OF FIRST COMM. ? .03
ENTER RATE OF EXCESS COMM. ? .05


THERE ARE NO EXCESS SALES
TOTAL COMMISSION IS 120



ENTER TOTAL SALES ?
```

For the following problem, flowchart and planned output, code a program to solve the problem.

7. **Problem:** Enter an amount of money (A) and two interest rates ($R1$ and $R2$). Calculate the interest using the greater interest rate. Calculate the interest even if both interest rates are the same.

Flowchart

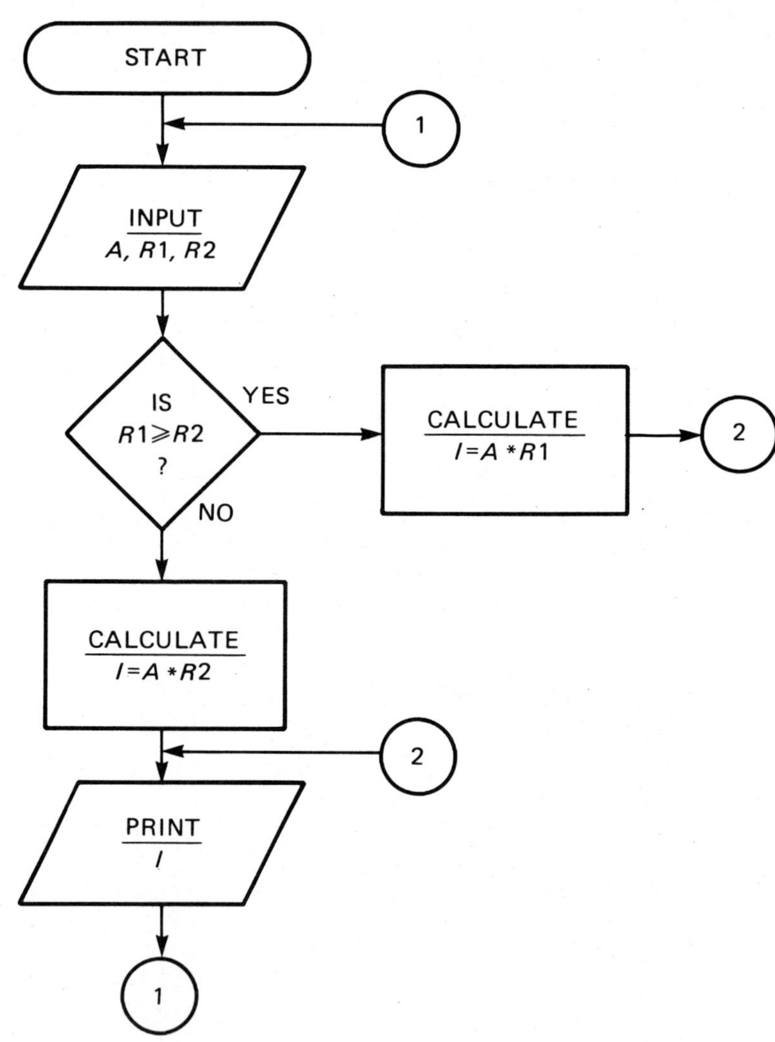

Planned Output

```
GETTING THE HIGHER INTEREST RATE


ENTER AMOUNT AND INTEREST RATES 1 AND 2
? 1000, .05, .07
THE RATE IS .07
THE INTEREST ON 1000 AT .07
IS $ 70


ENTER AMOUNT AND INTEREST RATES 1 AND 2
? 1000, .08, .02
THE RATE IS .08
THE INTEREST ON 1000 AT .08
IS $ 80


ENTER AMOUNT AND INTEREST RATES 1 AND 2
?
```

For each of the following problems:
a. Write an analysis which includes the development of an algorithm.
b. Write a plan for a computer solution to the problem.
c. Develop a flowchart.
d. Design a planned output.
e. Code a program to solve the problem and RUN it on the computer.

8. **Problem:** Enter two amounts of money ($A1$ and $A2$) and a rate of interest (R). Calculate the interest on the greater amount. Calculate the interest even if both amounts are the same.

9. **Problem:** Revise the graduated commissions program so that if someone has excess sales they get a $45 bonus added to their total commission.

10. **Problem:** The employees at McDonald's Pizza House are paid for a 40-hour week with double-time for overtime. Enter the number of hours worked by an employee and the hourly wage. Display the amount earned for the week including overtime (if any).

Section 4.5 DIVISIBILITY BY 2, 3, 4, 5, 9 AND 11 (INT(X))

Does 3 divide 16 without a remainder?

```
Quotient ──────► 5 R1 ◄── Remainder
                 ______
Divisor ───► 3 / 16
                  ▲
Dividend ─────────┘
```

Since the remainder is 1, we can say that 3 does not divide 16, or, 16 *is not divisible* by 3.

Is 4161 divisible by 3?

```
      1387  R0
    ________
  3 / 4161
```

This time the remainder is 0. Therefore, 4161 *is divisible* by 3.

We can always test for divisibility by dividing and examining the remainder. Whenever the remainder is 0, the dividend is divisible by the divisor.

How can we tell if a number is even?

We know that if a number is even it is divisible by 2. But instead of dividing by 2 and looking for a remainder of 0, we know a shortcut.

Rule for Divisibility by 2:

If the last digit of a number is 0, 2, 4, 6 or 8, the number is divisible by 2.

Example 1 Is 7316 divisible by 2?

Solution: Yes, since the last digit is a 6.

How can we tell if a number is divisible by 5?

Of course we can divide the number by 5 and check for a 0 remainder. But we know a shortcut.

Rule for Divisibility by 5:

If the last digit of a number is a 0 or 5, the number is divisible by 5.

Example 2 Is 7415 divisible by 5?

Solution: Yes, since the last digit is 5.

Are there shortcuts to determine if numbers are divisible by 3, 4, 9 and 11? The answer is yes. But these rules are not as easy to see as the rules for 2 and 5. Let us examine them. They are not difficult and you should enjoy learning them.

Rule for Divisibility by 3:

A number is divisible by 3 if the sum of the digits of the number is divisible by 3.

Example 3 Is 5271 divisible by 3?

Solution: We find the sum of the digits: $5 + 2 + 7 + 1 = 15$.
Since 15 is divisible by 3, 5271 is divisible by 3.

Rule for Divisibility by 4:

A number is divisible by 4 if the number formed by the last two digits is divisible by 4.

Example 4 Is 5283 divisible by 4?

Solution: Since 83 is not divisible by 4, 5283 is not divisible by 4.

Class Exercise 1 Use the divisibility rules previously stated, to determine if each of the numbers is

1. divisible by 2:	a. 3185	b. 1358	c. 5813
2. divisible by 3:	a. 7164	b. 6471	c. 1746
3. divisible by 4:	a. 9238	b. 3928	c. 8329
4. divisible by 5:	a. 7105	b. 5107	c. 7510

Rule for Divisibility by 9:

A number is divisible by 9 if the sum of the digits of the number is divisible by 9.

Example 5 Is 5283 divisible by 9?

Solution: Since the sum of the digits, 5 + 2 + 8 + 3 = 18, is divisible by 9, 5283 is divisible by 9.

Rule for Divisibility by 11:

> A number is divisible by 11 if the difference of the sums of the alternate digits is divisible by 11.

Example 6 Is 7183 divisible by 11?

Solution: Find the sums of the alternate digits.

```
  1 + 3   = 4
  ┌───┐
7 1 8 3
└───┘
7 + 8     = 15
```

Calculate the difference of the sums:

15 − 4 = 11

Since 11 is divisible by 11, 7183 is also divisible by 11.

Class Exercise 2 Determine if the numbers are (use the divisibility rules)

1. divisible by 9:	a. 7812	b. 1287	c. 2939
2. divisible by 11:	a. 8294	b. 132891	c. 41357
3. divisible by 3:	a. 21582	b. 51282	c. 13333
4. divisible by 4:	a. 5616	b. 6561	c. 31008

Class Exercise 3 Test each of the following for divisibility by 2, 3, 4, 5, 9 and 11.

1. 1024 2. 90937 3. 6876 4. 5434

Class Exercise 4 Using the divisibility rules, find the value of d which will make the numbers

1. divisible by 9: $53d87$
2. divisible by 4: $682d$ (give three values for d)
3. divisible by 3: $782d1$ (give four values for d)
4. divisible by 11: $2d563$

Suppose we want to determine if 4182 is divisible by 5. We can either:

a. check to see if the last digit is either a 0 or 5, or
b. divide the entire number by 5 and check for a 0 remainder.

How can we get the computer to determine if a number is divisible by 5?

To do this we will introduce the idea of a BASIC *function.*

The BASIC language offers (as an added feature) BASIC functions, a group of small programs that are stored inside the computer. The programmer can use these functions by writing the code name for the function in the correct place in the program.

The first BASIC function we will discuss is called the INTEGER function.

> The INTEGER function, written INT(X) where X is called the argument of the function, returns the largest integer not greater than the argument X.

For example,

INT(4.7) = 4; INT(19.8) = 19; INT(12.1) = 12 and INT(−2.1) = −3.

(It seems that the answer should be −2, but $-2 > -2.1$.)

Class Exercise 5 Find the value of each.

1. INT(2.6)
2. INT(14.1)
3. INT(23.7)
4. INT(−5.2)

Find the value assigned to L after executing the instruction.

5. 30 L = INT(7.5)
6. 20 L = INT(−7.3)
7. 90 L = 10 * INT(7.8)
8. 70 L = INT(3.6 + .5)

Consider the statement:

INT $(X/5) * 5 = X$

When $X = 20$, is the statement true or false?
By putting in 20 for X, we get:

INT $(20/5) * 5 = 20$?

INT $(4) * 5 = 20$?

$4 * 5 = 20$?

$20 = 20$

This statement is true when $X = 20$.
Is the statement true when $X = 12$?
This time we replace X by 12 and get:

INT $(12/5) * 5 = 12$?

INT $(2.4) * 5 = 12$?

$2 * 5 = 12$?

$10 \neq 12$

This statement is false when $X = 12$.
In fact, whenever X is divisible by 5, the statement INT $(X/5) * 5 = X$ is true. When X is not divisible by 5, the statement is false.
Let us now use this statement to develop a computer solution to the following problem.

Problem

Enter a number. Determine if the number is divisible by 5.

Solution:
a. *Analysis:* To determine if a number is divisible by 5 we will use the algorithm

INT $(X/5) * 5 = X$

b. *Plan:*
1. Put in a value for X.
2. Is INT $(X/5) * 5 = X$?
   a) If the answer is yes, then print X is divisible by 5. Go to 1.
   b) If the answer is no, then print X is not divisible by 5. Go to 1.

c. *Flowchart*

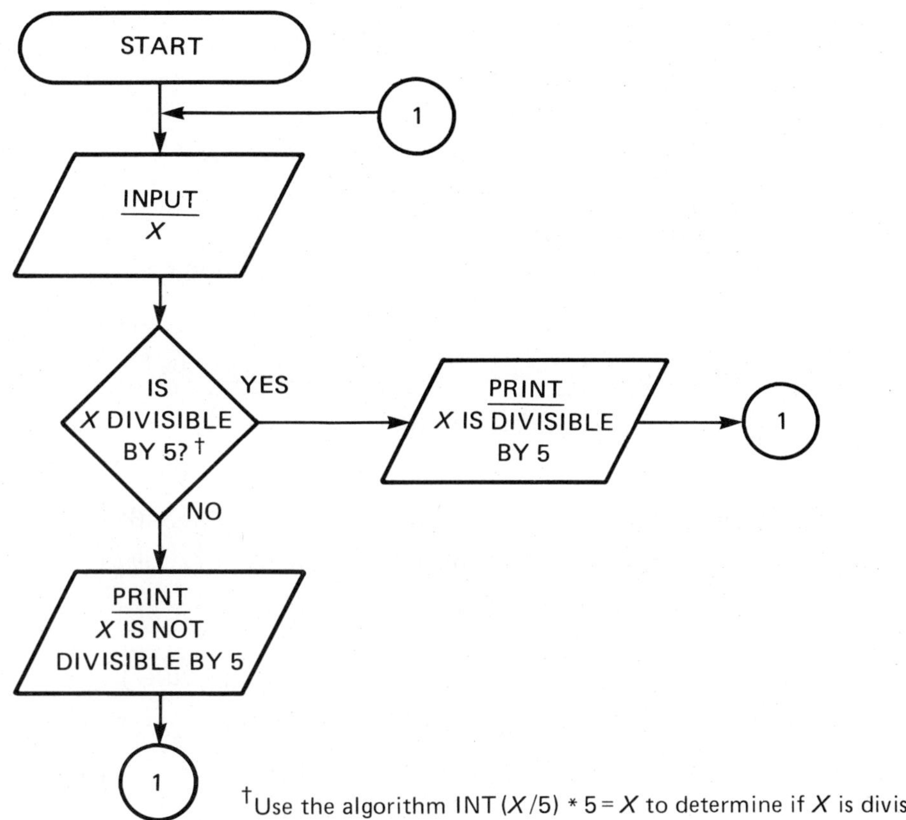

†Use the algorithm INT(X/5) * 5 = X to determine if X is divisible by 5.

d. *Planned Output*

```
IS THE NUMBER DIVISIBLE BY 5?



ENTER A NUMBER ? 361
361 IS NOT DIVISIBLE BY 5.


ENTER A NUMBER ? 450
450 IS DIVISIBLE BY 5.


ENTER A NUMBER ?
```

e. *The Program*

```
10  PRINT "IS THE NUMBER DIVISIBLE BY 5?"
20  PRINT: PRINT: PRINT
30  INPUT "ENTER A NUMBER"; X
40  IF INT(X/5)*5=X THEN GOTO 70
50  PRINT X; "ƀIS NOT DIVISIBLE BY 5."
60  PRINT: PRINT: GOTO 30
70  PRINT X; "ƀIS DIVISIBLE BY 5."
80  GOTO 60
90  END
```

Exercise 4.5

Determine if the numbers are (use the rules)

1. divisible by 2: a. 9267 b. 7296 c. 6972
2. divisible by 3: a. 8466 b. 6481 c. 4686
3. divisible by 4: a. 1924 b. 4912 c. 9214
4. divisible by 5: a. 6203 b. 2035 c. 6010

Determine if the following numbers are (use the rules)

5. divisible by 9: a. 4536 b. 6453 c. 8121
6. divisible by 11: a. 73876 b. 20801 c. 4162
7. divisible by 3: a. 5142 b. 4251 c. 3176
8. divisible by 4: a. 8004 b. 4008 c. 4125

Test each of the following for divisibility by 2, 3, 4, 5, 9 and 11:

9. 2016   10. 9130   11. 54810   12. 638112

Using the divisibility rules, find the value of C which will make the numbers

13. divisible by 9: $80C43$
14. divisible by 4: $713C$ (give two values for C)
15. divisible by 3: $684C2$ (give three values for C)
16. divisible by 11: $248C9$

Find the value of each of the following:

17. INT (8.2)
18. INT (13.9)
19. INT (6)
20. INT (−4.2)

Find the value assigned to M after executing the instruction.

21. 60 M = INT (36.1)
22. 10 M = INT (7.2 + .5)
23. 40 M = 100 * INT (7.6)
24. 20 M = INT (10.7) / 5

Revise the program that was written in this section to produce the following outputs.

25.

```
DIVISIBILITY BY 5
ENTER A NUMBER ? 361
THE NUMBER IS NOT DIVISIBLE BY 5



DIVISIBILITY BY 5
ENTER A NUMBER ? 450
THE NUMBER IS DIVISIBLE BY 5



DIVISIBILITY BY 5
ENTER A NUMBER ?
```

26.

```
IS IT DIVISIBLE BY 5?


WHAT IS THE NUMBER? 361
361 IS NOT DIVISIBLE BY 5


WHAT IS THE NUMBER? 450
450 IS DIVISIBLE BY 5


WHAT IS THE NUMBER?
```

For the following problem, flowchart and planned output, code a program to solve the problem.

27. **Problem:** Enter two numbers (F, S). Determine if the first number, F, is divisible by the second number, S. (*Special Case:* What value cannot be assigned to S? See ♦ in the flowchart.)

Flowchart

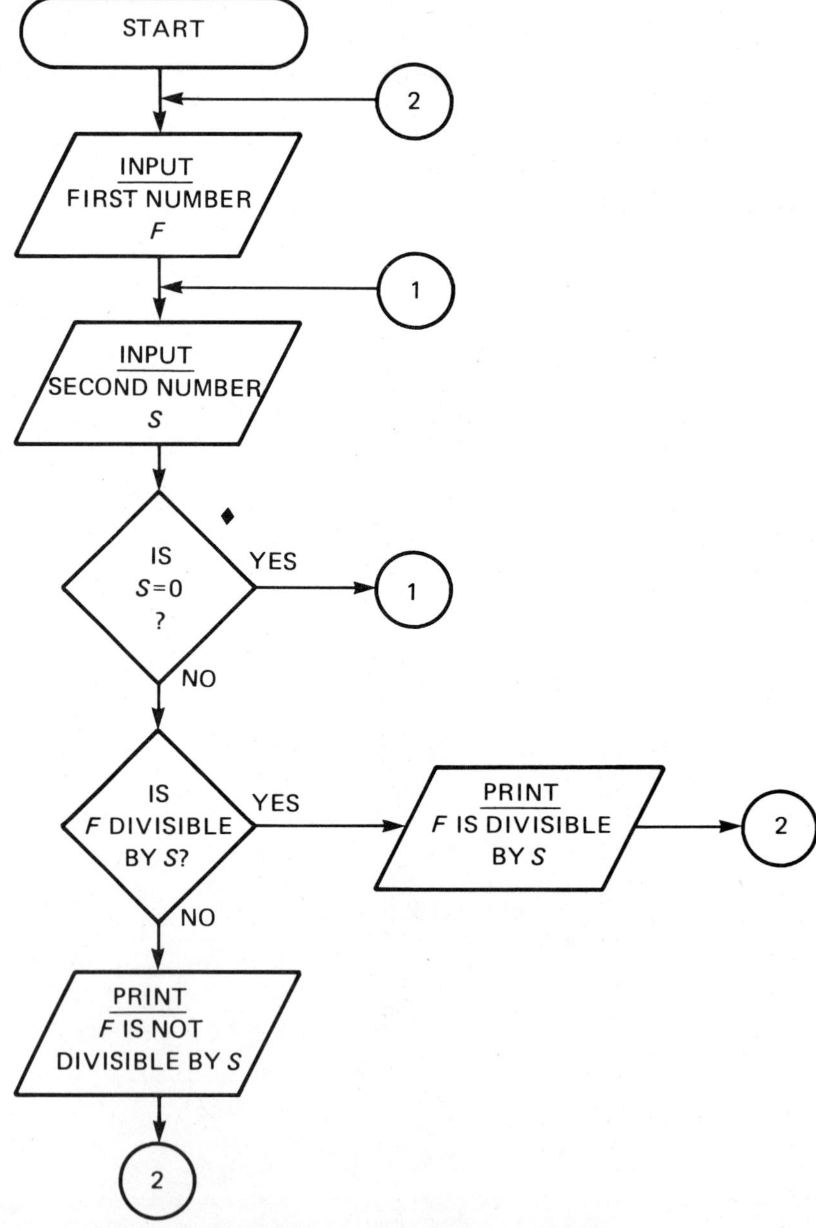

Planned Output

```
IS F DIVISIBLE BY S?
ENTER FIRST NUMBER ? 42
ENTER SECOND NUMBER ? 5
FIRST IS NOT DIVISIBLE BY SECOND


ENTER FIRST NUMBER ? 42
ENTER SECOND NUMBER ? 0
SECOND NUMBER CANNOT BE ZERO!
ENTER SECOND NUMBER ? 14
FIRST NUMBER IS DIVISIBLE BY SECOND


ENTER FIRST NUMBER ?
```

For each of the following problems:
a. Write an analysis which includes the development of an algorithm.
b. Write a plan for a computer solution to the problem.
c. Develop a flowchart.
d. Design a planned output.
e. Code a program to solve the problem and RUN it on the computer.

28. **Problem:** Enter a number. Determine if it is divisible by 6.

29. **Problem:** Enter a number. Determine if it is even.

30. **Problem:** Enter two nonzero numbers. Determine if the larger number is divisible by the smaller number. (As a bonus, revise the program to allow the entry of any numbers.)

END OF CHAPTER EXERCISES

Section 4.1 Translate into symbols:

1. two increased by five
2. eight decreased by six
3. the product of two and five
4. five times three increased by one
5. ℓ decreased by seven
6. *d* increased by seventeen
7. two times *s* increased by four
8. twelve times *e* decreased by nine

Translate into symbols. Use X to represent the number.

9. a number decreased by nineteen
10. a number increased by fifteen
11. three times a number
12. two times a number decreased by five
13. Six times a number increased by one is equal to forty-three.
14. Four times a number decreased by eight is equal to twelve.

Solve the equations. Check each result.

15. $2a + 7 = 9$
16. $4x - 72 = 28$
17. $3c + 75 = 99$
18. $9b - 104 = 67$

Solve the following problems. Check your results.

19. Three times a number increased by 47 is equal to 92. Find the number.

20. Six times a number decreased by 102 is 36. Find the number.

21. Five times a number increased by 76 is 261. Find the number.

22. Double a number decreased by 19 is 107. Find the number.

Use the algorithm $n = \frac{c - b}{a}$ to solve the following equations. Check each result.

23. $8n + 4 = 172$
24. $3n + 2 = 227$
25. $14n + 15 = 127$
26. $6n + 17 = 167$

For the given problem, flowchart and planned output, code a program to solve the problem.

27. **Problem:** Solve $\frac{x}{a} - b = c$ for x.

Flowchart

START
1
INPUT
A
IS
A=0
?
YES
PRINT
A CANNOT
BE ZERO
1
NO
INPUT
B, C
CALCULATE
X=A•(C+B)
PRINT
X
1
Planned Output
SOLVING X/A-B=C.
PUT IN A VALUE FOR A
? 2
PUT IN VALUES FOR B AND C
? 3, 1
FOR X/2-3=1
X=8
PUT IN A VALUE FOR A
? 0
A CANNOT BE ZERO
PUT IN A VALUE FOR A
?

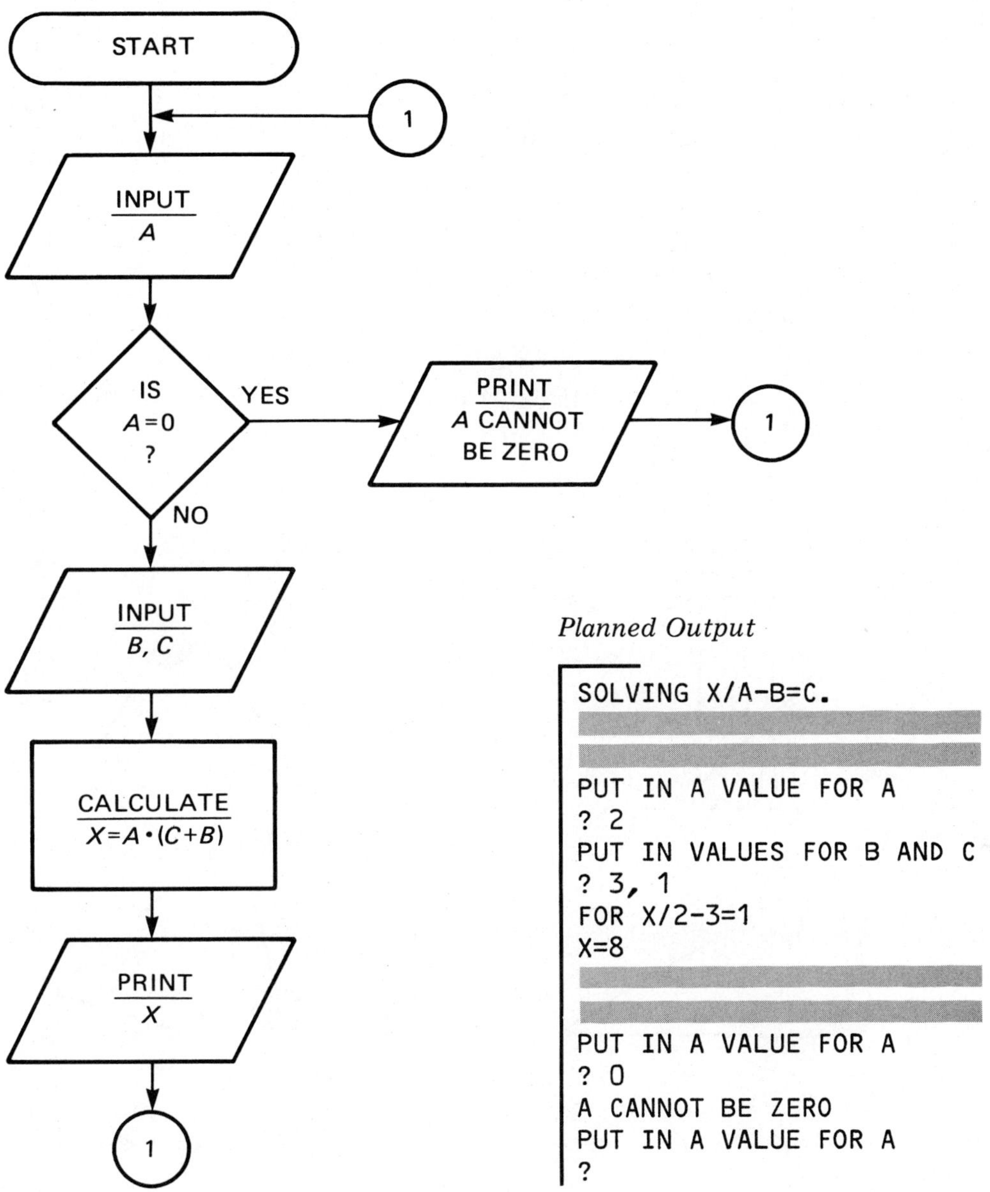

For each of the following problems:
a. Write an analysis which includes the development of an algorithm.
b. Write a plan for a computer solution to the problem.
c. Develop a flowchart.
d. Design a planned output.
e. Code a program to solve the problem and RUN it on the computer.

28. **Problem:** Solve $\frac{x}{a}=b$ for x.

29. **Problem:** Solve $\frac{x}{a}+b=c$ for x.

30. **Problem:** Enter the base (length) and height (width) of a rectangle. Display the perimeter and area.

Section 4.2

Classify the following triangles as acute, right or obtuse

31.

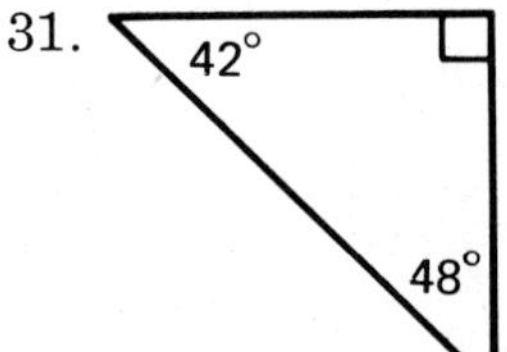

32.

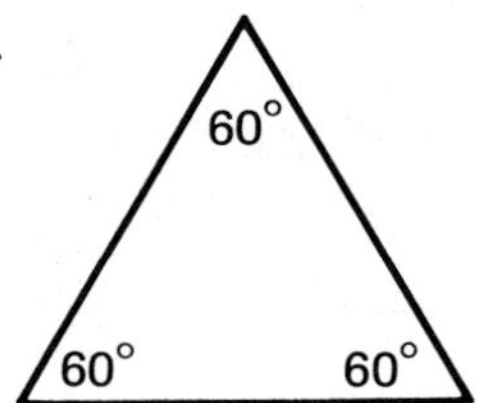

33.

48°
100°
32°

34.

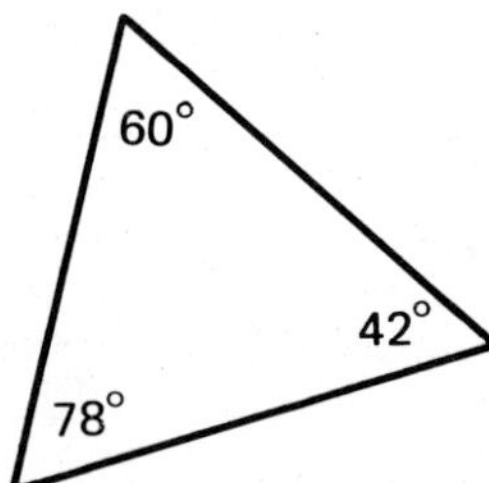

Each of the following are measures of two angles of a triangle. State whether the triangle is acute, right or obtuse.

35. 68°, 55°   36. 72°, 18°   37. 30°, 30°   38. 41°, 49°

For the following problem, flowchart and planned output, code a program that will solve the problem.

39. **Problem:** An isosceles triangle has two congruent (equal) sides. It also has two congruent angles (called base angles). Enter the measure of a base angle of an isosceles triangle. Find the measure of the vertex (non-base) angle. (Assume that the base angle is acute.)

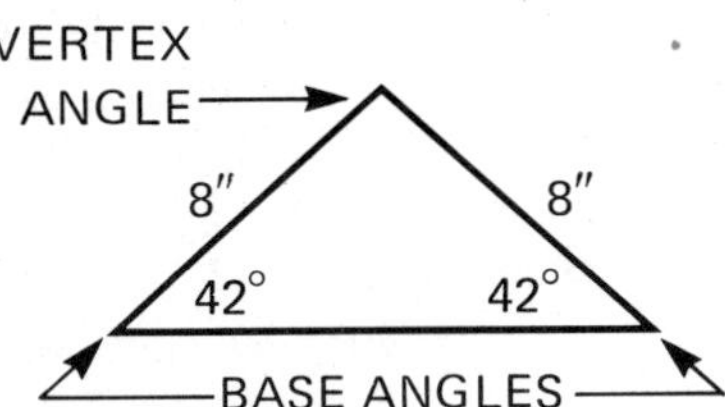

Flowchart

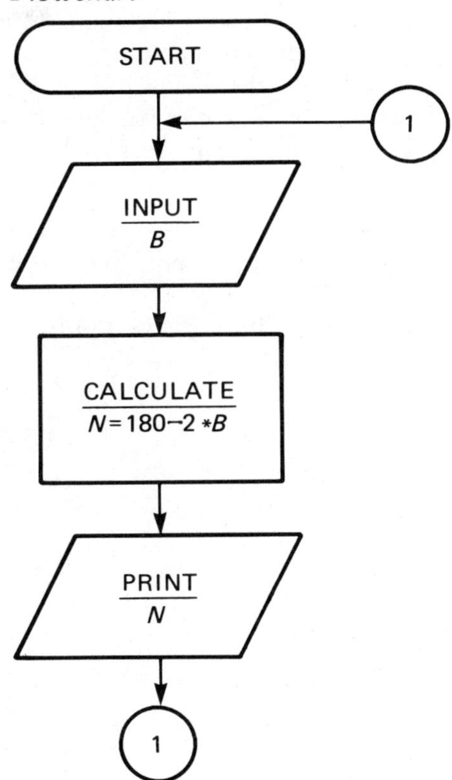

Planned Output

```
FINDING THE VERTEX ANGLE


ENTER THE BASE ANGLE
? 42
THE BASE ANGLES MEASURE 42
THE VERTEX ANGLE MEASURES 96 DEGREES


ENTER THE BASE ANGLE
?
```

For each of the following problems:
a. Write an analysis which includes the development of an algorithm.
b. Write a plan for a computer solution to the problem.
c. Develop a flowchart.
d. Design a planned output.
e. Code a program to solve the problem and RUN it on the computer.

40. **Problem:** A straight angle is an angle whose measure is 180°. Enter a number. Determine if that number may represent the measure of a straight angle.

41. **Problem:** Enter the measure of a base angle of an isosceles triangle. Find the measure of the vertex angle. (Do not assume that the angle entered is acute.)

42. **Problem:** Enter the measure of a base angle of an isosceles triangle. Determine if the triangle is obtuse. (Do not assume that the angle entered is acute.)

Section 4.3

For the problem, flowchart and planned output, code a program to solve the problem.

43. **Problem:** Enter A and B. Have the user choose to have the computer solve either $AX=B$ or $X/A=B$, for X.

Flowchart

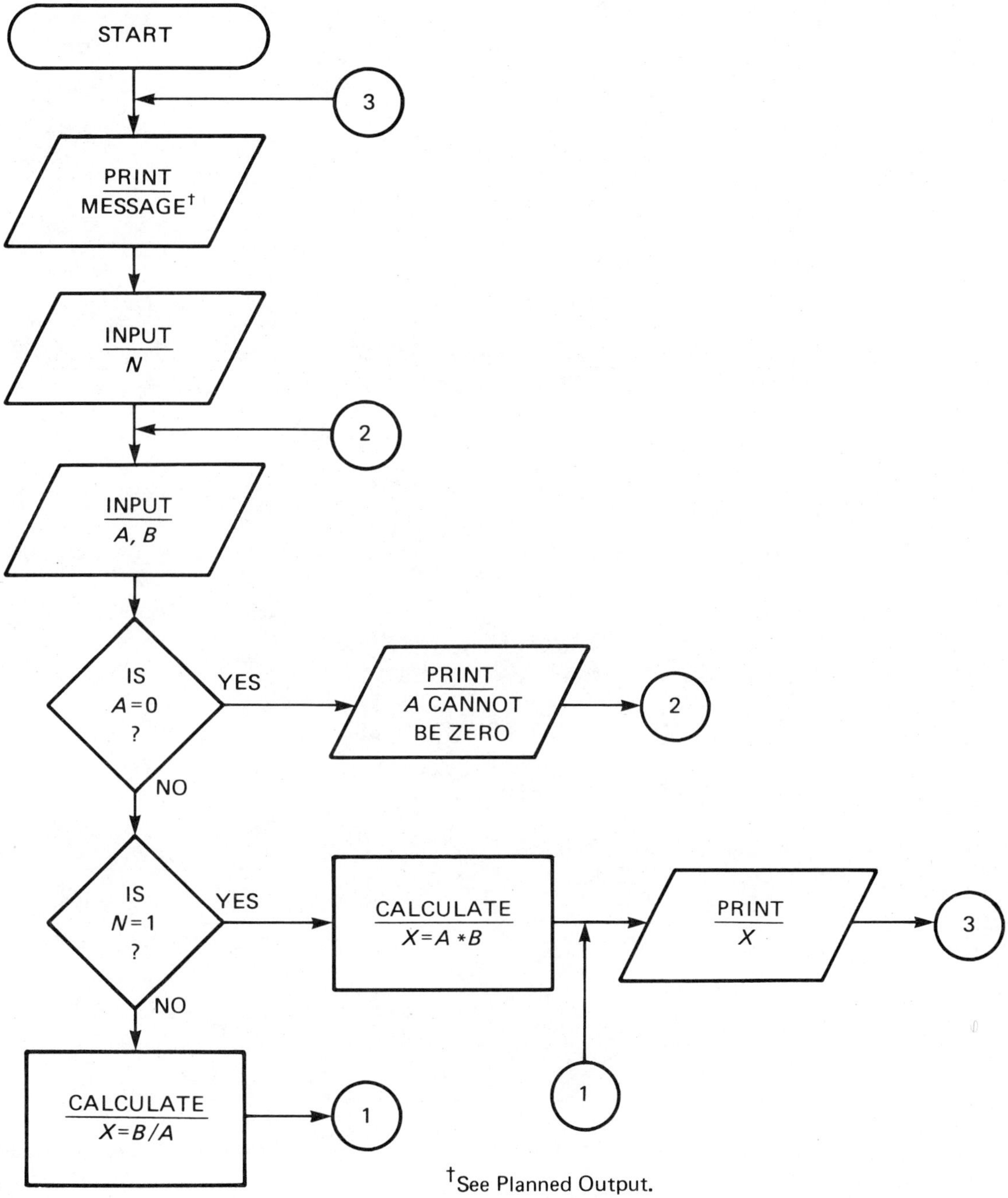

START

3

PRINT MESSAGE†

INPUT N

2

INPUT A, B

IS A=0 ?

YES

PRINT A CANNOT BE ZERO

2

NO

IS N=1 ?

YES

CALCULATE X=A*B

PRINT X

3

NO

CALCULATE X=B/A

1

1

†See Planned Output.

Planned Output

```
SOLVING AX=B OR X/A=B



YOU CAN HAVE EITHER EQUATION SOLVED
ENTER 1 TO SOLVE X/A=B
ENTER 2 TO SOLVE AX=B
? 2
ENTER A AND B ? 2, 6
FOR 2X=6
X=3



YOU CAN HAVE EITHER EQUATION SOLVED
ENTER 1 TO SOLVE X/A=B
ENTER 2 TO SOLVE AX=B
? 1
ENTER A AND B ? 2, 6
FOR X/2=6
X=12



YOU CAN HAVE EITHER EQUATION SOLVED
ENTER 1 TO SOLVE X/A=B
ENTER 2 TO SOLVE AX=B
? 1
ENTER A AND B ? 0, 6
A CANNOT BE ZERO
ENTER A AND B ?
```

For each of the following problems:
a. Write an analysis which includes the development of an algorithm.
b. Write a plan for a computer solution to the problem.
c. Develop a flowchart.
d. Design a planned output.
e. Code a program to solve the problem and RUN it on the computer.

44. **Problem:** Enter a number of inches. Convert the inches to feet and then to yards. (Be sure the number entered is positive.)

45. **Problem:** If a number is entered in inches, convert it to yards. If a number is entered in yards, have the computer convert it to inches.

46. **Problem:** Enter a number of feet. Have the user decide whether the computer converts the number to inches or yards.

Section 4.4

Solve the following problems.

47. Diane is paid a graduated commission. She is paid 4% on all sales up to $7500 and 6% on all sales in excess of $7500. If Diane's sales are $10,000, what is her commission?

48. William is paid 3% commission on the first $8000 of his monthly sales and 5% on all sales in excess of $8000. Last month his sales were $9400. What was his commission?

Find the total commission for each of the following.

	Sales	*Commission*
49.	$11,000	2% on the first $7400; 5% on the excess over $7400
50.	$13,500	3% on the first $8500; 6% on the excess over $8500
51.	$7800	5% on the first $4000; 8% on the excess over $4000
52.	$16,000	4% on the first $14,000; 5% on the excess over $14,000

For the following problem, flowchart and planned output, code a program to solve the problem.

53. **Problem:** Eleanor is paid a commission as follows:

a. 7% if the sales are more than $12,000.
b. 5% if the sales are from $8500 to $12,000.
c. 4% for sales under $8500.

Enter her sales. Display her commission.

Flowchart

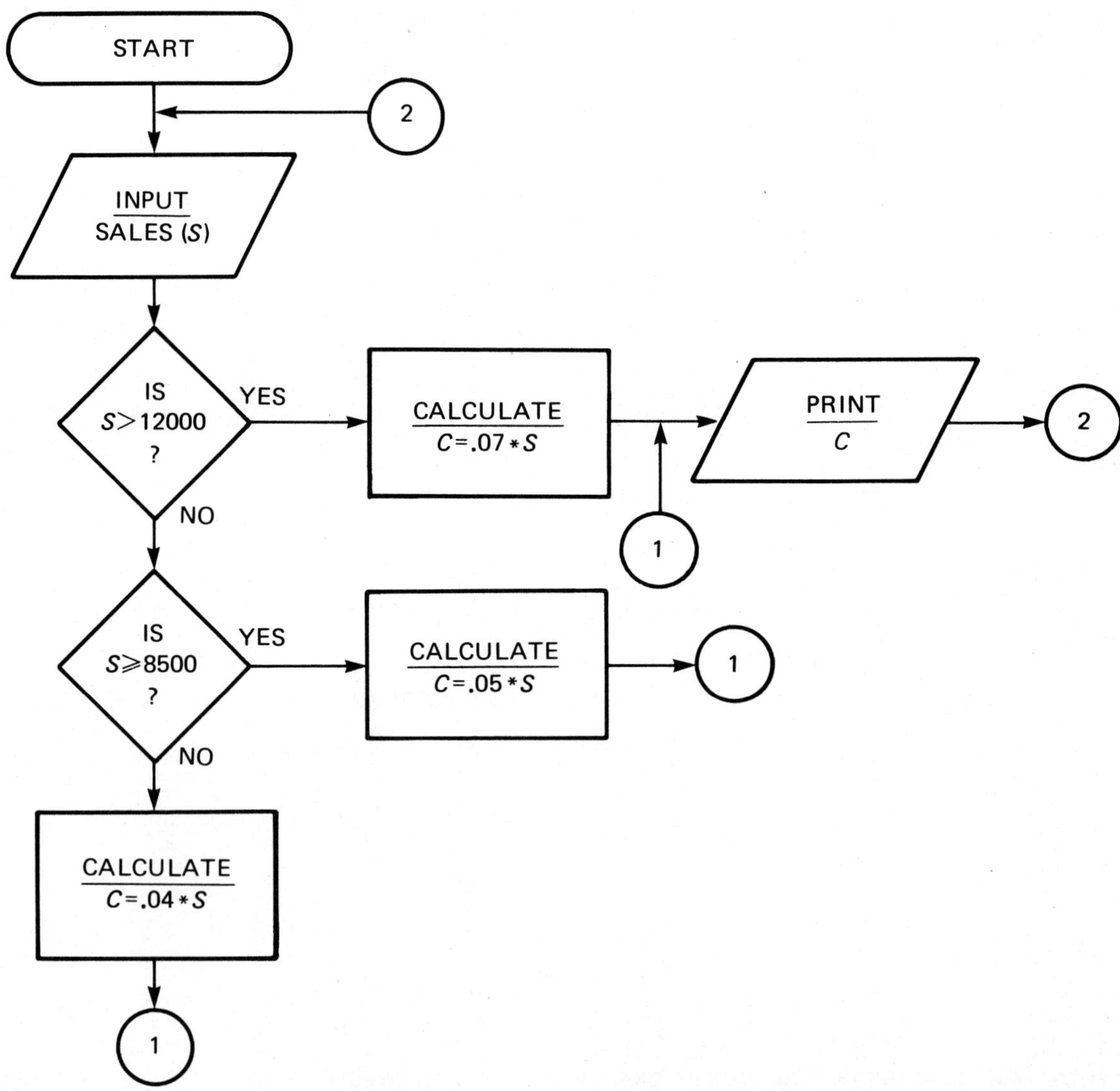

Planned Output

```
ELEANOR'S SALES COMMISSION


ENTER SALES ? 5000
FOR 5000 COMMISSION IS 200 DOLLARS


ENTER SALES ? 14000
FOR 14000 COMMISSION IS 980 DOLLARS


ENTER SALES ? 9000
FOR 9000 COMMISSION IS 450 DOLLARS


ENTER SALES ?
```

For each of the following problems:
a. Write an analysis which includes the development of an algorithm.
b. Write a plan for a computer solution to the problem.
c. Develop a flowchart.
d. Design a planned output.
e. Code a program to solve the problem and RUN it on the computer.

54. **Problem:** Robert is paid a commission as follows:

a. 8% if his sales are more than $5000.
b. 5% if his sales are less than or equal to $5000.

Enter his sales. Display his commission.

55. **Problem:** Ada is paid a commission as follows:

a. 9% if the sales are more than $10,000.
b. 6% if the sales are from $4500 to $10,000.
c. 3% for sales under $4500.

Enter her sales. Display her commission.

56. **Problem:** The employees at Alice's Restaurant are paid for a 40-hour week with time-and-one-half for overtime. Enter the number of hours worked by an employee and his hourly wage. Display the amount earned for the week including overtime (if any).

Section 4.5

Determine if the numbers are

57. divisible by 2: a. 8136 b. 6183 c. 3618
58. divisible by 3: a. 5523 b. 3552 c. 3251
59. divisible by 4: a. 8435 b. 3548 c. 5384
60. divisible by 5: a. 9203 b. 3290 c. 4195

Determine if the following numbers are

61. divisible by 9: a. 8876 b. 7868 c. 9994
62. divisible by 11: a. 7480 b. 61824 c. 46618
63. divisible by 3: a. 33233 b. 5202 c. 65142
64. divisible by 4: a. 9112 b. 1804 c. 5914

Test each of the following for divisibility by 2, 3, 4, 5, 9 and 11:

65. 8211 66. 5418 67. 5220 68. 7073

Using the divisibility rules, find the value of e which will make the number

69. divisible by 9: $241e2$ (give two values for e)
70. divisible by 4: $702e6$ (give five values for e)
71. divisible by 3: $61e72$ (give three values for e)
72. divisible by 11: $6e705$

Find the value of each of the following:

73. INT (6.1) 74. INT (6.7) 75. INT (18.2) 76. INT (−18.2)

Find the value assigned to P after executing the instruction.

77. 20 P = INT (8.2 + .5)
78. 30 P = INT (8.6 + .5)
79. 70 P = INT (100 * (4.26 + .5)) / 100
80. 60 P = INT (100 * (4.24 + .5)) / 100

For the given problem, flowchart, and planned output, code a program to solve the problem.

81. **Problem:** Enter a two-digit number. Display the ten's digit on one line and the unit's digit on the next line. (Assume a positive two-digit number is entered.)

Flowchart

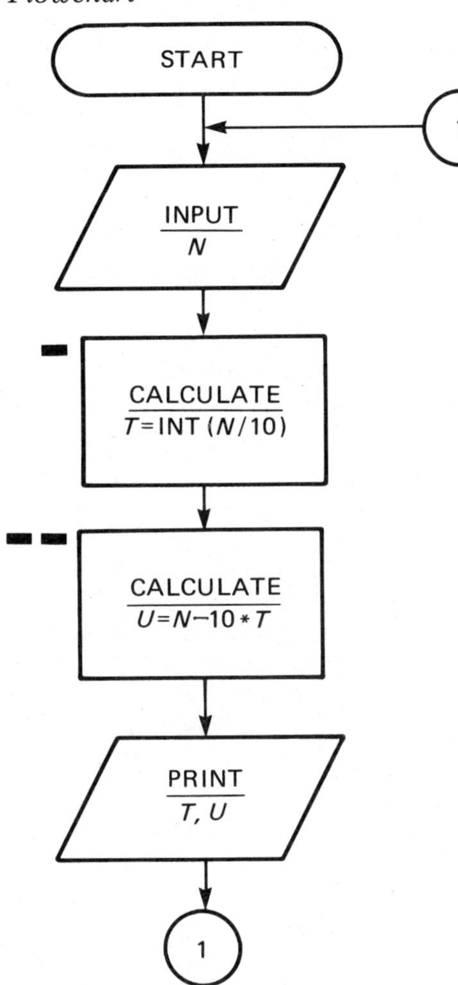

Discussion of the Flowchart:

▬ This algorithm isolates the ten's digit. Let us see how it works with the number 45. (We should get a 4, which is the ten's digit.)

$T = \text{INT}(N/10)$

$T = \text{INT}(45/10)$

$T = \text{INT}(4.5)$

$T = 4$

▬▬ This algorithm will give us the unit's digit. Using the same number, $N=45$ and $T=4$, we get:

$U = N - 10 * T$

$U = 45 - 10 * 4$

$U = 45 - 40$

$U = 5$, which is the unit's digit.

Planned Output

```
SEPARATING THE DIGITS


ENTER A POSITIVE TWO-DIGIT NUMBER ? 46
THE TEN'S DIGIT IS 4
THE UNIT'S DIGIT IS 6


ENTER A POSITIVE TWO-DIGIT NUMBER ?
```

For each of the following problems:
a. Write an analysis which includes the development of an algorithm.
b. Write a plan for a computer solution to the problem.
c. Develop a flowchart.
d. Design a planned output.
e. Code a program to solve the problem and RUN it on the computer.

82. **Problem:** Enter a number. Determine if the number is odd.

83. **Problem:** Enter a positive two-digit number. Separate the digits. (Do not assume that the number entered has two digits.)

84. **Problem:** Enter a two-digit number. Display the number with the digits reversed. (Do not assume that the positive number entered has two digits.)

Chapter 5

Applications in Algebra and Business

Section 5.1 PRINTING CHARTS (READ-DATA)

Preparing for an increase in prices, the B. Nice Toy Company asked their Data Processing Division (the part of the company that works with a computer) for a chart listing the item number and current price for their three best-selling toys. This chart was produced.

```
B. NICE TOY COMPANY
ITEM #          PRICE
1782            7.65
7630            8.17
9491            4.03
```

What makes this output different from any of our previous outputs?

There is no "?" indicating an INPUT statement.

In business, companies do not want a "?" appearing on bills, reports, charts or tables. We therefore use a different method to INPUT data. It is called READ from a DATA statement.

> *READ-DATA Statement:* This statement instructs the computer to READ a value from a DATA statement and assign the value to a specified variable in the READ statement.

Following are examples of how READ-DATA statements are used in a program.

Example 1 *Program #1*

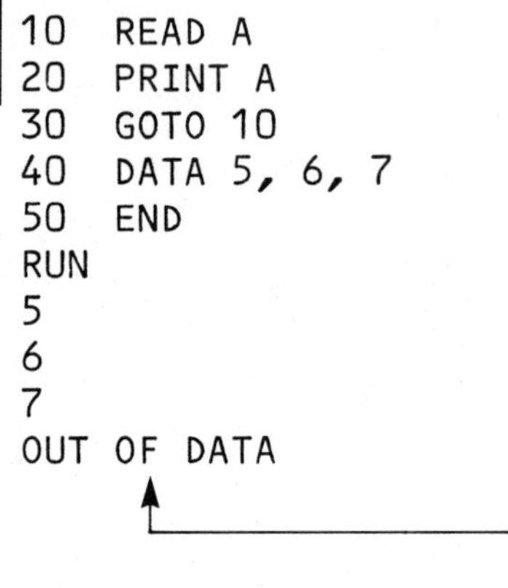

```
10  READ A
20  PRINT A
30  GOTO 10
40  DATA 5, 6, 7
50  END
RUN
5
6
7
OUT OF DATA
```

Each value from the DATA statement is assigned to A. Line 20 prints the values. (*Note:* No "?" appears in the output.)

After the computer is instructed to READ all the DATA, this message appears if you ask the computer to keep reading. The DATA will only be read once. (The message "OUT OF DATA" may take different forms on various computers.)

Example 2 *Program #2*

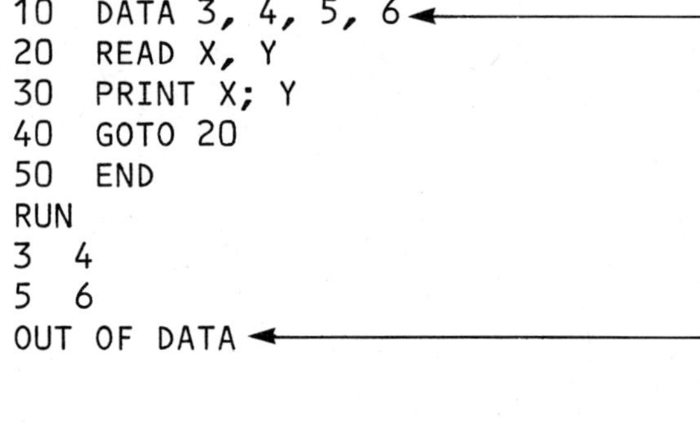

```
10  DATA 3, 4, 5, 6
20  READ X, Y
30  PRINT X; Y
40  GOTO 20
50  END
RUN
3  4
5  6
OUT OF DATA
```

The DATA statement may appear anywhere in the program.

We get the same message because we are again asking the computer to READ after all the values have been used.

Example 3 *Program #3*

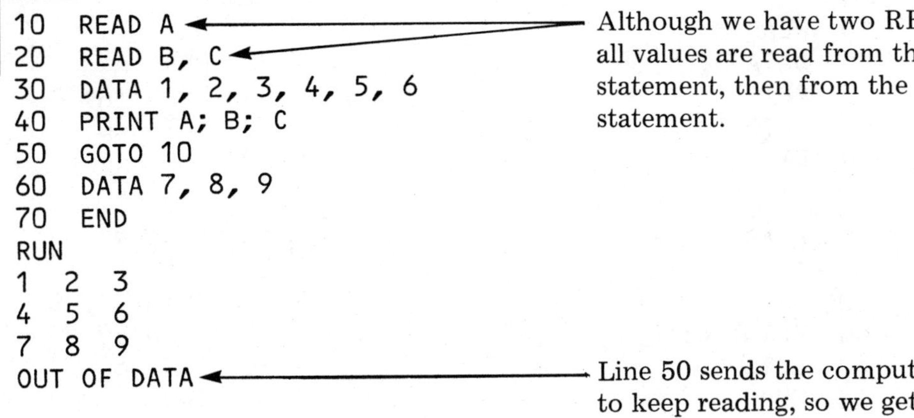

```
10  READ A
20  READ B, C
30  DATA 1, 2, 3, 4, 5, 6
40  PRINT A; B; C
50  GOTO 10
60  DATA 7, 8, 9
70  END
RUN
1  2  3
4  5  6
7  8  9
OUT OF DATA
```


Now let us look at the following table.

Several Things to Remember when Using READ-DATA Statements

1. Punctuation:
   a. Variables used in the READ statement are separated by commas.
   b. Numbers used in the DATA statement are separated by commas.
2. The DATA statement may appear anywhere in the program.
3. Each value in a DATA statement is assigned only once to a variable in the READ statement.†
4. When all values in a DATA statement are READ, the computer searches for another DATA statement. If no more DATA is available, an OUT OF DATA message appears.
5. Notice that in the outputs of the three programs, no question mark appears on the display when data is entered using the READ-DATA statement.

†Values may be read again with a RESTORE command. (See your manual.)

Class Exercise 1 Show the output for each program.

1.
```
10  READ X
20  DATA 3, 5, 7
30  PRINT X
40  GOTO 10
50  END
```

2.
```
10  DATA 7, 6, 5
20  READ A
30  PRINT A;
40  GOTO 20
50  END
```

3.
```
10  READ C
20  DATA 4, 5, 6, 7
30  READ E
40  PRINT C; E
50  GOTO 10
60  END
```

4.
```
10  READ R
20  READ S
30  PRINT R; S
40  GOTO 20
50  DATA 6, 5, 4, 3
60  END
```

5.
```
10  READ X, Y
20  P=X*Y
30  PRINT P
40  DATA 2, 3, 4, 5
50  GOTO 10
60  END
```

6.
```
10  READ C, N
20  PRINT C;
30  C=C+1
40  IF C>N THEN GOTO 70
50  GOTO 20
60  DATA 1, 5
70  END
```

Let us now use the READ-DATA statement to solve this problem.

Problem

Have the computer print the following chart.

```
B. NICE TOY COMPANY
ITEM#           PRICE
1782            7.65
7630            8.17
9491            4.03
```

Solution:

a. *Analysis:* Since there are no "?"s, there are no INPUT statements in the program. We will use READ-DATA statements to enter the data.

b. *Plan:*

1. Read in values for Item # (I) and Price (P).
2. Print I, P.
3. Go to 1.
4. DATA 1782, 7.65, 7630, 8.17, 9491, 4.03.

   Note: The values are stored as item #, price, item #, price, etc. That is because the values are read in that order.

5. End.

c. *Flowchart*

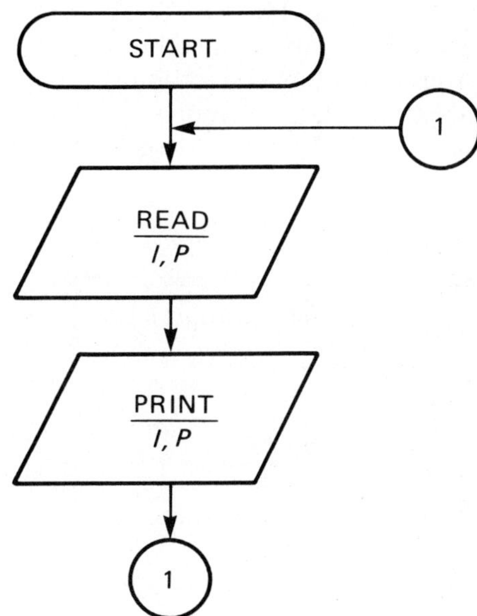

d. *Planned Output*

The planned output is in the statement of the problem.

e. *The Program*

```
  10  PRINT "B. NICE TOY COMPANY"
■ 20  PRINT "ITEM#", "PRICE"
  30  READ I, P
■ 40  PRINT I, P
  50  DATA 1782, 7.65, 7630, 8.17, 9491, 4.03
■■ 60  GOTO 30
  70  END
```

Discussion of the Program:

■ Your computer may line up ITEM # and the numbers under it or PRICE and the numbers under it differently. Make the adjustments in your planned outputs.

■■ This program will produce the planned output. However, line 60 sends the computer back to keep reading more values. After the three pairs of values are read, we get an OUT OF DATA message. In the next section, we will learn how to avoid getting this unwanted message.

Exercise 5.1 Show the output for each program.

1.
```
10  READ S
20  DATA 2, 3, 4
30  P=5*S
40  PRINT S, P
50  GOTO 10
60  END
```

2.
```
10  READ A, B
20  C=A+2*B
30  PRINT A, B, C
40  DATA 8, 7, 6, 5
50  GOTO 10
60  END
```

3.
```
10  READ Y
20  X=INT(Y)
30  PRINT Y; X
40  DATA 5.8, 7.2
50  GOTO 10
60  END
```

4.
```
10  DATA 5.8, 7.2
20  READ A
30  R=INT(A+.5)
40  PRINT A, R
50  GOTO 20
60  END
```

5.
```
10  READ E
20  DATA 10, 9, 8, 7
30  READ F
40  PRINT E, F
50  GOTO 30
60  END
```

6.
```
10  DATA 2, 5
20  READ K, N
30  PRINT K;
40  K=K+1
50  IF K>N THEN GOTO 70
60  GOTO 30
70  END
```

Revise the program that was written in this section to produce the following outputs.

7.

```
B. NICE TOY COMPANY
ITEM #          PRICE
4183            6.34
9200            7.26
1791            1.07
OUT OF DATA
```

8.

```
B. GOOD FRIENDS INC.
ITEM#               PRICE
6364                7.21
9007                8.06
2345                9.32
5816                2.59
OUT OF DATA
```

For the given problem, flowchart and planned output, code a program to solve the problem.

9. **Problem:** Read in ten numbers. Print only the positive numbers. (Use the following data: 8, 7, −4, −3, 6, 0, 5, −11, −10, 4.)

Flowchart *Planned Output*

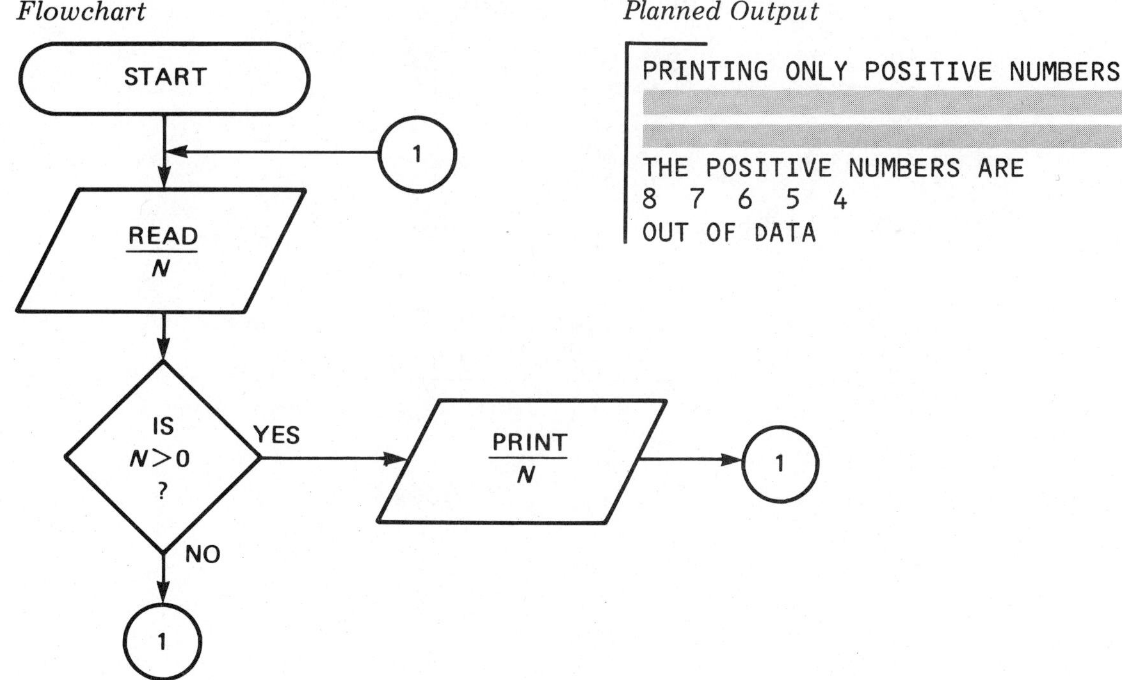


```
PRINTING ONLY POSITIVE NUMBERS


THE POSITIVE NUMBERS ARE
8  7  6  5  4
OUT OF DATA
```

For each of the following problems:
a. Write an analysis which includes the development of an algorithm.
b. Write a plan for a computer solution to the problem.
c. Develop a flowchart.
d. Design a planned output.
e. Code a program to solve the problem and RUN it on the computer.

10. **Problem:** Print the following chart:

```
RADIO SHACK TRS-80 COMPUTERS
ITEM #              PRICE
4001                41.36
7110                82.14
5136                19.03
1218                26.75
OUT OF DATA
```

11. **Problem:** Print the following chart:

```
APPLE COMPUTER SALES
QUANT.              COST
2                   10.46
3                   15.69
4                   20.92
OUT OF DATA
```

12. **Problem:** Read ten numbers. Print all numbers that are divisible by 8. (Use the following data: 5, 6, 7, 8, 32, 706, 2024, 4116, 8001, 373.)

Section 5.2 USING A COUNTER IN A PROGRAM

The International Record Shop is having a sale. All tapes are selling for $6.53 each. The store manager developed this chart for the cashiers.

```
INTERNATIONAL RECORD SHOP
NUMBER          COST
1               6.53
2               13.06
3               19.59
4               26.12
5               32.65
```

The following program will print the chart.

```
10  PRINT "INTERNATIONAL RECORD SHOP"
20  PRINT "NUMBER", "COST"
30  READ N, C
40  PRINT N, C
50  DATA 1, 6.53, 2, 13.06, 3, 19.59, 4, 26.12, 5, 32.65
60  GOTO 30
70  END
```

When we run the program, we get the output below.

```
INTERNATIONAL RECORD SHOP
NUMBER          COST
1               6.53
2               13.06
3               19.59
4               26.12
5               32.65

OUT OF DATA
```

What appears on our output that does not appear on the manager's?

Our output has an OUT OF DATA message.

How can we avoid having that message printed?

We must program the computer to *count* the number of times that it prints a number and cost. If the count is greater than five, we end the program.

Let us write a plan to eliminate the OUT OF DATA message.

Plan to Include a Count in our Program:
We will use memory cell K to keep the count.

1. We start our count at 1. Initialize: $K=1$.
2. Read N, C.
3. Print N, C.
4. Increase the count by 1.
5. Is $K>5$?
   a) If yes, go to 6.
   b) If no, go to 2.
6. End.

Next, we make a flowchart for the plan and a tracing of the flowchart.

Flowchart

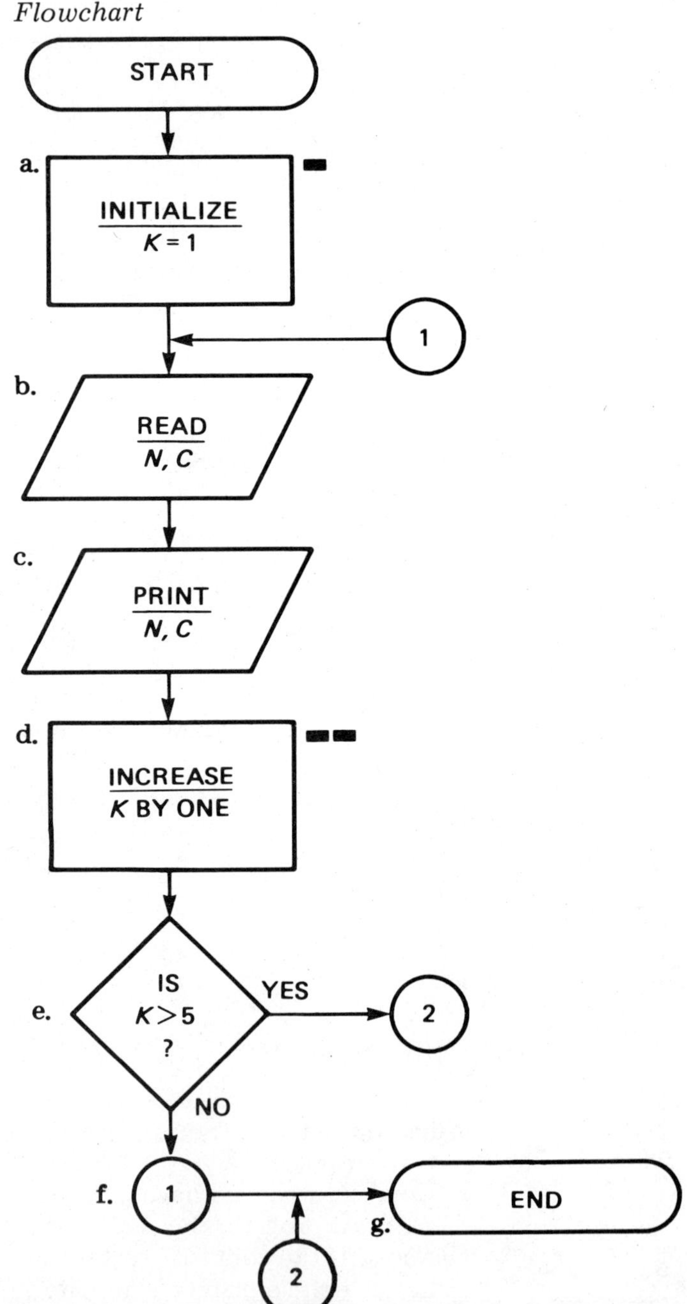

Discussion of the Flowchart:

■ Means to give initial value.

■■ How do we increase K by one and keep the result in memory cell K?

The computer instruction $K=K+1$ increases K by one and stores the new value back in K again. For example, if $K=4$, after executing the instruction $K=K+1$, K becomes $4+1$ or 5. If we execute the instruction again, we get $K=5+1$ or 6.

Tracing the Flowchart

These steps repeat until $K>5$. Then the program ends.

	Place in Flowchart	*K*	*N*	*C*
	a.	1		
	b.		1	6.53
	c.		[1]	[6.53]
	d.	2		
	e. Is 2> 5? No. Go to b.			
	b.		2	13.06
	c.		[2]	[13.06]
	d.	3		
	e. Is 3> 5? No. Go to b.			
	b.		3	19.59
	c.		[3]	[19.59]
	d.	4		
	e. Is 4> 5? No. Go to b.			
	b.		4	26.12
	c.		[4]	[26.12]
	d.	5		
	e. Is 5> 5? No. Go to b.			
	b.		5	32.65
	c.		[5]	[32.65]
$K>5$	d.	6		
	e. Is 6> 5. Yes.			
→	g. END			

Note that the program ended after the fifth set of data was printed. The computer did not go back to READ again. Therefore, it does not print the OUT OF DATA message.

Class Exercise 1 Show the value of C after executing the instructions once.

1. 10 $C=1$
   20 $C=C+1$
2. 30 $C=2$
   40 $C=C+2$
3. 10 $C=3$
   20 $C=C+3$
4. 20 $C=10$
   30 $C=C-1$
5. 40 $C=6$
   50 $C=2*C$
6. 30 $C=6$
   40 $C=3*C$
7. 50 $C=4$
   60 $C=C\uparrow 2$
8. 20 $C=8$
   30 $C=C-2$

Now we code the program.

```
  10   PRINT "INTERNATIONAL RECORD SHOP"
  20   PRINT "NUMBER", "COST"
■ 30   K=1
  40   READ N, C
  50   PRINT N, C
■ 60   K=K+1
■ 70   IF K>5 THEN GOTO 100
■ 80   GOTO 40
  90   DATA 1, 6.53, 2, 13.06, 3, 19.59, 4, 26.12, 5, 32.65
  100  END
```

Discussion of the Program:
■ Lines 30, 60, 70 and 80, working together, form an important algorithm called a *counter*. We will place these statements in a program to see why, together, they form a counter.

```
10  K=1 ◄──────────────────── (previously line 30)
20  PRINT K; ◄─────────────── (inserted to see what is happening to K)
30  K=K+1 ◄────────────────── (previously line 60)
40  IF K>5 THEN GOTO 60 ◄──── (previously line 70 with change in GOTO #)
50  GOTO 20 ◄──────────────── (previously line 80 with change in GOTO #)
60  END
RUN
1  2  3  4  5 ◄────────────── (because of the semi-colon in line 20, the output
                               is printed on the same line)
```

Do you see why this algorithm is called a counter?
It is a procedure that has the computer *count* from 1 to 5.

What change must we make in this program to have the computer count to 10?
Line 40 must be changed to:

```
40 IF K>10 THEN GOTO 60
```

Must we always count by ones?
Look at the solution to the next problem.

Problem

Write a program to count from 2 to 12 by twos.

Solution:

a. *Analysis:* The following changes must be made in our previous program.
1. Since we begin counting from 2, we initialize, $K=2$.
2. Our check (IF-THEN statement) is to see if $K>12$.
3. Since we wish to count by twos, we increase K by 2 after each printing.

b. *Plan:*
1. Initialize: $K=2$.
2. Print K.
3. Increase K by 2.
4. Check: Is $K>12$?
   a) If yes, go to 5.
   b) If no, go to 2.
5. End.

c. *Flowchart*

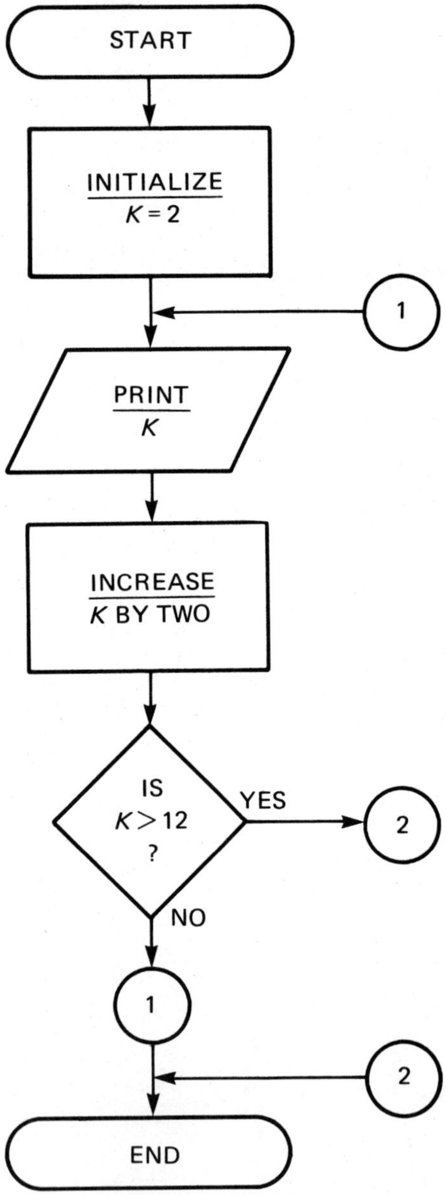

START

INITIALIZE
$K = 2$

PRINT
K

INCREASE
K BY TWO

IS
$K > 12$
?

YES

2

NO

1

2

END

d. *Planned Output*

```
COUNTING BY TWOS TO TWELVE

2  4  6  8  10  12
```

e. *The Program*

```
10  PRINT "COUNTING BY TWOS TO TWELVE"
20  PRINT
30  K=2
40  PRINT K;
50  K=K+2
60  IF K>12 THEN GOTO 80
70  GOTO 40
80  END
```

The next exercise will give you practice using counters.

Exercise 5.2 Show the value of T after executing the instructions once.

1. 40 $T=1$
   50 $T=T+1$
2. 60 $T=3$
   70 $T=T+2$
3. 30 $T=2$
   40 $T=T+1$
4. 50 $T=5$
   60 $T=2*T$
5. 60 $T=60$
   70 $T=T+5$
6. 70 $T=12$
   80 $T=T-1$
7. 20 $T=14$
   30 $T=T-2$
8. 10 $T=20$
   20 $T=T-5$

Show the outputs of the following programs.

9.
```
10  X=1
20  PRINT X
30  X=X+1
40  IF X>6 THEN GOTO 60
50  GOTO 20
60  END
```

10.
```
10  Y=5
20  PRINT Y;
30  Y=Y+2
40  IF Y>17 THEN 60
50  GOTO 20
60  END
```

11.
```
10  L=3
20  PRINT L;
30  L=L+3
40  IF L>18 THEN GOTO 60
50  GOTO 20
60  END
```

12.
```
10  R=6
20  PRINT R;
30  R=R-1
40  IF R<1 THEN 60
50  GOTO 20
60  END
```

13.
```
10  C=2
20  PRINT C;
30  C=2*C
40  IF C>20 THEN 60
50  GOTO 20
60  END
```

14.
```
10  N=1
20  PRINT "HELLO"
30  N=N+1
40  IF N>4 THEN 60
50  GOTO 20
60  END
```

15.
```
10  C=1
20  READ N
30  PRINT C;
40  C=C+1
50  IF C>N THEN 80
60  GOTO 30
70  DATA 5
80  END
```

16.
```
10  C=1
20  READ N
30  READ A, B
40  PRINT A, B
50  C=C+1
60  IF C>N THEN 90
70  GOTO 30
80  DATA 3, 1, 2, 4, 5, 7, 8
90  END
```

For the given problem, flowchart and planned output, code a program to solve the problem.

17. **Problem:** Print the counting numbers from 1 to 5 along with their squares.

Planned Output

```
NUMBERS AND SQUARES
NUMBER          SQUARE
1               1
2               4
3               9
4               16
5               25
```

Flowchart

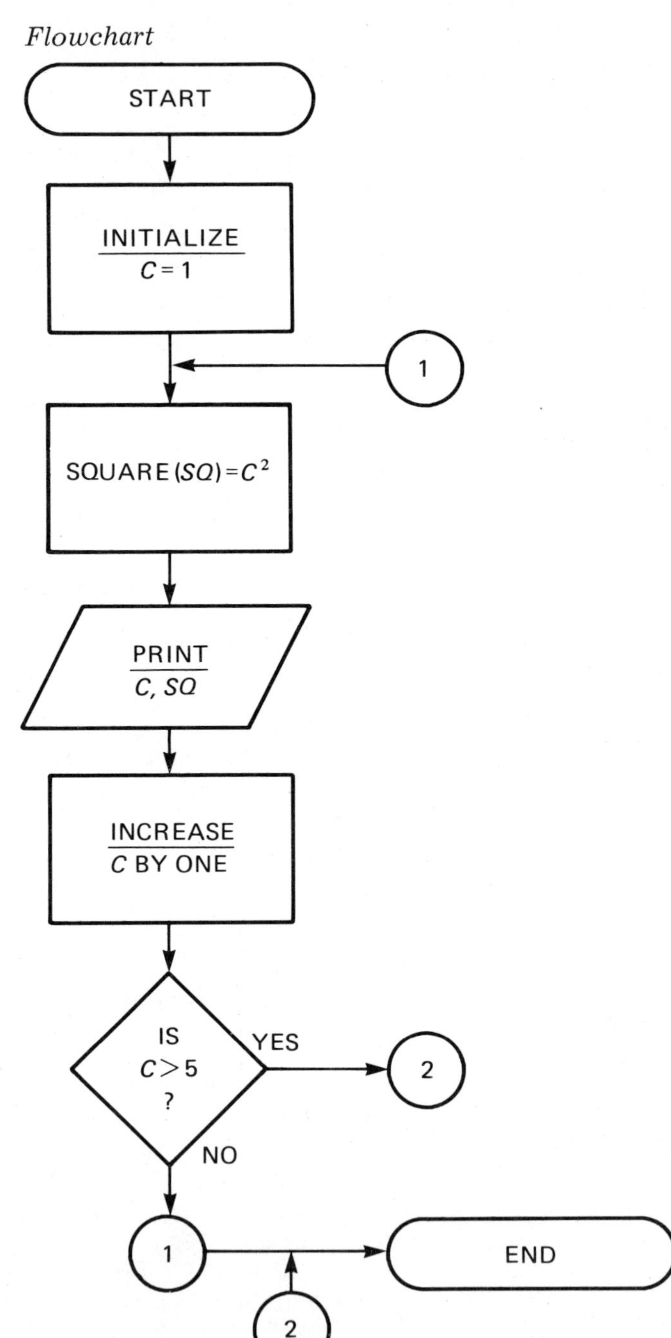

For each of the following problems:
a. Write an analysis which includes the development of an algorithm.
b. Write a plan for a computer solution to the problem.
c. Develop a flowchart.
d. Design a planned output.
e. Code a program to solve the problem and RUN it on the computer.

18. **Problem:** Count from 1 to 20 by ones.

19. **Problem:** Print the whole numbers from 1 to 10 along with their cubes.

20. **Problem:** The International Record Shop is having a sale. All tapes are selling for $6.53. Develop the following chart without using READ-DATA.

```
INTERNATIONAL RECORD SHOP
NUMBER            COST
1                 6.53
2                 13.06
3                 19.59
4                 26.12
5                 32.65
```

Section 5.3 MOTION PROBLEMS (PRINT TAB(X))

Suppose you are driving in a car at a speed (rate) of 30 miles per hour (mph). How far will you travel in 2 hours?

At 30 mph for 2 hours, you will travel 60 miles.

How far will you travel in 4 hours?

At 30 mph for 4 hours, you will travel 120 miles.

What relationship do you see between distance, rate and time?

distance = rate × time or

$$D = R \cdot T$$

Using this formula, we can solve many problems.

Example 1 How far can Lisa travel in 7 hours at the rate of 35 miles per hour?

Solution: We use the formula $D = R \cdot T$. The rate is 35 mph and the time is 7 hours.

$D = 35 \cdot 7$
$D = 245$

She can travel 245 miles.

Example 2 How long will it take Lara to drive 270 miles at a rate of 45 mph?

Solution: We use the formula $D = R \cdot T$. The distance is 270 miles and the rate 45 mph.

$$D = R \cdot T$$

$$270 = 45\,T$$

Divide both sides by 45: $\dfrac{270}{45} = \dfrac{45}{45}\,T$

$$6 = T \quad \text{or} \quad T = 6$$

Therefore, Lara needs 6 hours for her trip.

Example 3 As a jet takes off, the pilot tells the passengers that the trip of 3000 miles will take 5 hours. What is the speed (rate) of the jet?

Solution: In this example we are given the distance, 3000 miles, and the time, 5 hours. We again use the formula:

$$D = R \cdot T$$

$$3000 = R \cdot 5 \quad \text{or} \quad 3000 = 5R$$

Divide both sides by 5: $\dfrac{3000}{5} = \dfrac{5R}{5}$

$$600 = R$$

The jet flies at a speed (an average) of 600 mph.

Class Exercise 1 Solve the following problems.

1. How long will it take Carol to drive 360 miles at a rate of 40 mph?
2. How far can Al travel in 5 hours at the rate of 48 mph?
3. What is the rate of a plane that travels 2400 miles in 8 hours?

Using the formula $D = R \cdot T$, find the missing value for each of the following:

4. $R = 25$ mph
$T = 3$ hours
$D = ?$
5. $R = 36$ mph
$T = ?$
$D = 108$ miles
6. $R = ?$
$T = 4\frac{1}{2}$ hours
$D = 180$ miles

Now we turn to the computer to solve the following problem.

Problem

Have the computer print a chart for the distance traveled by a car going at a rate of 50 mph for the first 6 hours.

Chart

```
DISTANCE A CAR TRAVELS AT 50 MPH
RATE          TIME          DISTANCE
50            1             50
50            2             100
50            3             150
50            4             200
50            5             250
50            6             300
```

Solution:

a. *Analysis:*

1. Note that the columns above are separated into fields but not the fields that are defined by using a comma within a PRINT statement. We must therefore use the BASIC function, PRINT TAB(X).

> A PRINT TAB(X) instruction moves the cursor to the specified position, X, on the current line.† This is similar to tabulating on a typewriter.

†This may vary somewhat on your computer. Check your computer manual.

For example, the instruction

10 PRINT TAB(5); "A"; TAB(15); "B"

will cause the computer to display an A in position 5 on a line and then display B in position 15 on the same line. (Note that on most computers no punctuation is required after a PRINT TAB(X); however, the specified number must be in parentheses. Also note that some computers leave a lead blank when displaying numbers. Therefore, when using the PRINT TAB(X) to display a number, you might have to adjust the number in the parentheses.)

2. Notice that there is no "?" in the chart. We will write this program without the use of INPUT or READ-DATA. We will have the computer generate (count) the values for the time. Then we will use the formula $D=R \cdot T$ to get the distance.

b. *Plan:*

1. Initialize both the rate and time: $R=50$ and $T=1$.
2. Calculate: $D=R \cdot T$.
3. Print R, T, D.
4. Increase T by one.
5. Is $T>5$?
   a) If yes, then go to 6.
   b) If no, then go to 2.
6. End.

c. *Flowchart*

START
INITIALIZE
R = 50, T = 1
1
CALCULATE
D = R • T
PRINT
R, T, D
INCREASE
T BY ONE
IS
T > 5
?
YES
2
NO
1
END
2

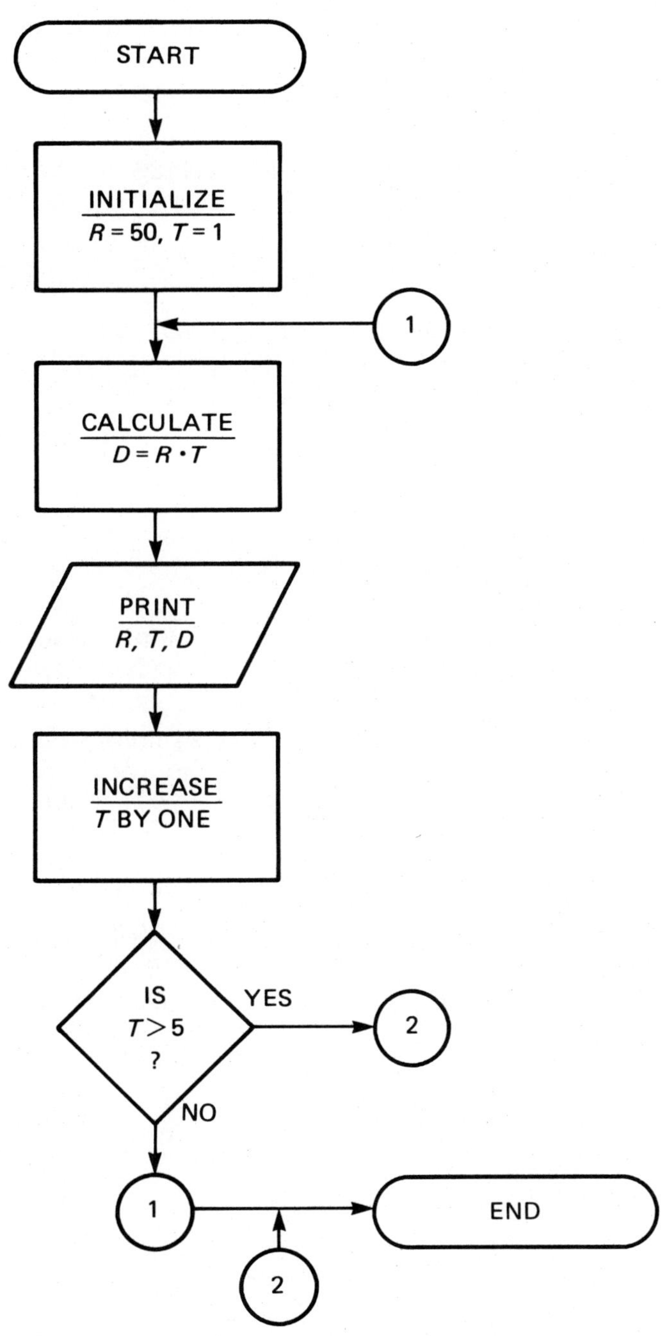

d. *Planned Output*

See the chart under the statement of the problem on page 253.

e. *The Program*

```
10    PRINT "DISTANCE A CAR TRAVELS AT 50 MPH"
20    PRINT "RATE"; TAB(13); "TIME"; TAB(25); "DISTANCE"
30    R=50
40    T=1
50    D=R*T
60    PRINT R; TAB(13); T; TAB(25); D
70    T=T+1
80    IF T>5 THEN GOTO 100
90    GOTO 50
100   END
```

Exercise 5.3

Solve the following:

1. If Jeremy rides his bicycle 45 miles from Plainview to Demarest in 3 hours, how fast will he be riding?

2. How long will it take Lynda to drive 24 miles from home to school at a speed of 30 mph?

3. David can cross a 36 mile lake by boat in 1½ hours. How fast is he traveling?

Use the formula $D = R \cdot T$ to find the missing value for each of the following:

4. $R = 46$ mph
   $T = 7$ hours
   $D = ?$

5. $R = 32$ mph
   $T = ?$
   $D = 256$ miles

6. $R = ?$
   $T = 3\frac{1}{2}$ hours
   $D = 350$ miles

Revise the program that was written in this section to produce the following outputs.

7.
```
TRAVELING AT 50 MPH
RATE        TIME        DIST.
50          1           50
50          2           100
50          3           150
50          4           200
50          5           250
50          6           300
```

8.
```
USING D=50×T
R       T       D
50      1       50
50      2       100
50      3       150
50      4       200
50      5       250
```

For the given problem, flowchart and planned output, code a program to solve the problem.

9. **Problem:** Read in six values to represent the rate in mph of a car. Print the following chart (see Planned Output).

Flowchart

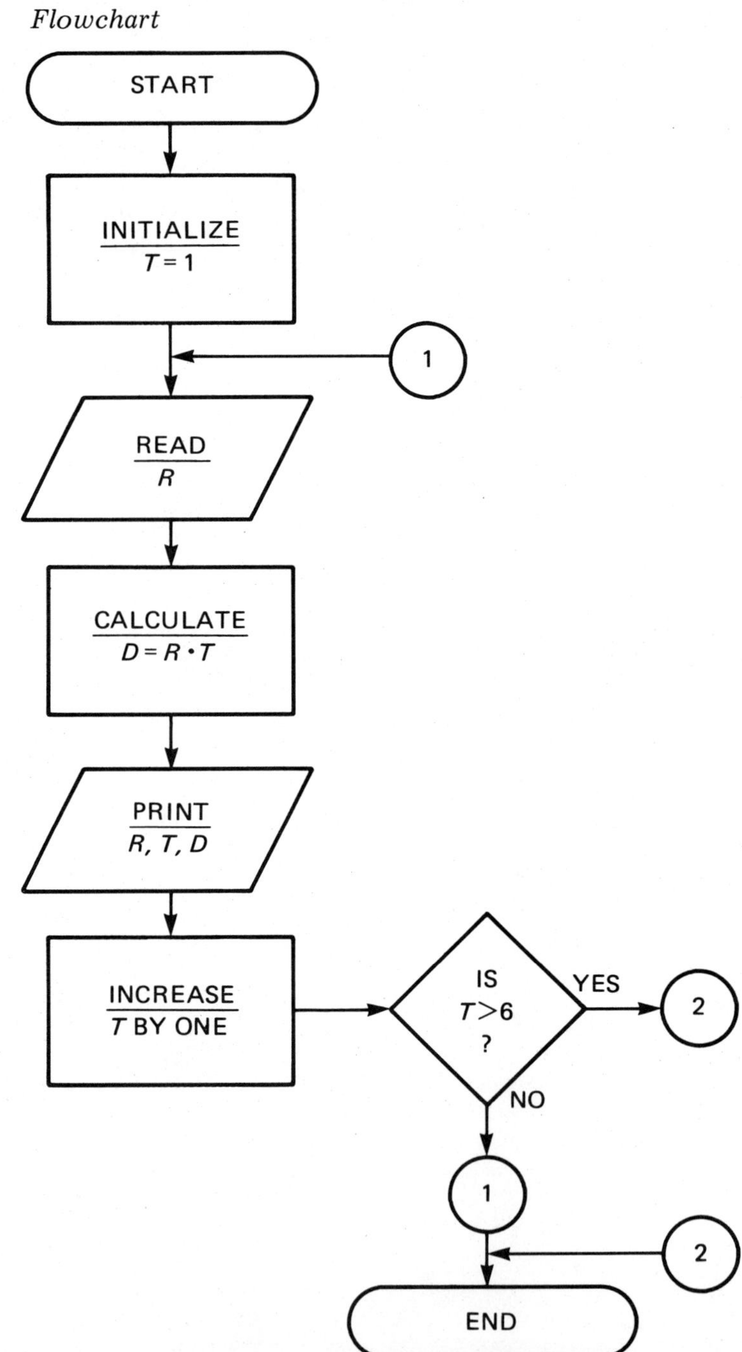

Planned Output

```
VARYING THE RATE AND TIME
RATE        TIME        DISTANCE
15          1           15
22          2           44
25          3           75
30          4           120
34          5           170
39          6           234
```

For each of the following problems:
a. Write an analysis which includes the development of an algorithm.
b. Write a plan for a computer solution to the problem.
c. Develop a flowchart.
d. Design a planned output.
e. Code a program to solve the problem and RUN it on the computer.

10. **Problem:** Have the computer print a chart showing the areas of parallelograms each of whose bases has length 25, and whose altitudes are integers from 10 to 15 inclusive.

11. **Problem:** Have the computer print a chart showing the areas of triangles each of whose bases has length 16, and whose heights are integers from 6 to 12 inclusive.

12. **Problem:** If a rate and time are entered, find the distance. If a rate and distance are entered, find the time.

Section 5.4 SEQUENCES AND SERIES (ACCUMULATORS IN A PROGRAM)

What do the following sequences of numbers have in common?

Sequence #1:

1, 4, 7, 10, 13, 16, . . .

Sequence #2:

3, 6, 9, 12, 15, 18, . . .

Both sequences have the same (common) difference between consecutive terms. The numbers in a sequence are called terms.

Sequence #1: 1, 4, 7, 10, 13, 16, . . .

Difference between terms: 4−1 (3), 7−4 (3), 16−13 (3)

Sequence #2: 3, 6, 9, 12, 15, 18, . . .

Difference between terms: 6−3 (3), 9−6 (3), 18−15 (3)

> A sequence that has a common difference between consecutive terms is called an *arithmetic sequence.*

For example: 6, 10, 14, 18, 22, 26, . . .

10−6 (4), 14−10 (4), 26−22 (4)

is an arithmetic sequence.

Notice that the difference between any two consecutive (following in order) terms is 4.

Sequences #1 and #2 are also arithmetic sequences. Each has a common difference of 3 between all pairs of consecutive terms.

See if you can identify arithmetic sequences.

Class Exercise 1 Which of the following are arithmetic sequences? State the reason.

1. 2, 5, 8, 11, 14, . . .
2. 1, 3, 5, 7, 9, . . .
3. 3, 9, 15, 21, 27, . . .
4. 1, 1, 2, 3, 4, 8, . . .

Fill in the blanks for the following arithmetic sequences.

5. 7, _, _, 28, 35, 42, . . .
6. 8, _, _, _, 28, 33, 38, . . .
7. _, _, 16, 22, 28, 34, . . .
8. _, 16, _, 30, 37, . . .

How can we find the sum of the terms of the following arithmetic sequence?

1, 2, 3, 4, 5, . . . , 100 (the numbers from 1 to 100)

We can add the numbers:

$S = 1 + 2 + 3 + \ldots + 100$

Not a simple task. There must be an easier way!

> The sum of the numbers of an arithmetic sequence is called an *arithmetic series.*

To find the sum of an arithmetic series, we use the formula:

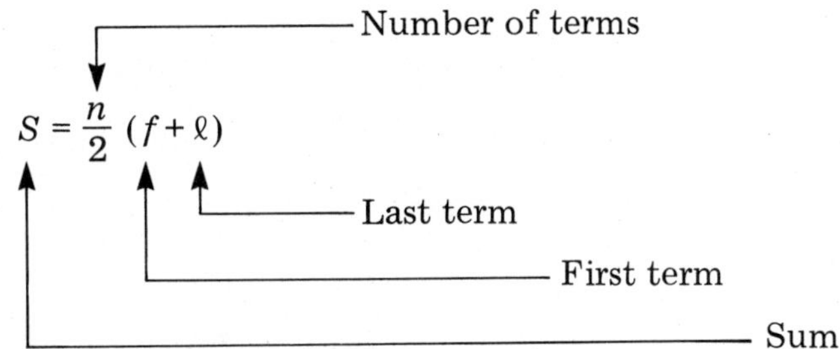

Let us use this formula to find the sum of the first 100 counting numbers:

$1 + 2 + 3 + \ldots + 100$ (100 terms)

Because this is an arithmetic series (the common difference between terms is 1), we use the formula:

$$S = \frac{n}{2}(f + \ell)$$

Since the number of terms is 100 ($n = 100$), the first term is 1 ($f = 1$) and the last term is 100 ($\ell = 100$), we get:

$$S = \frac{100}{2}(1 + 100)$$

$$S = 50(101)$$

$$S = 5050$$

This is much easier than actually adding each term.

Class Exercise 2 Use the formula $S = \frac{n}{2}(f + \ell)$ to find the sum of the following arithmetic series:

1. $2 + 4 + 6 + 8 + 10 + 12 + 14 + 16$ ($n = 8$, $f = 2$ and $\ell = 16$)
2. $7 + 10 + 13 + 16 + 19 + 22 + 25 + 28$
3. $6 + 13 + 20 + 27 + 34$
4. $37 + 45 + 53 + 61 + 69 + 77 + 85 + 93 + 101$

Is the following an arithmetic sequence?

$$1, \quad 4, \quad 16, \quad 64, \quad 256, \ldots$$

$$4-1 \quad 16-4 \quad 64-16 \quad 256-64$$

$$(3) \quad (12) \quad (48) \quad (192)$$

Examine the differences.

Since there is no common difference between consecutive terms, it is not an arithmetic sequence.

Let us look at the ratios between consecutive terms:

$$1, \quad 4, \quad 16, \quad 64, \quad 256, \ldots$$

$$\frac{4}{1} = 4 \qquad \frac{64}{16} = 4 \quad \frac{256}{64} = 4$$

This sequence has a common ratio of 4.

A sequence with a common ratio between consecutive terms is called a *geometric sequence.*

For example:

$$2, \quad 6, \quad 18, \quad 54, \quad 162, \ldots$$ is also a geometric sequence.

$$\frac{6}{2} = 3 \quad \frac{18}{6} = 3 \qquad \frac{162}{54} = 3$$

It has a common ratio of 3.

See if you can identify geometric sequences.

Class Exercise 3 Which of the following are geometric sequences? State the reason.

1. 1, 3, 9, 27, 81, . . .
2. 3, 12, 48, 192, . . .
3. 1, 2, 4, 8, 16, . . .
4. 5, 25, 125, 625, . . .

Fill in the blanks for the following geometric sequences.

5. _, 6, 12, _, _, . . .
6. 4, 24, _, _, . . .
7. 7, _, 175, 875, _, . . .
8. 100, 50, _, _, . . .

The sum of the numbers of a geometric sequence is called a *geometric series.*

We can find the sum of a geometric series by using the formula:

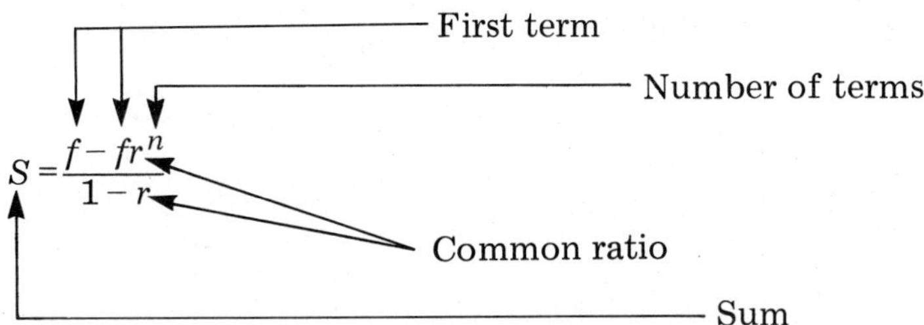

Example 1 How much money do you earn in ten days if you are paid \$1 the first day, \$2 the second day, \$4 the third day, \$8 the fourth day, etc.?

Solution: We must find the sum of the geometric series:

$S = 1 + 2 + 4 + 8 + 16 + \ldots$ (10 terms)

In this series the number of terms is 10 ($n = 10$), the first term is 1 ($f = 1$) and the common ratio is 2 ($r = 2$). (You may wish to use a calculator.)

$$S = \frac{f - f \cdot r^n}{1 - r}$$

$$S = \frac{1 - 1 \cdot 2^{10}}{1 - 2} = \frac{1 - 1024}{-1}$$

$$S = \frac{-1023}{-1} = 1023$$

You would earn \$1023 in ten days!

Class Exercise 4 Use the formula $S = \frac{f - f \cdot r^n}{1 - r}$ to find the sum of the following geometric series.

1. $2 + 4 + 8 + 16 + 32 + 64 + 128 + 256$
2. $4 + 12 + 36 + 108 + \ldots$ (7 terms)
3. $5 + 10 + 20 + 40 + 80 + 160 + 320 + 640$
4. $2 + 6 + 18 + 54 + 162 + 486 + 1458$

Computers can do approximately one million calculations per second. They, therefore, do not need formulas to find the sum of a series. Let us see how the computer can solve the following problem without a series sum formula.

Problem

> Have the computer find the sum of an arithmetic series where we are given the first term, F, the common difference, D, and the number of terms, N. (All numbers entered are positive.)

Solution:

a. *Analysis:* If we know the first and last terms of an arithmetic sequence, we can count from the first number to the last in increments of the common difference. By adding these numbers, we would get the sum of the sequence.

We have the first term in our problem. How can we get the last term?

In an arithmetic sequence, the last term (L) is found using this formula:

$$L = f + (n-1) \cdot d$$ †

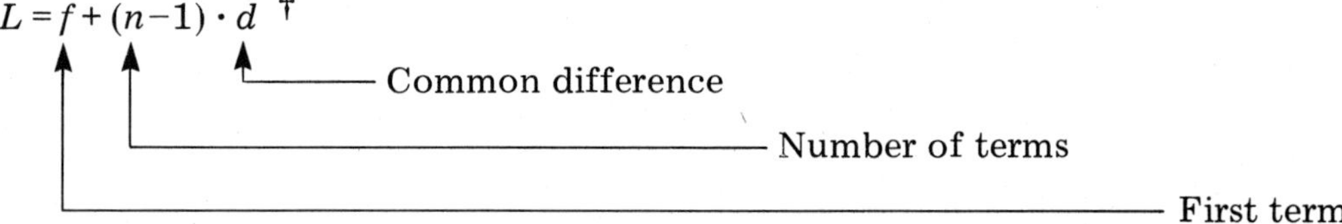

For example, to find the last term of an arithmetic sequence where $f=1$, $d=3$ and $n=6$ we get:

$L = 1 + (6-1) \cdot 3$

$L = 1 + 5 \cdot 3$

$L = 1 + 15 = 16$

We can check our result by actually listing the 6 terms of the sequence:

1, 4, 7, 10, 13, 16

Last term is the same result as we calculated with the formula.

Now we can use this algorithm in our plan.

---

†A development of this algorithm may be found on page 245 in Elgarten, et al., *Using Computers in Mathematics*, Addison-Wesley Publishing Company, Menlo Park, California, 1983.

b. *Plan:*

1. Put in values for the first term (F), the common difference (D), and the number of terms (N).
2. Find the last term (L): $L=F+(N-1)\cdot D$.
3. A memory cell, S, will be reserved to add (accumulate) all terms. We initialize the sum (S) at 0: $S=0$.
4. We will count beginning with the first term. So we initialize the count (C) at F: $C=F$.
5. Accumulate the count (the terms of the sequence) in S.
6. Increase the count by the common difference (D): $C=C+D$.
7. Is $C>L$? (Have we reached the last term yet?)
   a) If yes, then print S. Go to 1.
   b) If no, then go to 5.

c. *Flowchart*

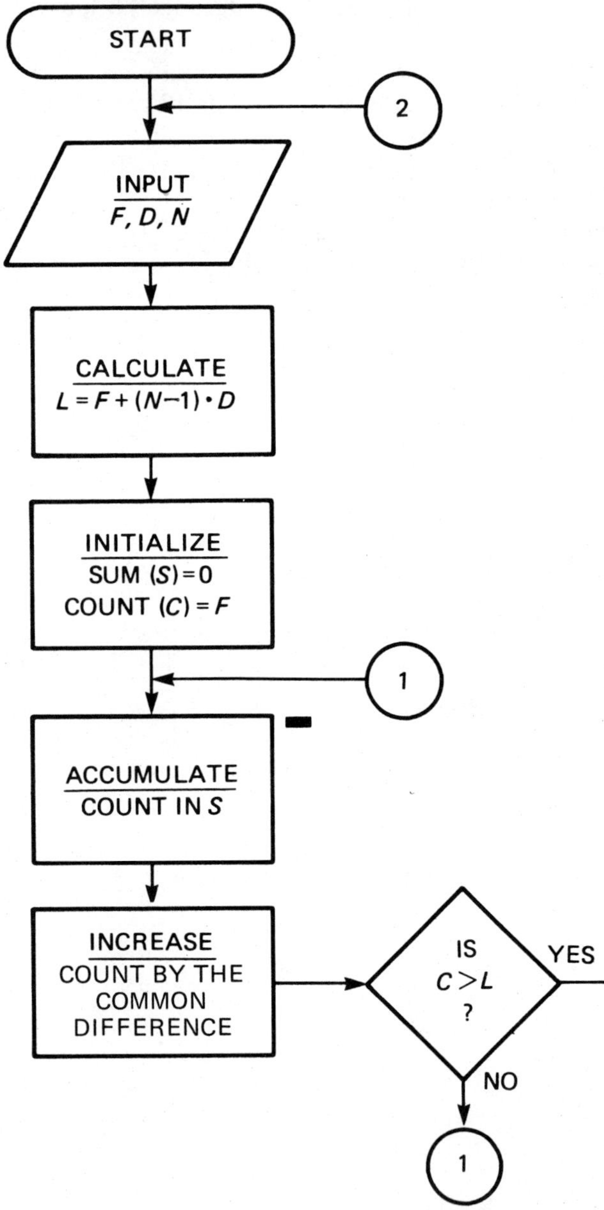

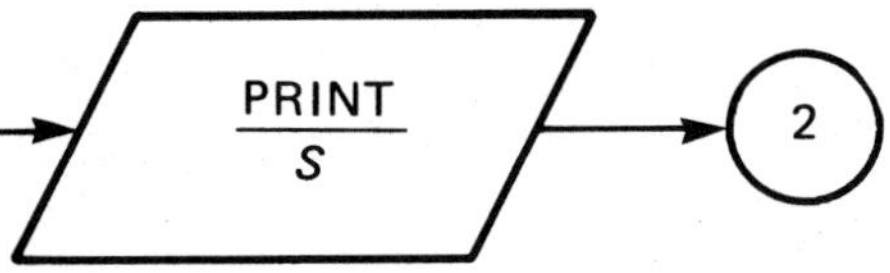

START

INPUT
F, D, N

CALCULATE
$L = F + (N-1) \cdot D$

INITIALIZE
SUM $(S) = 0$
COUNT $(C) = F$

ACCUMULATE
COUNT IN S

INCREASE
COUNT BY THE
COMMON
DIFFERENCE

IS
$C > L$
?

YES

PRINT
S

NO

1

2

Discussion of the Flowchart:

▬ In this assignment box we ask the computer to accumulate (or sum) the contents of memory cell C in memory cell S. That is, we wish to add C to S and store the new result back in S. The instruction $S=S+C$ will accomplish this.

As you trace the flowchart, you may consider the following:

F	D	N	S	C	
1	3	6	0	1	You see the way C enumerates the sequence 1, 4, 7, 10, 13, 16.
			1	4	
			5	7	
			12	10	
See how the computer adds term after term.			22	13	
			35	16	
			51	19>16 (last term)	
			51		

d. *Planned Output*

```
SUM OF AN ARITHMETIC SERIES


ENTER FIRST TERM ? 1
ENTER COMMON DIFFERENCE ? 3
ENTER NUMBER OF TERMS ? 6
THE SUM OF THE SERIES IS 51


ENTER FIRST TERM ?
```

e. *The Program*

```
    10  PRINT "SUM OF AN ARITHMETIC SERIES"
    20  PRINT : PRINT
    30  INPUT "ENTER FIRST NUMBER"; F
    40  INPUT "ENTER COMMON DIFFERENCE"; D
    50  INPUT "ENTER NUMBER OF TERMS"; N
    60  L=F+(N-1)*D
    70  S=0
    80  C=F
■   90  S=S+C
■■ 100  C=C+D
   110  IF C>L THEN GOTO 130
   120  GOTO 90
   130  PRINT "THE SUM OF THE SERIES IS"; S
   140  GOTO 20
   150  END
```

Discussion of the Program:

■ Line 90 contains the *accumulator.* The contents of memory cell C are added to memory cell S and this new result is stored back in memory cell S.

■■ Line 100 is generating each term of the series. Remember each new term is just D (the common difference) more than the previous term.

Exercise 5.4

Which of the following are arithmetic sequences? State the reason.

1. 10, 13, 14, 15, . . .
2. 10, 13, 16, 19, . . .
3. 40, 35, 30, 25, . . .
4. 2, 2.5, 3, 3.5, 4, . . .

Fill in the blanks for the following arithmetic sequences.

5. 16, _, 24, 28, _
6. 50, _, _, _, 42, 40
7. _, _, _, 6.7, 7.2
8. _, 30, _, 28.2, 27.3

Use the formula $S = \frac{n}{2}(f + \ell)$ to find the sums of the following arithmetic series:

9. $3 + 6 + 9 + 12 + 15 + 18 + 21 + 24 + 27$
10. $8 + 13 + 18 + 23 + 28 + 33 + 38 + 43 + 48 + 53 + 58$
11. the sum of the even numbers from 2 to 100
12. the sum of the counting numbers from 1 to 150

Which of the following are geometric sequences?

13. 7, 14, 21, 28, ...
14. 7, 14, 28, 56, ...
15. 3, 12, 15, 132, ...
16. 4, 8, 16, 32, 64, ...

Fill in the blanks for the following geometric sequences.

17. _, 9, 27, _, _
18. 7, 28, _, _
19. 50, 5, _, _
20. 64, _, _, 8, 4, _

Use the formula $S = \frac{f - f \cdot r^n}{1 - r}$ to find the sum of the following geometric series.

21. $8 + 16 + 32 + 64 + 128$
22. $18 + 54 + 162 + \ldots$ (6 terms)
23. $1 + 3 + 9 + 27 + \ldots$ (8 terms)
24. $16 + 32 + 64 + 128 + \ldots$ (7 terms)

Revise the program that was written in this section to produce the following outputs.

25.

```
SUM AN ARITHMETIC SERIES



ENTER F, D AND N ? 1, 3, 6
SUM OF THE SERIES IS 51



ENTER F, D AND N ?
```

26.

```
FIND THE SUM OF THE SERIES


THE FIRST TERM IS 1
THE DIFFERENCE IS 3
THE NUMBER OF TERMS IS 6
THE SUM OF THE SERIES IS 51
```

For the given problem, flowchart and planned output, code a program to solve the problem

27. **Problem:** Enter a counting number, N. Find the sum of the squares and the sum of the cubes of the counting numbers from 2 to N.

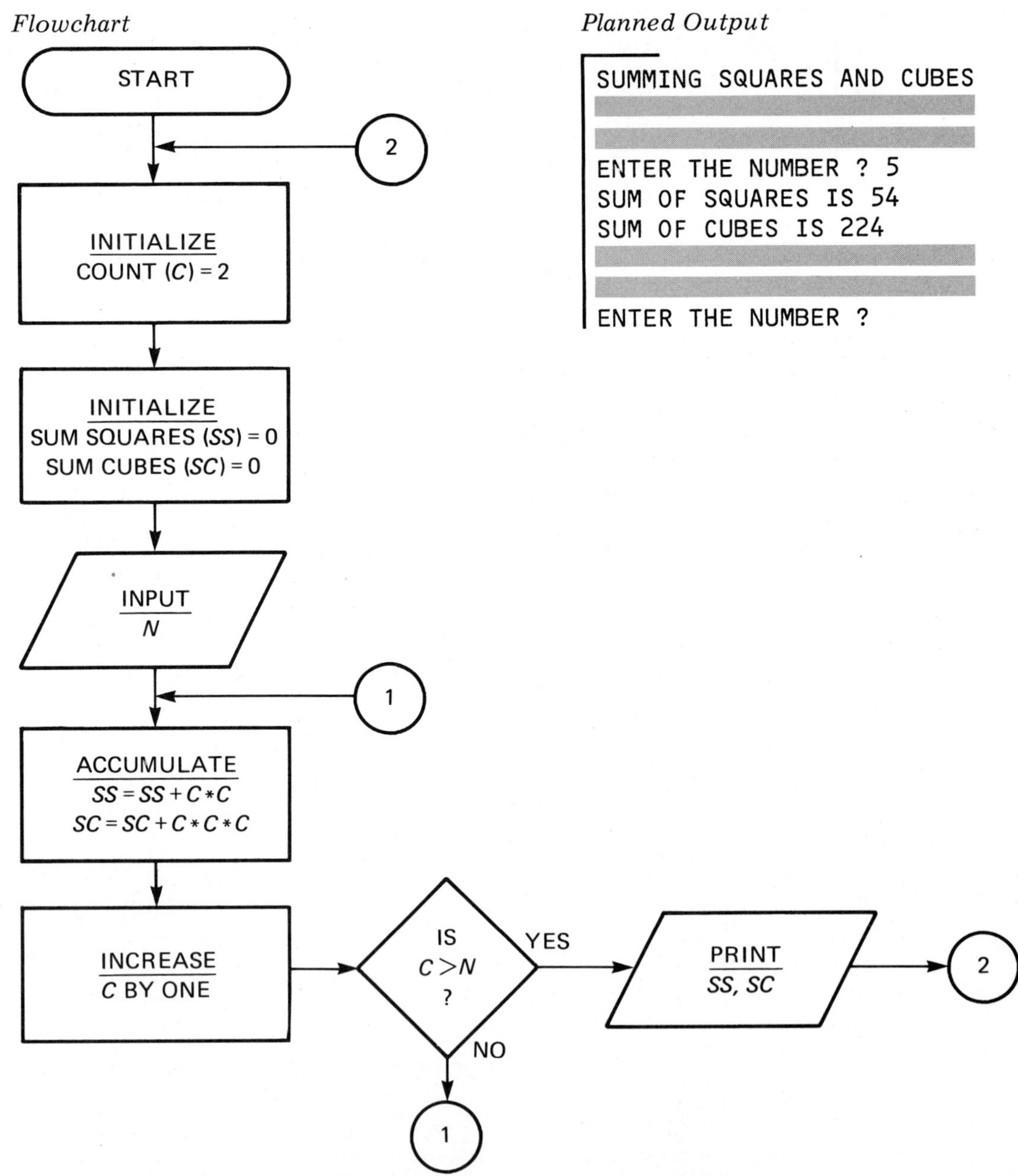

For each of the following problems:
a. Write an analysis which includes the development of an algorithm.
b. Write a plan for a computer solution to the problem.
c. Develop a flowchart.
d. Design a planned output.
e. Code a program to solve the problem and RUN it on the computer.

28. **Problem:** Enter a counting number, N, which is greater than 1. Have the computer find the sum of the counting numbers from 1 to N.

29. **Problem:** Enter a counting number, N, which is greater than 1. Have the computer find the sum of the squares of the counting numbers from 1 to N.

30. **Problem:** Have the computer find the sum of an arithmetic series where we are given the first term, F, the common difference, D, and the number of terms, N. If the number of terms is not more than ten, have the computer list all the terms in the series.

Section 5.5 FINDING THE FACTORS OF NUMBERS (FOR-NEXT LOOP)

We know that since:

$1 \times 24 = 24$,

$2 \times 12 = 24$,

$3 \times 8 = 24$, and

$4 \times 6 = 24$,

we can say that the *factors* of 24 are:

1, 2, 3, 4, 6, 8, 12 and 24.

Example 1 What are the factors of 15?

Solution: Since $1 \times 15 = 15$ and
$3 \times 5 = 15$,
the factors of 15 are 1, 3, 5 and 15.

We notice that each of the factors of 15 divides 15 with a remainder of zero.

> The *factors* of a whole number are those counting numbers which divide the given whole number with a remainder of zero.

Class Exercise 1 Find the factors of each number:

1. 12   2. 48   3. 25   4. 23

Now let us see how the computer finds factors of numbers.

Problem

Enter a counting number. Have the computer list all the factors of the number.

Solution:

a. *Analysis:* A number, C, is a factor of N if C divides N leaving a remainder of zero. To test this, we used the algorithm:

INT $(N/C) * C = N$ (see Section 4.5).

If the above is true, C is a factor of N. If it is false, then C is not a factor of N.

To get all the factors of N, we will try all the counting numbers from 1 to N.

b. *Plan:*

1. Initialize a counter (C): $C=1$.
2. Put in a value for N.
3. Is INT $(N/C)*C=N$?
   a) If yes, print C. Go to 4.
   b) If no, continue.
4. Increase C by one.
5. Is $C>N$?
   a) If yes, go to 1.
   b) If no, go to 3.

c. *Flowchart*

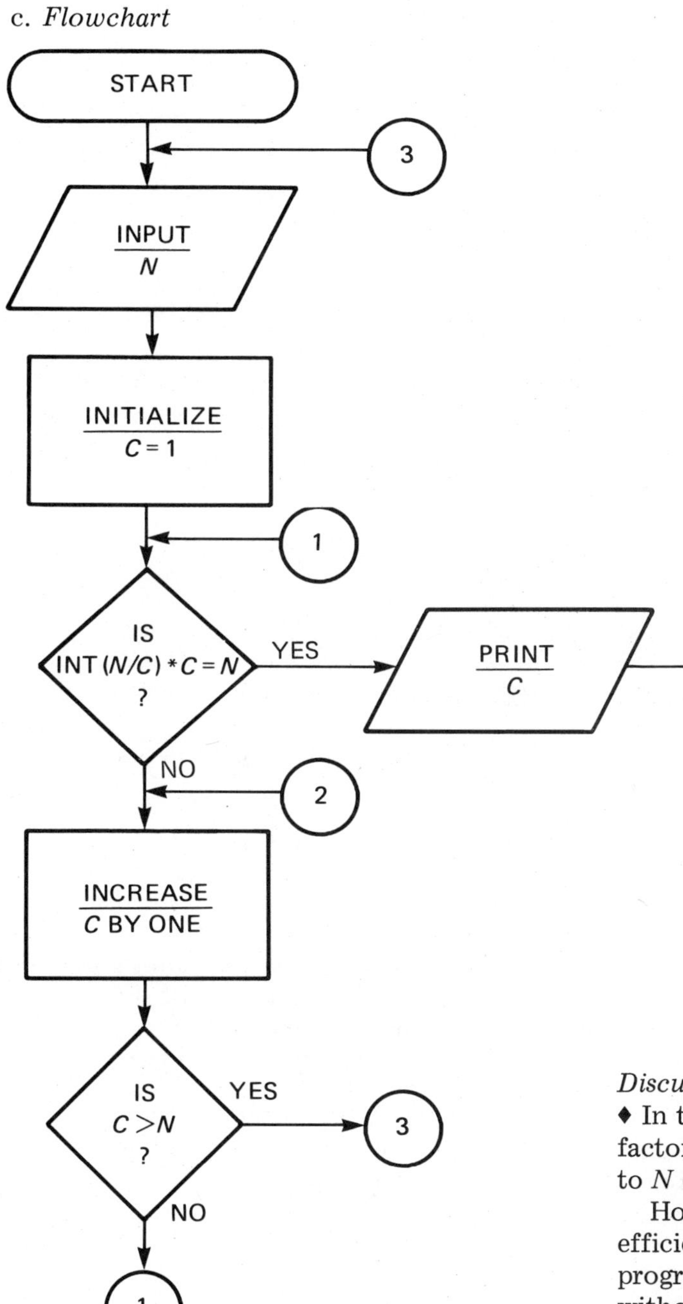

Discussion of the Flowchart:

♦ In this approach, we try as possible factors all the counting numbers from 1 to N (we check to see if $C>N$).

How can you make this program more efficient? See if you can revise the program so that we get the same result without trying *all* the counting numbers from 1 to N.

d. *Planned Output*

```
LIST FACTORS OF A NUMBER


ENTER THE NUMBER ? 24
THE FACTORS ARE
1  2  3  4  6  8  12  24


ENTER THE NUMBER ?
```

e. *The Program*

```
    10   PRINT "LIST FACTORS OF A NUMBER"
    20   PRINT : PRINT
    30   INPUT "ENTER THE NUMBER"; N
    40   PRINT "THE FACTORS ARE"
    50   C=1
■   60   IF INT(N/C)*C=N THEN PRINT C;
    70   C=C+1
■■  80   IF C>N THEN GOTO 100
    90   GOTO 60
■■  100  PRINT
■■  110  GOTO 20
    120  END
```

Discussion of the Program:

■ The semi-colon at the end of line 60 will keep the cursor on the same line while displaying the factors.

■■ After all the factors are displayed (when $C > N$), the computer goes to line 100. This PRINT statement is used to release the cursor from the hold of the semi-colon. Line 110 then sends the computer back to skip two lines (at line 20).

Many programs require the use of a counter. A computer construction called a FOR-NEXT loop can perform the same function as the counter in our previous program. We can rewrite lines 50-90 using the FOR-NEXT loop as follows.

From Our Program

```
  :
30    INPUT "ENTER THE NUMBER"; N
40    PRINT "THE FACTORS ARE"
50    C=1
60    IF INT(N/C)*C=N THEN PRINT C;
70    C=C+1
80    IF C>N THEN GOTO 100
90    GOTO 60
100   PRINT
```

Revision Using a FOR-NEXT Loop

```
      :
■    50    FOR C=1 TO N
     60    IF INT(N/C)*C=N THEN PRINT C;
■■   70    NEXT C
     80    PRINT
```

Discussion of the Revised Program:

■ *C* is initialized at 1. Control then proceeds to line 60, etc.

■■ On line 70, NEXT *C* increases *C* by one and then checks to see if $C>N$. If $C>N$, then control passes to the next line (in this case, line 80). Otherwise, control goes back (loops) to the first instruction following the FOR statement (line 60).

Let us look at the outputs of several programs that use a FOR-NEXT loop.

Program #1

■
```
10  FOR X=1 TO 3
20  PRINT "SMILE"
30  NEXT X
40  END
RUN
SMILE
SMILE
SMILE
```

Program #2

■■
```
10  FOR Y=1 TO 9 STEP 2
20  PRINT Y;
30  NEXT Y
40  END
RUN
1  3  5  7  9
```

Program #3

■■■
```
10  FOR L=5 TO 1 STEP -1
20  PRINT L;
30  NEXT L
40  END
RUN
5  4  3  2  1
```

Discussion of Programs #1, 2 and 3:

■ Line 10 of Program #1 indicates that whatever is between lines 10 and 30 will be executed three times.

■■ Line 10 of Program #2 uses STEP 2 to indicate that instead of Y going from 1 to 9 by ones (stepping-up by one), two should be added to Y each time line 30 is executed. We can also increase Y by a decimal fraction (STEP .3, etc.).

■■■ Line 10 of Program #3 indicates that -1 should be added to L each time line 30 is executed. (This has the effect of decreasing the value of L.) Whenever a negative number follows the word STEP, the NEXT part of the loop checks to see if the value of the change is less than the number following the word "TO." For example, in Program #3, line 30 checks to see if $L<1$.

The next set of exercises will give you practice working with FOR-NEXT loops.

Exercise 5.5 Find the factors of each number.

1. 6
2. 36
3. 14
4. 19

Show the output for the following programs.

5.
```
10  FOR X=1 TO 4
20  PRINT X
30  NEXT X
40  END
```

6.
```
10  FOR A=1 TO 6
20  PRINT A;
30  NEXT A
40  END
```

7.
```
10  FOR J=1 TO 4
20  PRINT "HELLO"
30  NEXT J
40  END
```

8.
```
10  FOR C=1 TO 3
20  PRINT C, "COMPUTER"
30  NEXT C
40  END
```

9.
```
10  FOR K=1 TO 4
20  T=2*K
30  PRINT T;
40  NEXT K
50  END
```

10.
```
10  FOR L=1 TO 4
20  S=3*L
30  PRINT L, S
40  NEXT L
50  END
```

11.
```
10  FOR B=2 TO 10 STEP 2
20  PRINT B
30  NEXT B
40  END
```

12.
```
10  FOR B=3 TO 15 STEP 3
20  PRINT B;
30  NEXT B
40  END
```

13.
```
10  FOR D=0 TO 1.1 STEP .3
20  PRINT D
30  NEXT D
40  END
```

14.
```
10  FOR E=6 TO 1 STEP -1
20  PRINT E;
30  NEXT E
40  END
```

Revise the program that was written in this section to produce the following outputs.

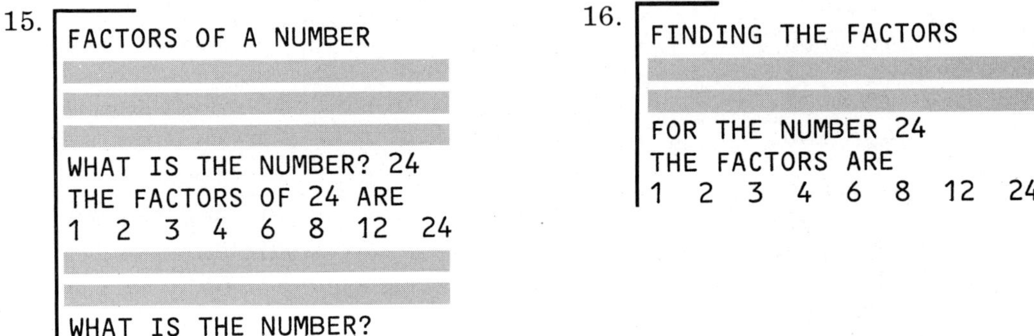

For the given problem, flowchart and planned output, code a program to solve the problem.

17. **Problem:** Enter a number *N*. Have the computer find the sum of the factors of *N*. (Assume *N* is a positive integer.)

Flowchart *Planned Output*

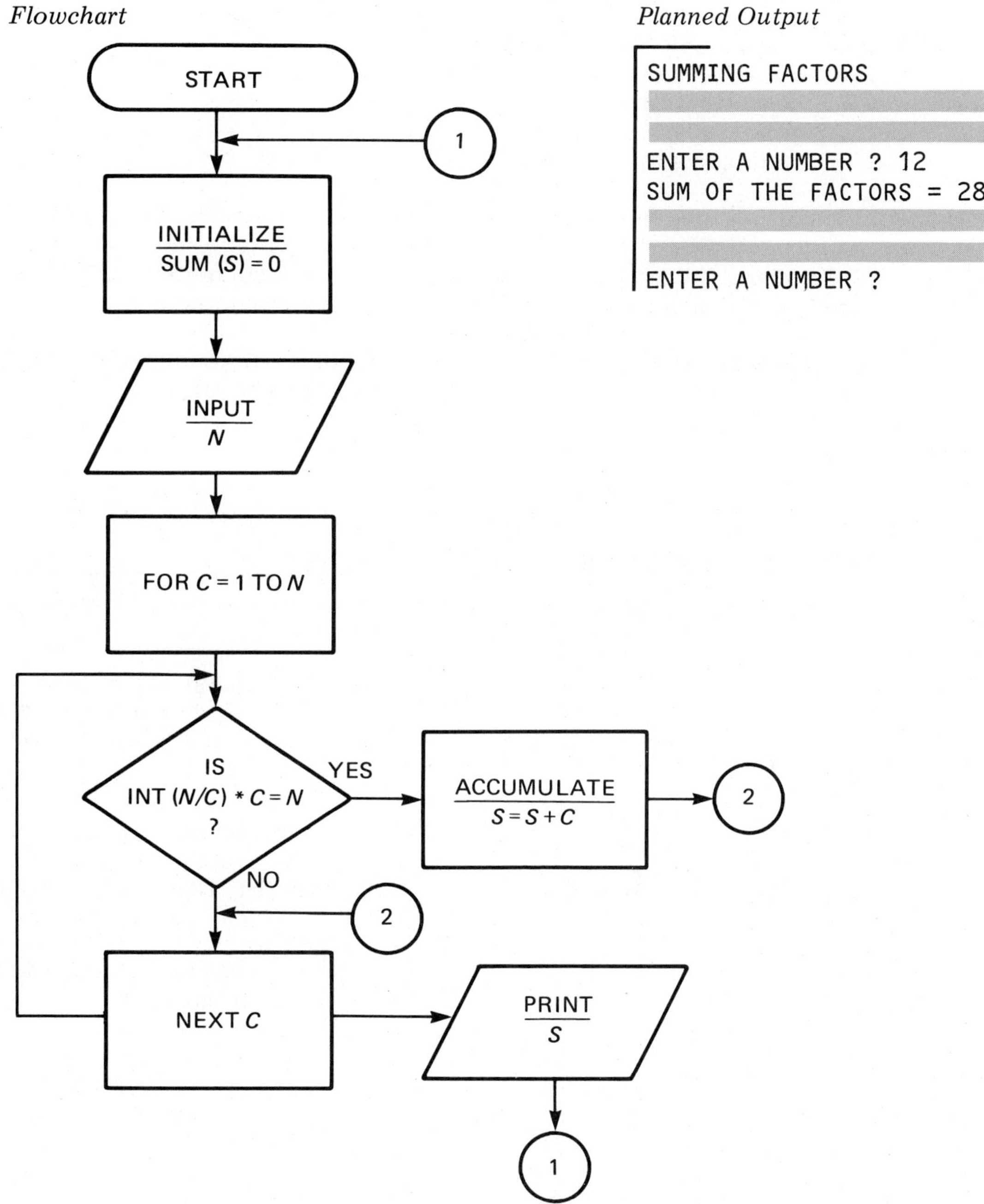

For each of the following problems:
a. Write an analysis which includes the development of an algorithm.
b. Write a plan for a computer solution to the problem.
c. Develop a flowchart.
d. Design a planned output.
e. Code a program to solve the problem and RUN it on the computer.

18. **Problem:** Revise the program written in this section using a FOR-NEXT loop.

19. **Problem:** Enter a counting number, N. Have the computer display the numbers from 1 to N in increments of one. Only accept positive integers for N.

20. **Problem:** Enter a number, N. If N is even, have the computer count from 2 to N by twos. Otherwise, have the computer count from 1 to N by ones. (N must be a positive integer.)

Section 5.6 PRIME NUMBERS (STRING VARIABLES)

Look at the factors of the numbers below.

The number 12:

$1 \times 12 = 12$

$2 \times 6 = 12$

$3 \times 4 = 12$

Factors of 12 are 1, 2, 3, 4, 6 and 12.

The number 13:

$1 \times 13 = 13$

Factors of 13 are 1 and 13.

The number 17:

$1 \times 17 = 17$

Factors of 17 are 1 and 17.

> When a positive whole number has exactly two different whole number factors, it is called a *prime number.*

The number 13 and the number 17 each have exactly two different factors. We can say that 13 and 17 are prime numbers.

> Except for 1, numbers that are not prime are called *composite.*

The number 12 is a composite number.

Example 1 Which of the following are prime numbers?

21, 29, 51

Solution: To determine if a number is prime, we must know its factors.

Number	*Factors*
21	1, 3, 7, 21
29	1, 29
51	1, 3, 17, 51

Since 29 has exactly two different factors, it is prime. The remaining numbers have more than two factors and therefore are not prime. They are composite.

Is 1 a prime number? To answer the question we ask ourselves, how many different factors does it have? Since it has only one factor, it is not prime.

Try the next class exercise to see if you can tell if a number is prime.

Class Exercise 1 Tell which of the following are prime numbers:

1. 11	2. 37	3. 42	4. 39
5. 7	6. 19	7. 9	8. 15

Now we turn to the computer to solve the following problem.

Problem Enter a number. Determine if the number is prime.

Solution:

a. *Analysis:*

1. We will consider a number prime if it has two factors. So for each number entered:
   a) Count the number of factors (F).
   b) If the number of factors (F) is two, the number is prime.
2. When testing possible factors of N we need only test numbers from 1 to $N/2$, since the only number greater than $N/2$ that is a factor is the number itself. For example, the factors of 12 are 1, 2, 3, 4, 6 and 12. Note that the only factor greater than $N/2$ $(12/2=6)$ is the number (12) itself. But we must remember in our program, to add 1 to the count so that we include the number itself as a factor.

b. *Plan:*

1. We must initialize to count the number of factors (F) of a number. F=1. (See *Analysis* 2.)
2. Put in a number, N.
3. Is $N = 1$?
   a) If yes, then go to 9.
   b) If no, continue.
4. We will now find all factors by trying the numbers from 1 to $N/2$. (See *Analysis* 2.)
5. For C=1 to $N/2$.
6. Is INT $(N/C) * C = N$?
   a) If yes, then increase the number of factors by one: F=F+1. Go to 7.
   b) If no, continue.
7. Next C.
8. Is $F = 2$?
   a) If yes, then print "prime." Go to 1.
   b) If no, then continue.
9. Print "not prime." Go to 1.

c. *Flowchart*

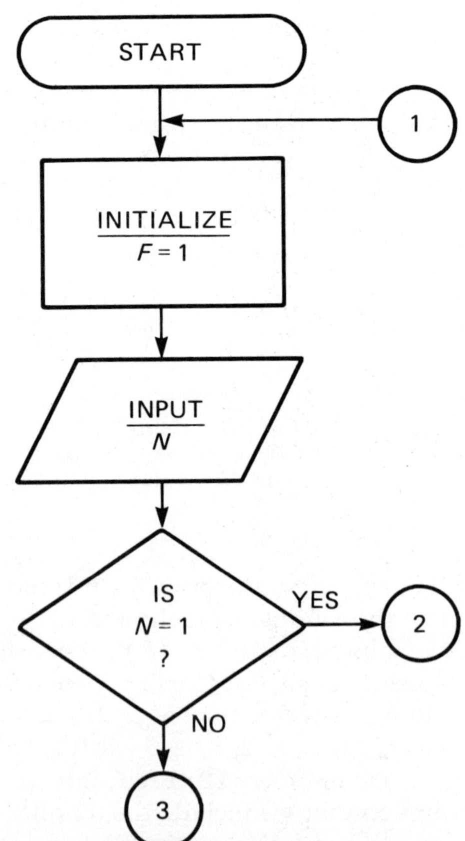

The "All-Purpose Machine"
The Many, Often Hidden, Uses of Computing

Computer pioneer John Von Neumann said the device should not be called the computer but rather the "all-purpose machine." It is not, after all, just a machine for doing calculations. The most striking thing about it is that it can be put to *any number of uses.* Though by no means all-encompassing, the uses presented in this double-length gallery include examples from government, education and training, medicine, transportation, space, and business. Some may surprise you.

1 King Cotton and the computer: Cotton growers use this IBM computer to check commodity prices.

35 Berlin's international convention center, which contains 80 meeting halls, uses a Honeywell Delta building management system, which automatically monitors 4000 points in the center's temperature, lighting, elevator, and fire alarm systems.

36 A poll taker for the *New York Daily News* taking a survey: Computers are used extensively in public opinion polling.

c. *Flowchart* (cont'd)

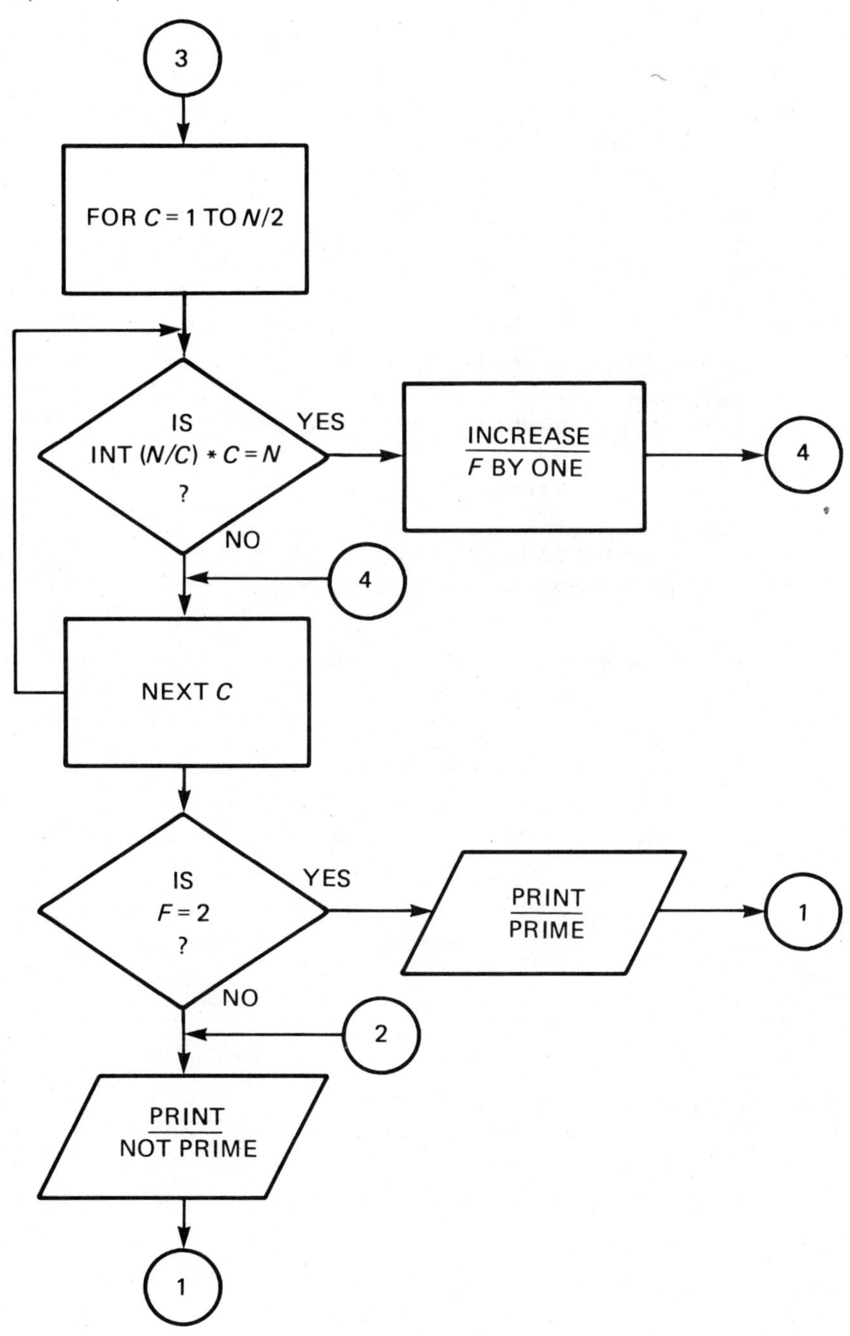

d. *Planned Output*

```
IS THE NUMBER PRIME?


WHAT IS THE NUMBER? 13
13 IS PRIME


■ DO YOU WISH TO CONTINUE?
TYPE IN 1 FOR YES, 2 FOR NO
? 1


WHAT IS THE NUMBER? 6
6 IS NOT PRIME


■ DO YOU WISH TO CONTINUE?
TYPE IN 1 FOR YES, 2 FOR NO
? 2


THE END
```

Discussion of the Planned Output:

■ We are giving the user of the program the option of continuing or ending the program. This was not included in the flowchart. (See Section 4.3 for a discussion of giving the user an option.)

e. *The Program*

```
10   PRINT "IS THE NUMBER PRIME?"
20   PRINT : PRINT
30   F=1
40   INPUT "WHAT IS THE NUMBER"; N
50   IF N=1 THEN GOTO 100
60   FOR C=1 TO N/2
70   IF INT(N/C)*C=N THEN F=F+1
80   NEXT C
90   IF F=2 THEN GOTO 110
100  PRINT N; "ƀIS NOT PRIME" : GOTO 120
110  PRINT N; "ƀIS PRIME"
120  PRINT : PRINT
130  PRINT "DO YOU WISH TO CONTINUE?"
140  PRINT "TYPE IN 1 FOR YES, 2 FOR NO"
150  INPUT R
160  IF R=1 THEN GOTO 20
170  PRINT : PRINT
180  PRINT "THE END"
190  END
```

Discussion of the Program:

■ Note the use of the colon to combine two statements on one line.

In our program we provided the user with the option of either entering more data or terminating the program. We set up a code so that a 1 meant yes and a 2 meant no.

Can the computer handle a response of "yes" or "no"?

The computer can handle not only numeric data (numbers) but also alphanumeric data (letters, characters and/or numbers). So a computer *can* handle an answer of "yes" or "no."

To do this we must introduce a new type of variable called a string variable.

> A *string variable* is a variable that names a memory cell that can store alphanumeric characters.

To form a string variable, place a $ after any numeric variable. The chart below compares some numeric and string variables.

Numeric Variables	String Variables
$A, B, C, \ldots, Z$	$A\$, B\$, C\$, \ldots, Z\$$
$AA, AB, \ldots, AZ$	$AA\$, AB\$, \ldots, AZ\$$
$A1, B1, \ldots, Z1$	$A1\$, B1\$, \ldots, Z1\$$

A string variable may hold up to 255 characters (depending upon the computer).

Below are some examples of how we can use string variables.

Example 2

```
10  INPUT "ENTER YOUR NAME"; A$
20  PRINT "HELLOb"; A$
30  END
RUN
ENTER YOUR NAME ? LARA
HELLO LARA
```

String variable used since we will enter a name (alphanumeric characters). (Line 10)

Computer will not leave a space when displaying the contents of a string. We must put in the space. (Line 20, "HELLOb")

Example 3

```
10  READ A$
20  PRINT "HELLOb"; A$
30  DATA LARA
40  END
RUN
HELLO LARA
```

How does Example 3 differ from Example 2? It shows how to use a string variable with a READ-DATA statement.

The next example will show how the computer can handle a yes or no response.

Example 4

```
10  PRINT "DO YOU WISH TO CONTINUE?"
20  PRINT "ANSWER EITHER YES OR NO."
30  INPUT R$
■ 40  IF R$="YES" THEN GOTO 10
50  PRINT "THE END"
60  END
RUN
DO YOU WISH TO CONTINUE?
ANSWER EITHER YES OR NO.
? YES
DO YOU WISH TO CONTINUE ?
ANSWER EITHER YES OR NO.
? NO
THE END
```

■ In line 40, when comparing a string variable to a word (or any alphanumeric characters), you must put the characters (word) in quotes.

The next program will alphabetize two names.

Example 5

```
10  READ A$, B$
20  IF A$<B$ THEN GOTO 50
30  PRINT B$, A$
40  GOTO 10
50  PRINT A$, B$
60  GOTO 10
70  DATA WILLIAM, AIMEE, JOHN, VICKIE
80  END
RUN
AIMEE           WILLIAM
JOHN            VICKIE
OUT OF DATA
```

The first time through the program, WILLIAM is assigned to $A\$$ and AIMEE is assigned to $B\$$. Line 20 wants to determine if WILLIAM is less than AIMEE ($A\$ < B\$$). One letter is less than another letter if it is closer to A in the alphabet.† ($A < B < C$ etc.) In this case, $A\$$ is not less than $B\$$; therefore, control passes to line 30.

The second time the READ is executed, $A\$$ = JOHN and $B\$$ = VICKIE. This time $A\$$ is less than $B\$$ and control passes to line 50.

---

†A more detailed discussion of how the computer assigns numerical values (ASCII codes) to letters will be presented in Section 7.4.

The next set of exercises will give you practice working with prime numbers and string variables.

Exercise 5.6 Tell which of the following are prime numbers.

1. 33
2. 31
3. 22
4. 1
5. 81
6. 2
7. 27
8. 59

Show the complete output for the following programs.

9.
```
10  PRINT "TELL ME YOUR NAME"
20  READ A$
30  DATA ED, EVA
40  PRINT "YOUR NAME ISƀ"; A$
50  PRINT : PRINT
60  GOTO 10
70  END
```

10.
```
10   PRINT "ARE YOU HAPPY?"
20   PRINT "ANSWER YES OR NO"
30   READ R$
40   IF R$="YES" THEN GOTO 70
50   PRINT "TOO BAD!"
60   PRINT : PRINT : GOTO 10
70   PRINT "I'M GLAD!"
80   GOTO 60
90   DATA YES, NO
100  END
```

11.
```
10  READ A, A$
20  PRINT "MY NAME ISƀ"; A$
30  PRINT "I AM"; A; "ƀYEARS OLD"
40  PRINT : PRINT : GOTO 10
50  DATA 5, JEREMY, 13, AMY
60  END
```

12.
```
10  FOR I=1 TO 3
20  READ N$
30  PRINT N$
40  NEXT I
50  DATA ANN, AVA, AL
60  END
```

Rewrite the program that was written in this section to produce the following outputs.

13.

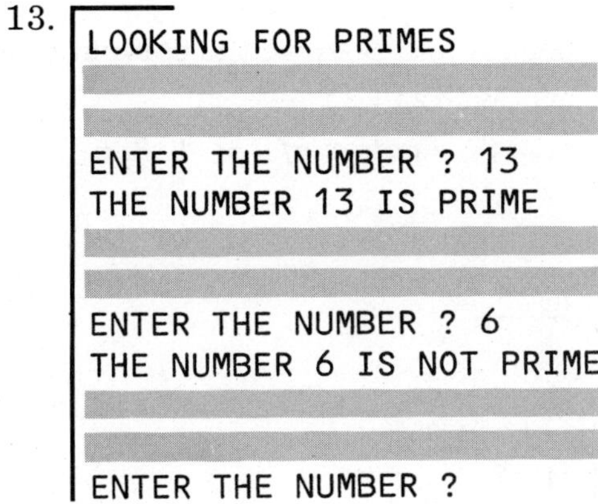

14.

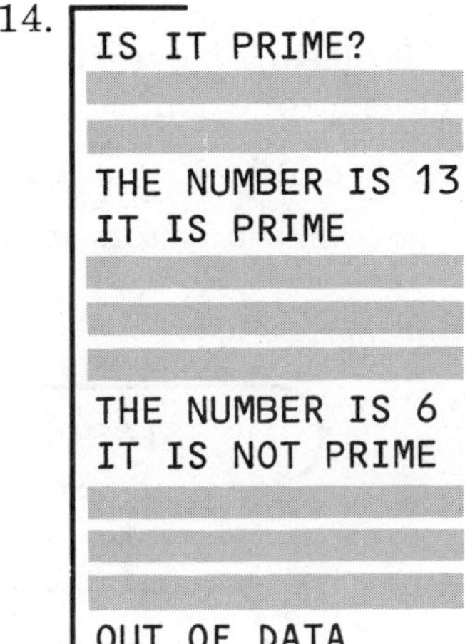

See if you can eliminate the OUT OF DATA message.

For the given problem, flowchart and planned output, code a program to solve the problem.

15. **Problem:** Enter two counting numbers, A and B. Have the computer list their common factors. (For example, suppose we list the factors of 10 and 15:

Factors of 10: 1, 2, 5, 10 *Factors of 15:* 1, 3, 5, 15

Their common factors are 1 and 5.)

Flowchart

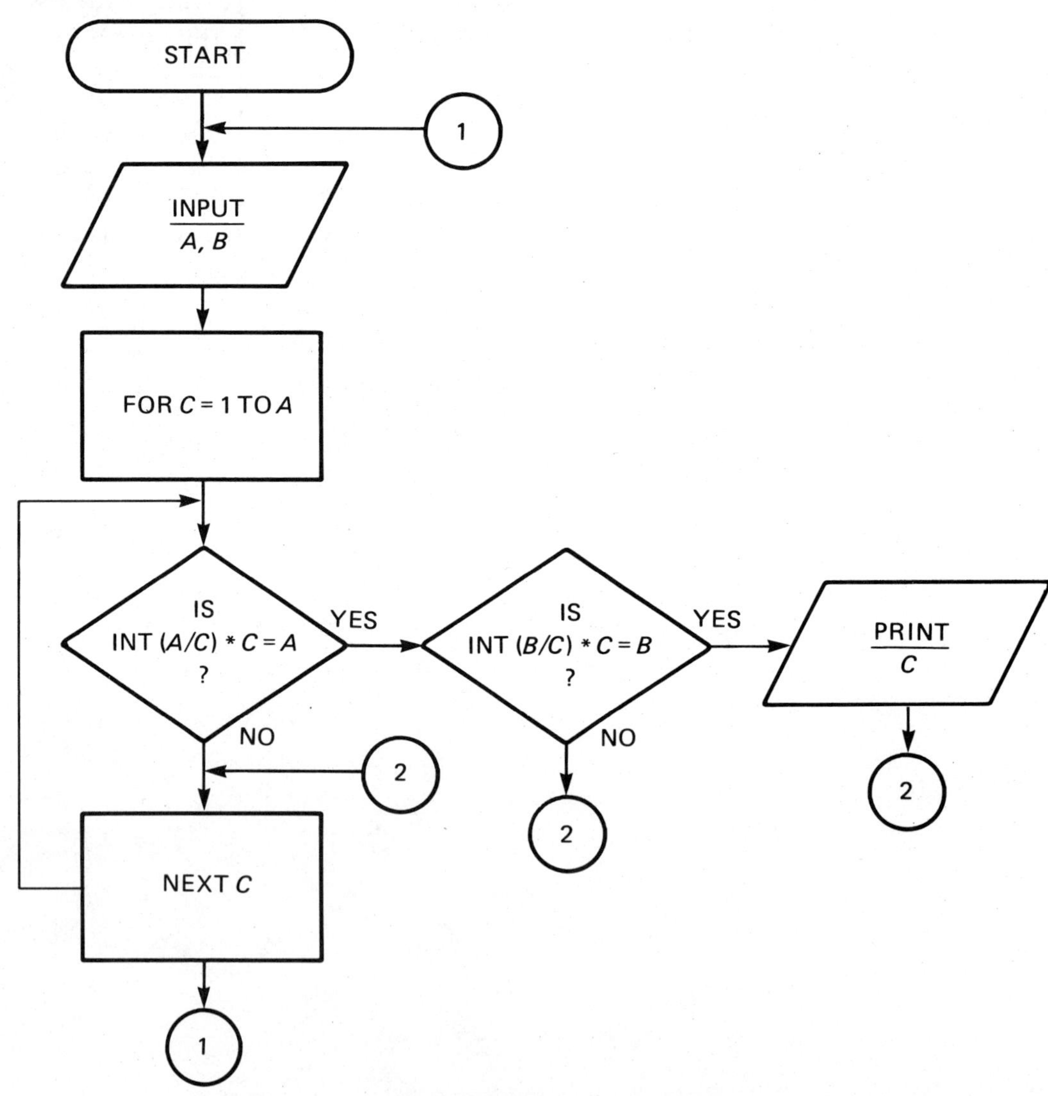

Planned Output

```
FINDING COMMON FACTORS


ENTER TWO NUMBERS ? 10, 15
THE COMMON FACTOR(S) FOR 10 AND 15
1  5


ENTER TWO NUMBERS ? 7, 8
THE COMMON FACTOR(S) FOR 7 AND 8
1


ENTER TWO NUMBERS ?
```

For each of the following problems:
a. Write an analysis which includes the development of an algorithm.
b. Write a plan for a computer solution to the problem.
c. Develop a flowchart.
d. Design a planned output.
e. Code a program to solve the problem and RUN it on the computer.

16. **Problem:** Revise the program that was written in this section to accept an answer of yes or no to either continue or end the program.

17. **Problem:** Have the computer read ten names. Have it determine and then display how many times the name TED appears. Use the following data list: JOE, ALICE, TED, BARBARA, TED, TED, TED, GERRY, JILL, TED.

18. **Problem:** Proper factors of a number are factors of the number that are less than the number. Enter a counting number. Have the computer display all the proper factors of the number. Give the user the option of either entering more data or ending the program.

END OF CHAPTER EXERCISES

Section 5.1 Show the output for each program.

1.
```
10  READ C
20  F=3*C
30  PRINT C, F
40  DATA 1, 2, 3
50  GOTO 10
60  END
```

2.
```
10  READ G
20  A=G+6
30  PRINT A
40  READ H
50  PRINT G, H
60  DATA 3, 4, 5, 6
70  GOTO 10
80  END
```

3.
```
10  READ A
20  READ B
30  PRINT A; B
40  GOTO 20
50  DATA 5, 6, 7, 8
60  END
```

4.
```
10  PRINT "FINDING PRODUCTS"
20  READ A, B
30  P=A*B
40  PRINT A, B, P
50  GOTO 20
60  DATA 3, 4, 4, 5
70  END
```

5.
```
10  PRINT "FINDING THE SUM"
20  PRINT "1ST", "2ND", "SUM"
30  READ A, B
40  S=A+B
50  PRINT A, B, S
60  GOTO 30
70  DATA 3.4, 5.7, 8.6, 7.9
80  END
```

6.
```
10  PRINT "COUNTING"
20  READ F, N
30  PRINT F;
40  F=F+1
50  IF F>N THEN 80
60  GOTO 30
70  DATA 1, 3, 4, 7
80  GOTO 20
90  END
```

For the given problem, flowchart and planned output, code a program to solve the problem.

7. **Problem:** Have the computer find the perimeter (where possible) of equilateral triangles whose sides have possible lengths selected from the following: 5, 6, 7, 8, −3, 0, 9 and 10. Have the computer display the sides and perimeters as shown in the planned output.

Flowchart

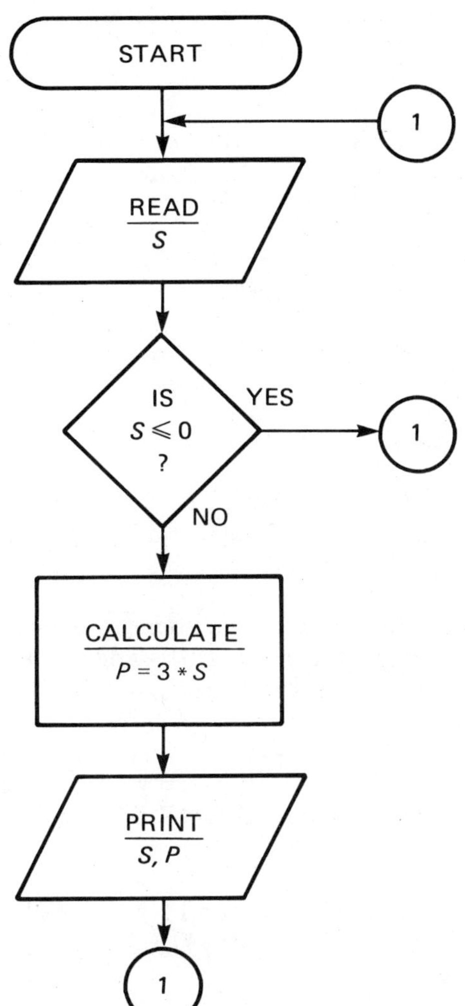

Planned Output

```
PERIMETERS OF EQUILATERAL TRIANGLES
SIDES           PERIMETERS
■ 5             15
6               18
7               21
8               24
9               27
10              30
OUT OF DATA
```

Discussion of the Planned Output:

■ Notice that 0 and the negative number were not included in the chart.

For each of the following problems:
a. Write an analysis which includes the development of an algorithm.
b. Write a plan for a computer solution to the problem.
c. Develop a flowchart.
d. Design a planned output.
e. Code a program to solve the problem and RUN it on the computer.

8. **Problem:** Have the computer print the following chart:

```
RECORD SALE APRIL 26-30
NUMBER            COST
1                 5.43
2                 10.86
3                 16.29
4                 21.72
OUT OF DATA
```

9. **Problem:** Have the computer read ten numbers. Have it then display all the numbers that are divisible by 3. (Use the following data: 5, 6, 7, 567, 657, 765, 843, 348, 438, 439.)

10. **Problem:** Have the computer read the measures of ten angles. Have it then display a chart listing the measure of the angle and its supplement. If the angle does not have a supplement, do not have it displayed. Use the following data: 70, 75, 83, 95, 105, 120, 138, 155, 185, −6.

Section 5.2

In each, determine the value that will be stored in memory cell D after both instructions are executed once.

11. 30 $D=1$
    40 $D=D+1$

12. 20 $D=5$
    30 $D=D+2$

13. 50 $D=11$
    60 $D=D+4$

14. 40 $D=12$
    50 $D=D-1$

15. 80 $D=10$
    90 $D=2*D$

16. 60 $D=7$
    70 $D=3*D$

17. 40 $D=8$
    50 $D=D-5$

18. 70 $D=2$
    80 $D=D\uparrow 3$

Show the output of the following programs.

19.
```
10  Y=2
20  PRINT Y
30  Y=Y+2
40  IF Y>10 THEN 60
50  GOTO 20
60  END
```

20.
```
10  PRINT "COUNTING BY 3"
20  Y=3
30  PRINT Y;
40  Y=Y+3
50  IF Y>21 THEN 70
60  GOTO 30
70  END
```

21.

```
10  PRINT "SAY BYE-BYE!"
20  C=1
30  PRINT "BYE-BYE!"
40  C=C+1
50  IF C>7 THEN 70
60  GOTO 30
70  END
```

22.

```
10  PRINT "READING 4 PAIRS"
20  K=1
30  READ A, B
40  PRINT A, B
50  K=K+1
60  IF K>4 THEN 90
70  GOTO 30
80  DATA 1, 2, 3, 4, 5, 6, 7, 8, 9, 10
90  END
```

23.

```
10  PRINT "COUNT DOWN"
20  D=5
30  PRINT D
40  D=D-1
50  IF D<1 THEN 70
60  GOTO 30
70  PRINT "BLAST-OFF!!"
80  END
```

24.

```
10  E=1
20  K=9-E
30  PRINT K;
40  E=E+1
50  IF E>8 THEN 70
60  GOTO 20
70  END
```

For the given problem, flowchart and planned output, code a program to solve the problem.

25. **Problem:** Enter two counting numbers, A and B. Have the computer count from the smaller number to the larger number in increments of one (by ones) and display the results.

Flowchart

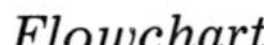

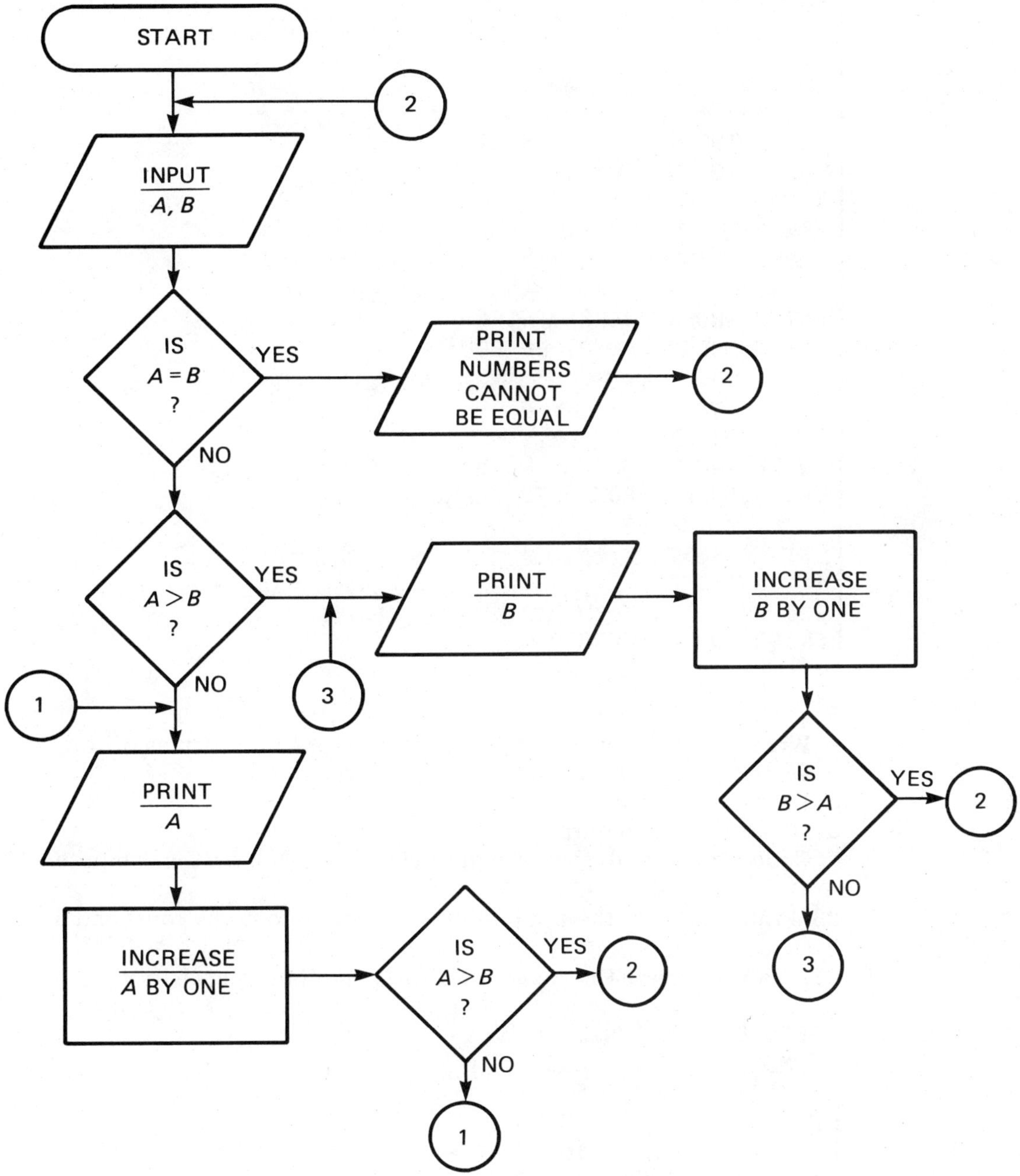
START
2
INPUT
A, B
IS
A = B
?
YES
NO
PRINT
NUMBERS
CANNOT
BE EQUAL
2
IS
A > B
?
YES
NO
PRINT
B
INCREASE
B BY ONE
3
1
PRINT
A
IS
B > A
?
YES
NO
2
3
INCREASE
A BY ONE
IS
A > B
?
YES
NO
2
1

Planned Output

```
COUNTING FROM THE SMALLER TO LARGER NO.



ENTER TWO NUMBERS ? 5, 9
THE NUMBERS FROM 5 TO 9 ARE
5  6  7  8  9



ENTER TWO NUMBERS ? 4, 4
THE NUMBERS CANNOT BE EQUAL



ENTER TWO NUMBERS ? 9, 5
THE NUMBERS FROM 5 TO 9 ARE
5  6  7  8  9



ENTER TWO NUMBERS ?
```

For each of the following problems:
a. Write an analysis which includes the development of an algorithm.
b. Write a plan for a computer solution to the problem.
c. Develop a flowchart.
d. Design a planned output.
e. Code a program to solve the problem and RUN it on the computer.

26. **Problem:** Have the computer count from 4 to 40 by fours and display the results.

27. **Problem:** Have the computer display the following chart without using READ-DATA.

```
RECORD SALE APRIL 26-30
NUMBER          COST
1               5.43
2               10.86
3               16.29
4               21.72
5               27.15
```

28. **Problem:** Enter two counting numbers, A and B. Have the computer count from A to B by ones and display the results. (Remember if A is the larger number, it will be counting down.)

Section 5.3

Solve the following:

29. How long will it take Noreen to drive 266 miles at a rate of 38 mph?

30. What is the rate of a car that travels 336 miles in 8 hours?

31. How far can Christopher travel in 3½ hours at the rate of 36 mph?

Using the formula $D = R \cdot T$, find the missing value for each of the following:

32. $R = 29$ mph
    $T = 5\frac{1}{2}$ hours
    $D = ?$

33. $R = 28$ mph
    $T = ?$
    $D = 168$ miles

34. $R = ?$
    $T = 5\frac{1}{2}$ hours
    $D = 165$ miles

For the given problem, flowchart and planned output, code a program to solve the problem.

35. **Problem:** When an object is dropped from a given height, we use the formula $s = 16t^2$ to find the distance, s, in feet it falls in t seconds. For example, if we want to know the distance an object falls after 3 seconds, we get:

$s = 16t^2$

$s = 16 \cdot 3^2$

$s = 16 \cdot 9 = 144$ feet

Have the computer print a chart for the distance an object falls for integral values from 1 to 5 seconds.

Flowchart

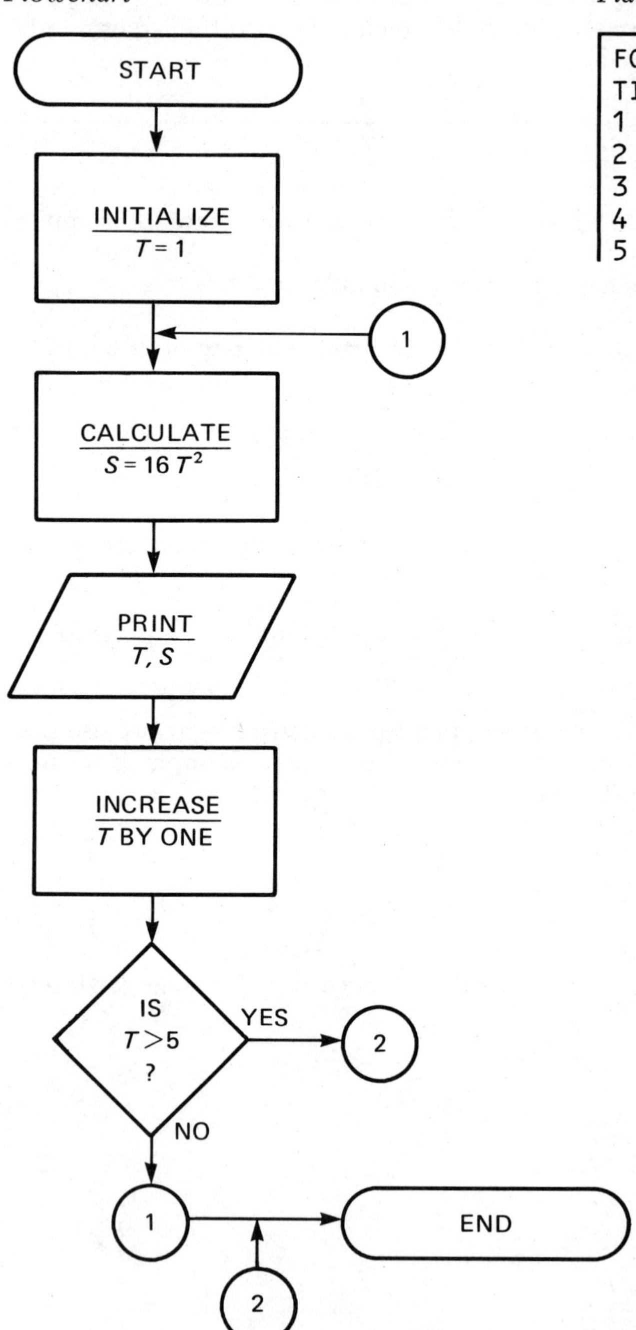

Planned Output

```
FOLLOWING A FALLING OBJECT
TIME          DISTANCE
1             16
2             64
3             144
4             256
5             400
```

For each of the following problems:
a. Write an analysis which includes the development of an algorithm.
b. Write a plan for a computer solution to the problem.
c. Develop a flowchart.
d. Design a planned output.
e. Code a program to solve the problem and RUN it on the computer.

36. **Problem:** Have the computer print a chart for the distance an object falls ($s=16t^2$) for integral values from 3 to 10 seconds. (Do not use READ-DATA.)

37. **Problem:** Have the computer read ten numbers. Then have it set up a chart to print the number, the square of the number, and the integer value of the number. Use the following data: 2.3, 4.5, 8, 9, −10.7, 11, 12, 20, −21.6, 30.

38. **Problem:** The Random Computer Company pays its salespersons a monthly salary of $800 plus 2% commission on monthly sales. Have the computer print the table below showing the salesperson number, sales and salary. (Read only the sales.)

```
SALESPERSON ƀƀƀƀƀƀ SALES ƀƀƀƀƀƀƀƀ SALARY
1                 13400          1068
2                 22000          1240
3                 8600           972
4                 32460          1449.2
```

Section 5.4

Which of the following are arithmetic sequences? State the reason.

39. 8, 12, 16, 20, . . .
40. 17, 24, 31, 38, . . .
41. 99, 94, 89, 84, . . .
42. 7, 9, 12, 16, . . .

Fill in the blanks for the following arithmetic sequences.

43. 12, _, _, 39, 48, . . .
44. 13, _, _, 25, . . .
45. 26, _, _, 50, . . .
46. 43, _, _, _, 23

Use the formula $S=\frac{n}{2}(f+\ell)$ to find the sum of the following arithmetic series.

47. $4+8+12+16+20+24+28$
48. $16+24+32+40+48+56+64+72+80$
49. $28+24+20+16+12+8+4$
50. $9+14+19+24+29+34+39+44+49+54+59$

Which of the following are geometric sequences? State the reason.

51. 2, 10, 50, 250, . . .

52. 3, 9, 27, 81, . . .

53. 64, 32, 16, 8, . . .

54. 3, 15, 75, 375, . . .

Use the formula $S = \frac{f - f \cdot r^n}{1 - r}$ to find the sum of the following geometric series.

55. $1 + 2 + 4 + 8 + 16 + \ldots$ (8 terms)

56. $1 + 4 + 16 + \ldots$ (5 terms)

57. $9 + 27 + 81 + \ldots$ (6 terms)

58. $3 + 18 + 108 + \ldots$ (7 terms)

Fill in the blanks for the following geometric sequences.

59. 5, 15, _, _, _, 1215

60. 3, _, _, 24, 48, . . .

61. 64, _, 16, 8, _, _,

62. 6, 24, _, _, _

For the given problem, flowchart and planned output, code a program to solve the problem.

63. **Problem:** Find the sum of a geometric series where we are given the first term, *F*, the common ratio, *R*, and the number of terms, *N*.

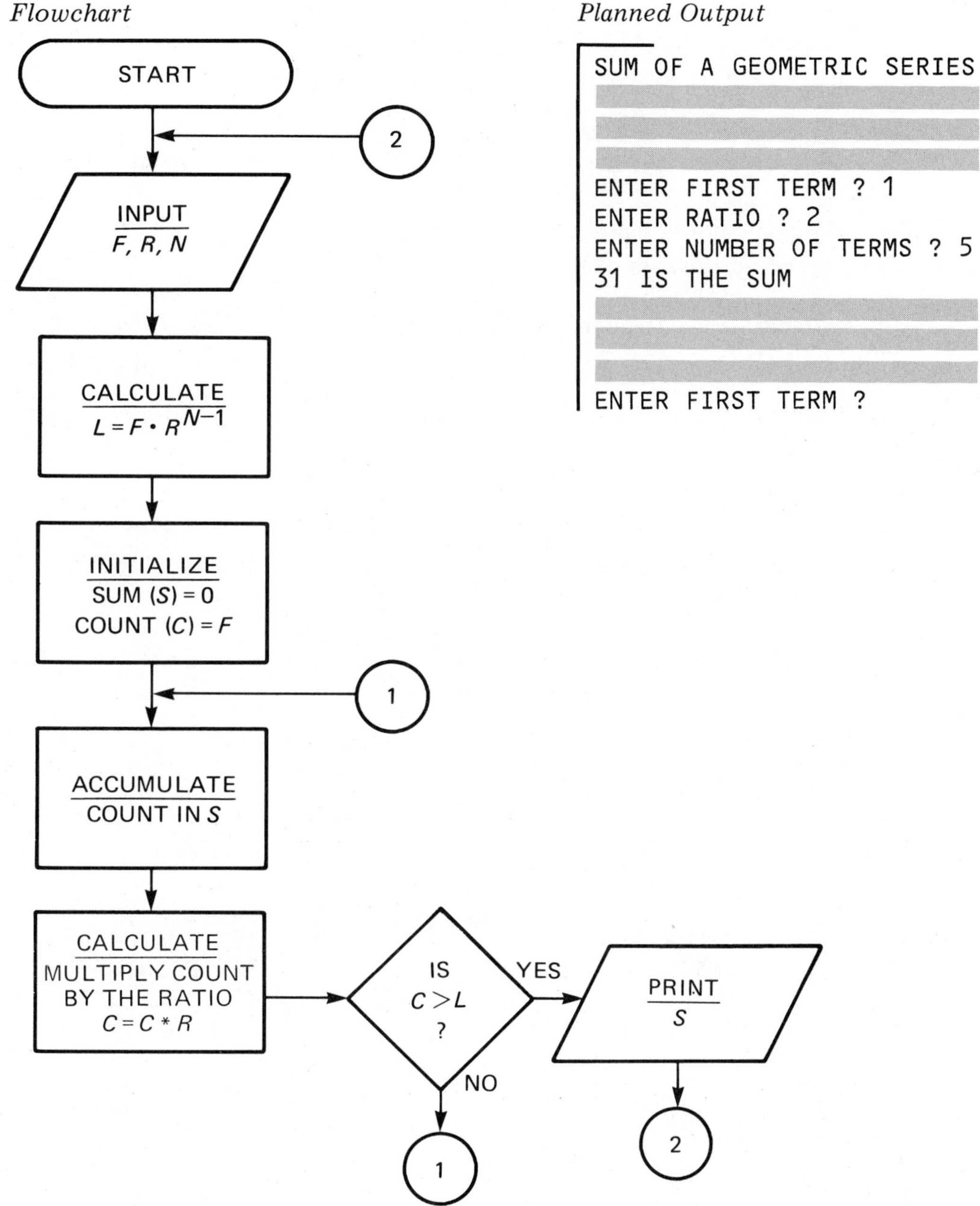
Flowchart
START
2
INPUT
F, R, N
CALCULATE
L = F • R^(N–1)
INITIALIZE
SUM (S) = 0
COUNT (C) = F
1
ACCUMULATE
COUNT IN S
CALCULATE
MULTIPLY COUNT
BY THE RATIO
C = C * R
IS
C > L
?
YES
NO
PRINT
S
1
2
Planned Output
SUM OF A GEOMETRIC SERIES
ENTER FIRST TERM ? 1
ENTER RATIO ? 2
ENTER NUMBER OF TERMS ? 5
31 IS THE SUM
ENTER FIRST TERM ?

For each of the following problems:
a. Write an analysis which includes the development of an algorithm.
b. Write a plan for a computer solution to the problem.
c. Develop a flowchart.
d. Design a planned output.
e. Code a program to solve the problem and RUN it on the computer.

64. **Problem:** Put in values for f, r and n. Use the formula $S = \frac{f - f \cdot r^n}{1-r}$ to have the computer find the sum of the geometric series.

65. **Problem:** Have the computer find the sum of the cubes of the counting numbers from 1 to 10.

66. **Problem:** Enter two counting numbers, A, B. Have the computer find the sum of the squares of the counting numbers from A to B (if $B>A$) or from B to A (if $A>B$). If $A=B$, have two different counting numbers entered.

Section 5.5

Find the factors of each number.

67. 45 68. 28 69. 17 70. 16

Show the output for the following programs.

71.
```
10  FOR C=3 TO 8
20  PRINT C
30  NEXT C
40  END
```

72.
```
10  FOR Q=5 TO 11
20  PRINT Q;
30  NEXT Q
40  END
```

73.
```
10  FOR L=2 TO 5
20  PRINT "GOOD"
30  NEXT L
40  END
```

74.
```
10  FOR N=1 TO 4
20  PRINT "TEST #"; N
30  NEXT N
40  END
```

75.
```
10  FOR P=5 TO 7
20  PRINT P, "APPLES"
30  NEXT P
40  END
```

76.
```
10  FOR X=2 TO 12 STEP 2
20  PRINT X;
30  NEXT X
40  END
```

77.

```
10  FOR A=0 TO 30 STEP 5
20  PRINT A;
30  NEXT A
40  END
```

78.

```
10  FOR S=10 TO 1 STEP -1
20  PRINT S
30  NEXT S
40  PRINT "BLAST-OFF!!"
50  END
```

79.

```
10  FOR I=1 TO 3
20  PRINT I;
30  NEXT I
40  FOR L=1 TO 4
50  PRINT L;
60  NEXT L
70  END
```

80.

```
  10  FOR I=1 TO 3
  20  PRINT I;
  30  NEXT I
■ 40  PRINT
  50  FOR L=1 TO 4
  60  PRINT L;
  70  NEXT L
  80  END
```

■ How does the PRINT statement make the output of #80 different than the output of #79?

For the given problem, flowchart and planned output, code a program to solve the problem.

81. **Problem:** Have the computer find X^N without using the exponentiation (↑) key.† (Recall that if $X=2$ and $N=3$, X^N is $2^3=2\cdot2\cdot2=8$.)

†This program does not handle the special case when $N=0$. As a bonus, see if you can handle this case in your program. As an added bonus, see if you can handle the case when N is a negative integer.

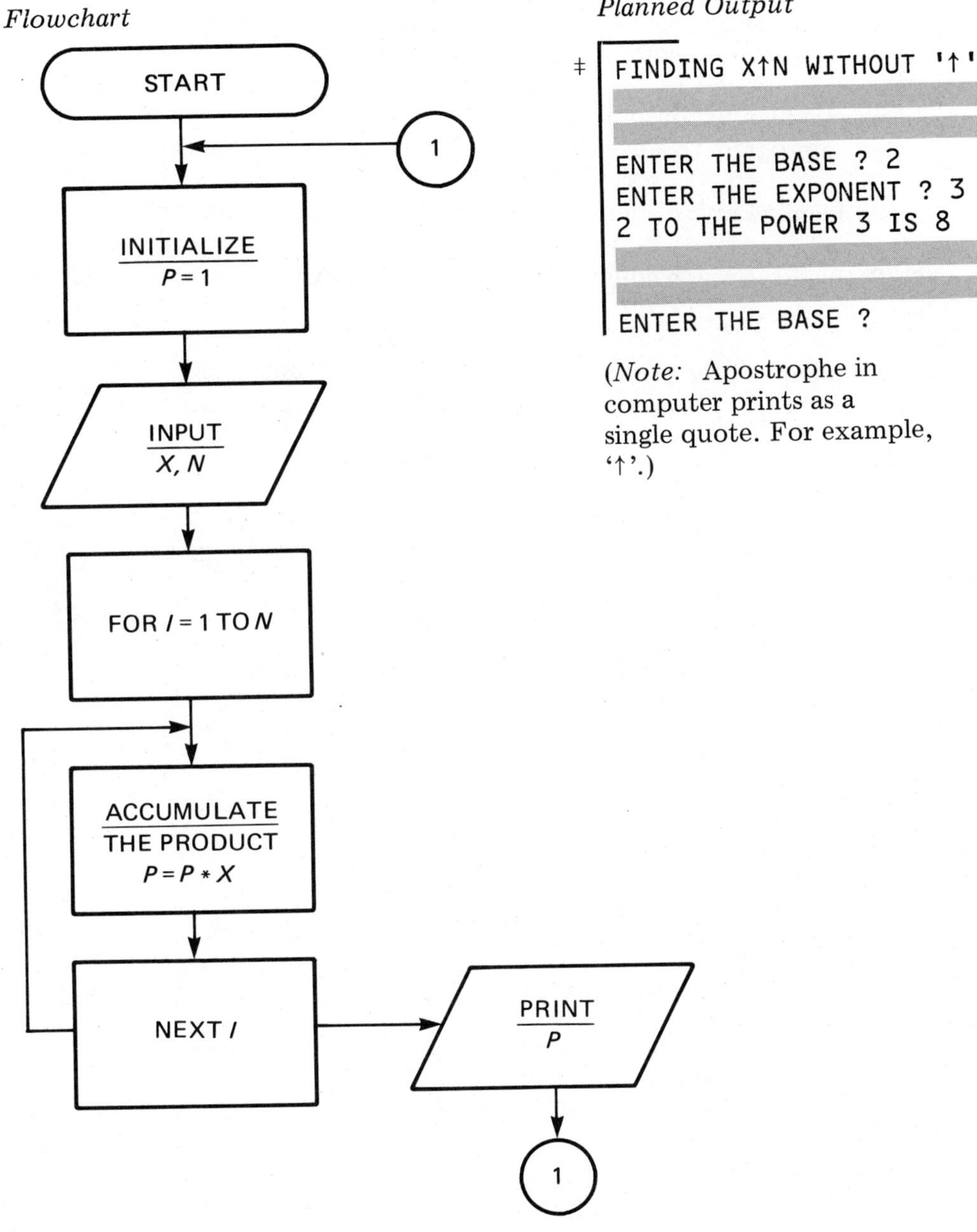

(*Note:* Apostrophe in computer prints as a single quote. For example, '↑'.)

‡On a Radio Shack Model III, the exponentiation symbol is [.

For each of the following problems:
a. Write an analysis which includes the development of an algorithm.
b. Write a plan for a computer solution to the problem.
c. Develop a flowchart.
d. Design a planned output.
e. Code a program to solve the problem and RUN it on the computer.

82. **Problem:** Enter a counting number, N. Print the number of factors of N.

83. **Problem:** Enter a counting number, N. Print the factors of N and the number of factors of N.

84. **Problem:** Enter counting numbers, F, N, I. Count from F to N in increments of I. (Be sure that $N > F$ and that I is less than the interval from F to N.)

Section 5.6

Tell which of the following are prime numbers.

85. 45	86. 3	87. 21	88. 57
89. 4	90. 23	91. 5	92. 25

Show the complete output for the following programs.

93.
```
10  READ X
20  FOR I=1 TO X
30  READ N$
40  PRINT "NAME"; I; "ƀISƀ"; N$
50  NEXT I
60  DATA 3, JILL, JANE, JOE
70  END
```

94.
```
10  FOR I=1 TO 2
20  READ C$, D$
30  IF C$<D$ THEN GOTO 60
40  PRINT D$, C$
50  GOTO 70
60  PRINT C$, D$
70  NEXT I
80  DATA GERRY, GERI, LANE, LOIS
90  END
```

95.
```
10  PRINT "LOOKING FOR ET"
20  C=0
30  FOR I=1 TO 5
40  READ B$
50  IF B$ = "ET" THEN C=C+1
60  NEXT I
70  DATA ET, SOS, UPS, ET, ET
80  PRINT "I FOUND ET"; C; "ƀTIMES"
90  END
```

96.
```
10  PRINT "HOW MANY A'S?"
20  C=0
30  FOR L=1 TO 8
40  READ E$
50  IF E$ = "A" THEN C=C+1
60  NEXT L
70  DATA C, A, E, A, F, A, R, A, A, A
80  PRINT "FOUND"; C; "ƀA'S"
90  END
```

For the given problem, flowchart and planned output, code a program to solve the problem.

97. **Problem:** There are many algorithms for determining if a number is prime. Use the flowchart to write a program to determine if a number is prime. (A discussion of this method for finding primes can be found on pages 299-303 in Elgarten, et al., *Using Computers in Mathematics*, Addison-Wesley Publishing Company, Menlo Park, California, 1983.)

Flowchart

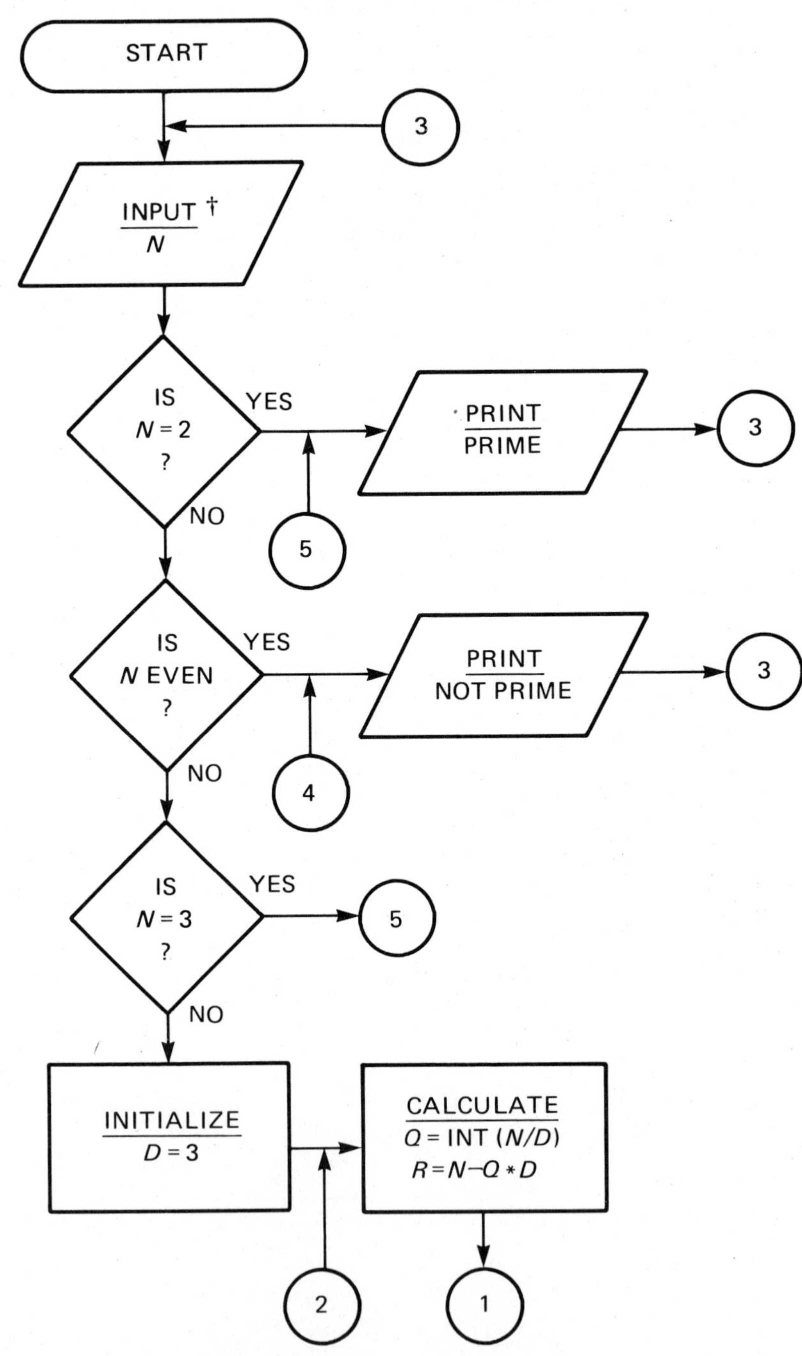

† N must be greater than 1.

Flowchart (cont'd)

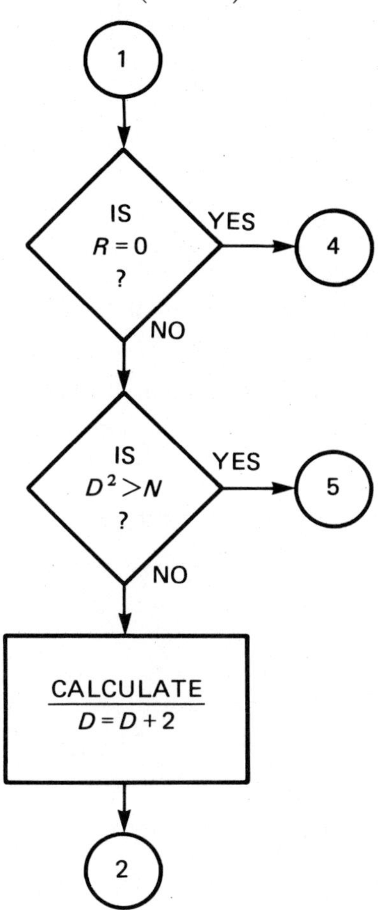

Planned Output

```
DETERMINE IF IT IS PRIME.


ENTER YOUR NUMBER ? 13
13 IS PRIME


ENTER YOUR NUMBER ? 6
6 IS NOT PRIME


ENTER YOUR NUMBER ?
```

For each of the following problems:
a. Write an analysis which includes the development of an algorithm.
b. Write a plan for a computer solution to the problem.
c. Develop a flowchart.
d. Design a planned output.
e. Code a program to solve the problem and RUN it on the computer.

98. **Problem:** Enter three counting numbers, A, B, and C. Have the computer list the common factors of the three numbers.

99. **Problem:** Enter a counting number. Have the computer display the sum of the proper factors of the number. (See Exercise 5.6, Problem 18.)

100. **Problem:** A perfect number is a counting number that is equal to the sum of its proper factors. Enter a counting number. Have the computer determine if the number is a perfect number. (For example, 6 is a perfect number because the sum of its proper factors $1+2+3$ equals 6.)

Chapter 6
Applications in Geometry

Section 6.1 APPROXIMATING SQUARE ROOTS (SQR(X) AND ROUNDING)

Can you fill in the correct number in each shape?

1. □ · □ = 36

2. △ · △ = 49

You probably had no trouble finding 6 for number 1, and 7 for number 2.

6 is a square root of 36.

7 is a square root of 49.

The square root of a number may be positive or negative. When we use the symbol "$\sqrt{}$" (called a radical) for a square root, we are only concerned with the positive square root. Therefore,

$\sqrt{36} = 6$, and $\sqrt{49} = 7$.

> The *square root of a number* is the factor of the number that when multiplied by itself produces the given number.

Example 1 Find $\sqrt{16}$.

Solution: We must look for two equal factors whose product is 16.

factor → 1	$1\overline{)16}$ with quotient 16 ← factor	Factors are not equal.
factor → 2	$2\overline{)16}$ with quotient 8 ← factor	Factors are not equal.
factor → 4	$4\overline{)16}$ with quotient 4 ← factor	Factors are equal.

Therefore, $\sqrt{16} = 4$.

Class Exercise 1 Find:

1. $\sqrt{25}$ 2. $\sqrt{1}$ 3. $\sqrt{121}$ 4. $\sqrt{144}$

So far all the square roots we have found are integers.
How can we find $\sqrt{40}$?

Looking at the possible factorings: $1 \cdot 40$

$2 \cdot 20$

$4 \cdot 10$

$5 \cdot 8$,

we see there is no pair of equal factors. So $\sqrt{40}$ is not an integer. We must approximate its value.

A good guess would be a number between 6 and 7, but closer to 6 [since 40 is closer to 6^2 (36) than to 7^2 (49)]. Let's guess 6.3.

If the exact answer is 6.3 then $6.3 \overline{)40}$ would have a quotient of 6.3. (You might wish to use your calculator at this point.)

$$6.3 \overline{)40} = 6.35$$

Our next guess would have to be between 6.3 and 6.35, so we will find the average of these numbers:

$$\frac{6.3 + 6.35}{2} = 6.325$$

Let us see if dividing by 6.325 will give us 6.325:

$$6.325 \overline{)40} = 6.324 \ldots$$

We may continue this process to get as close an approximation to $\sqrt{40}$ as we want. *The approximation is accurate to as many places as match in the divisor and quotient.*

Looking at the results above we can say

is approximately equal to

$\sqrt{40} \approx 6.32$ since 6.325 and 6.324 match to two decimal places.

We, therefore, are accurate to the nearest hundredth.

Class Exercise 2 Find the value of each of the following to the nearest tenth. (*Hint:* The divisor and quotient must match in the tenths place.)

1. $\sqrt{12}$ 2. $\sqrt{20}$ 3. $\sqrt{43}$ 4. $\sqrt{94}$

Find the value of each of the following to the nearest hundredth. (*Hint:* The divisor and quotient must match in the hundredths place.)

5. $\sqrt{27}$ 6. $\sqrt{75}$ 7. $\sqrt{2}$ 8. $\sqrt{110}$

Example 2 Find the positive value of X to the nearest hundredth.

1. $X^2 = 16$ 2. $X^2 = 25$ 3. $X^2 = 40$

Solution:

1. For $X^2 = 16$ we are looking for one of two equal factors of 16:

$$X = \sqrt{16}$$
$$X = 4.00$$

2. For $X^2 = 25$:

$$X = \sqrt{25}$$
$$X = 5.00$$

3. For $X^2 = 40$:

$X = \sqrt{40}$ We must approximate the answer (see above algorithm).

$$X = 6.32$$

Class Exercise 3 Find X to the nearest hundredth (only the positive value).

1. $X^2 = 4$ 2. $X^2 = 49$ 3. $X^2 = 70$ 4. $X^2 = 10$

How can we use the computer to find the square root of a number?
The computer has a square root function that will give us a square root.

> The *SQUARE ROOT function*, written SQR(X), where X is called the argument of the function, returns the square root of the argument X. The square root may also be found using $X\uparrow(1/2)$. X cannot be negative.

For example, if $X=4$, then SQR(X) becomes SQR(4)=2. When $X=40$ the SQR(X) becomes SQR(40)=6.32455532. (The number of decimal places depends on the computer that you are using.)

Class Exercise 4 Find the value assigned to S after executing the instructions.

1. ```
   10 X=16
   20 S=SQR(X)
   ```
2. ```
   20 A=64
   30 S=SQR(A)
   ```
3. ```
   40 T=2
   50 S=SQR(2*T)
   ```
4. ```
   60 L=10
   70 S=SQR(10*L)
   ```

Now we will use the square root function in a program.

Problem

> Enter a positive number. Have the computer display the square root rounded to the nearest hundredth.

Solution:

a. *Analysis:* We will use the function SQR(X) to find the square root of the number.

b. *Plan:*

1. Enter a number, X.
2. Is $X<0$?
   a) If yes, go to 1.
   b) If no, continue.
3. Use the square root function: N=SQR(X).
4. Round N to the nearest hundredth, NR (N rounded).
5. Print N, NR.
6. Go to 1.

c. *Flowchart*

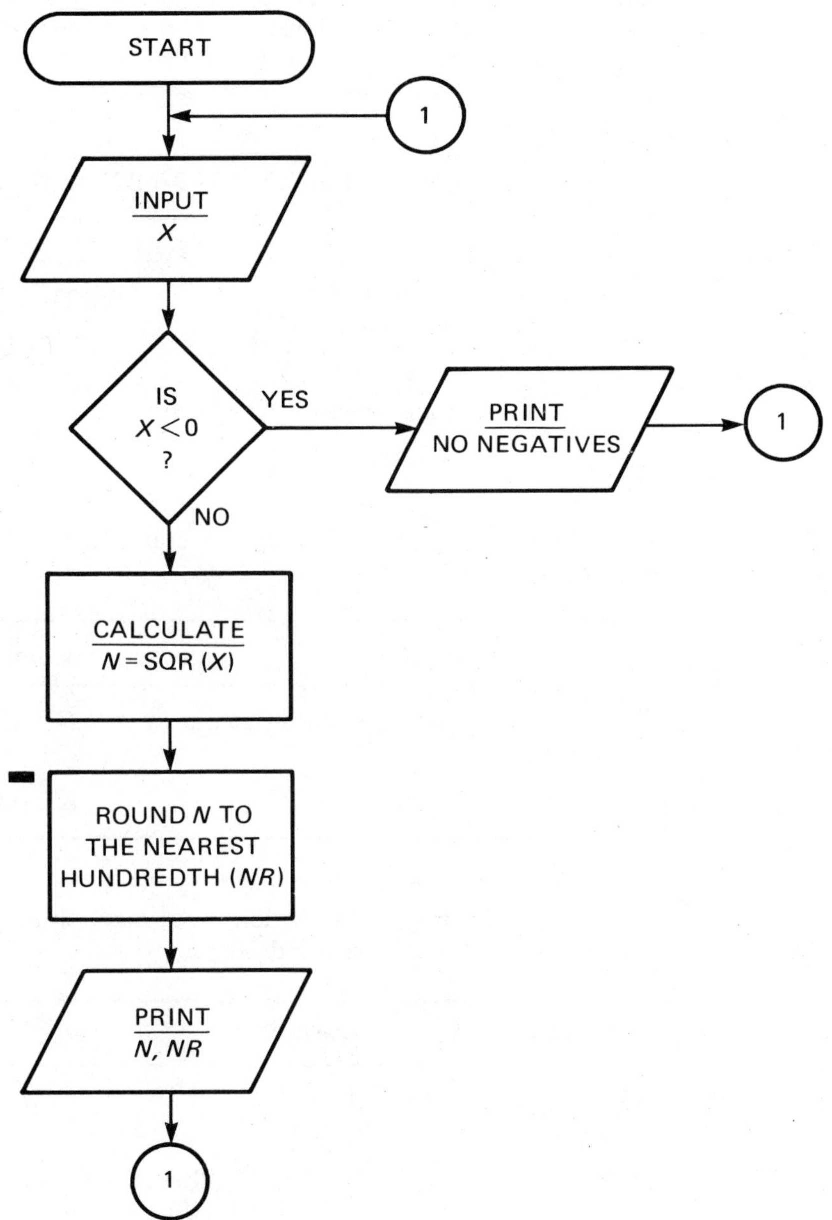

Discussion of the Flowchart:

▬ We must develop an algorithm to round a number to the nearest hundredth (two decimal places).

Let us review how to round to the nearest hundredth.

> *Rule:* If the thousandths (third decimal) place is 5 or more, raise the hundredths (second decimal) place one unit and cut off (truncate) the rest of the decimal fraction. Otherwise, just truncate the decimal fraction beginning at the third place.

Example 3 Use the rule to round 6.234 and 6.237 to the nearest hundredth.

Solution:

Number: 6.234 (4: Less than 5) 6.237 (7: 5 or more)

Rounded to the nearest hundredth:

6.23 6.24

For the computer, we will use the following procedure to round a number, N, to the nearest hundredth.

Procedure	Computer Translation
1. Add 5 to the thousandths place (that is, add .005 to the number.)	$R = N + .005$
2. Truncate R to two decimal places.	$NR = \text{INT}(100 * R) / 100$

NR is now N rounded to the nearest hundredth.

Let us use the computer translation, which is our algorithm, to see how it will round two different numbers to the nearest hundredth.

Computer Translation	$N = 6.234$	$N = 6.237$
1. $R = N + .005$	$R = 6.234 + .005$ $= 6.239$	$R = 6.237 + .005$ $= 6.242$
2. $NR = \text{INT}(100 * R)/100$	$NR = \text{INT}(100 * 6.239)/100$ $= \text{INT}(623.9)/100$ $= 623/100$ $= 6.23$	$NR = \text{INT}(100 * 6.242)/100$ $= \text{INT}(624.2)/100$ $= 624/100$ $= 6.24$

6.23 and 6.24 are the same results that we got from Example 3, where we used a different procedure. (For more practice working with the algorithm, see Exercise 6.1, Problems 21-24.)

We continue with our solution to the problem.

d. *Planned Output*

```
FINDING THE SQUARE ROOT OF A NUMBER


ENTER THE NUMBER ? 40
SQUARE ROOT IS 6.32455532
ROUNDED TO THE NEAREST HUNDREDTH IS 6.32


ENTER THE NUMBER ? -40
SORRY. NUMBER MUST BE POSITIVE


ENTER THE NUMBER ?
```

e. *The Program*

```
  10   PRINT "FINDING THE SQUARE ROOT OF A NUMBER"
  20   PRINT : PRINT
  30   INPUT "ENTER THE NUMBER"; X
  40   IF X<0 THEN GOTO 110
  50   N=SQR(X)
■ 60   R=N+.005
■ 70   NR=INT(100*R)/100
  80   PRINT "SQUARE ROOT IS"; N
  90   PRINT "ROUNDED TO THE NEAREST HUNDREDTH IS"; NR
  100  GOTO 20
  110  PRINT "SORRY. NUMBER MUST BE POSITIVE"
  120  GOTO 20
  130  END
```

Discussion of the Program:

- Lines 60 and 70 make up the algorithm to round N to the nearest hundredth.

The next exercises will give you practice using the square root function and rounding.

Exercise 6.1

Find:

1. $\sqrt{81}$ 2. $\sqrt{144}$ 3. $\sqrt{169}$ 4. $\sqrt{0}$

Find the value of each of the following to the nearest tenth.

5. $\sqrt{130}$ 6. $\sqrt{28}$ 7. $\sqrt{60}$ 8. $\sqrt{50}$

Find the value of each of the following to the nearest hundredth.

9. $\sqrt{38}$ 10. $\sqrt{7}$ 11. $\sqrt{89}$ 12. $\sqrt{60}$

Find X to the nearest hundredth (only the positive value).

13. $X^2=1$ 14. $X^2=36$ 15. $X^2=100$ 16. $X^2=85$

Find the value assigned to R after executing the instructions.

17.
```
20 B=25
30 R=SQR(B)
```

18.
```
10 T=81
20 R=SQR(T)
```

19.
```
50 V=8
60 R=SQR(2*V)
```

20.
```
80 M=20
90 R=INT(SQR(M))
```

For each of the following, use the algorithm to round the number N to the nearest hundredth. (The *Discussion of the Flowchart* in this section can serve as a model solution.)

Algorithm: $R = N + .005$

$NR = \text{INT}(100*R)/100$

↑ NR: N rounded to the nearest hundredth

21. 9.464 22. 3.069 23. 4.2345 24. 7.3168

For each of the following, use the algorithm to round the number N to the nearest tenth.

Algorithm: $R = N + .05$

$NR = \text{INT}(10*R)/10$

↑ NR: N rounded to the nearest tenth

25. 8.32 26. 9.68 27. 3.045 28. 6.152

Revise the program that was written in this section to produce the following outputs.

29.

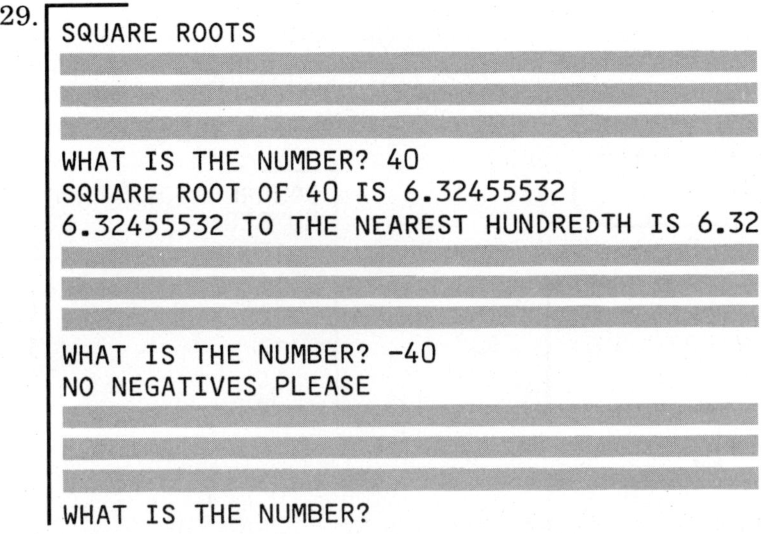

30.

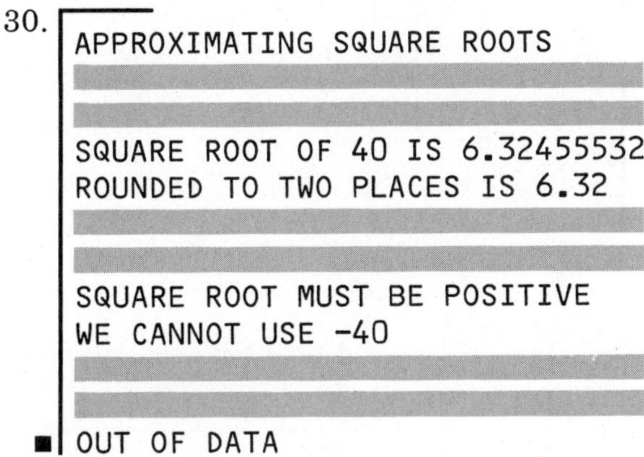

■ See if you can eliminate the OUT OF DATA message.

For the given problem, flowchart and planned output, code a program to solve the problem.

31. **Problem:** Have the computer display a chart of the square roots of all the whole numbers from 10 to 15. The square roots should be rounded to the nearest thousandth.

Flowchart

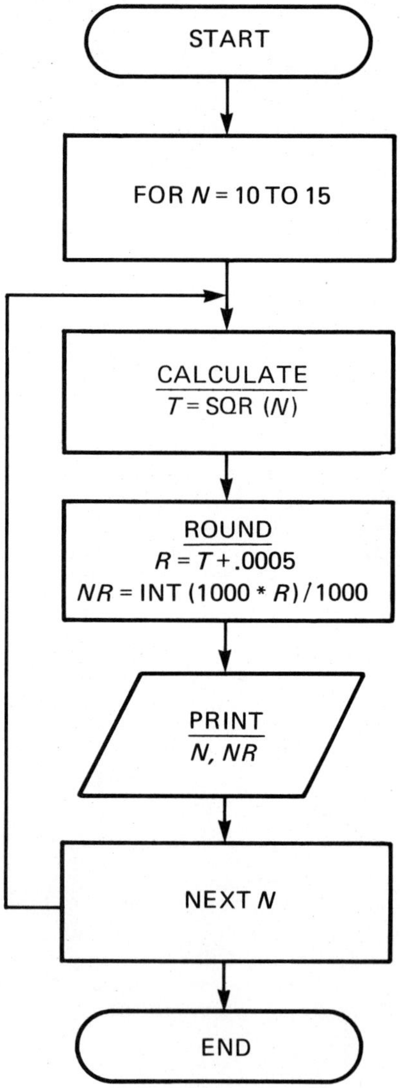

START

FOR N = 10 TO 15

CALCULATE
T = SQR (N)

ROUND
$R = T + .0005$
NR = INT (1000 * R)/1000

PRINT
N, NR

NEXT N

END

Planned Output

```
SQUARE ROOTS FROM 10-15
NUMBER            SQ. ROOT
  10                 3.162
  11                 3.317
  12                 3.464
  13                 3.606
  14                 3.742
  15                 3.873
```

For each of the following problems:
a. Write an analysis which includes the development of an algorithm.
b. Write a plan for a computer solution to the problem.
c. Develop a flowchart.
d. Design a planned output.
e. Code a program to solve the problem and RUN it on the computer.

32. **Problem:** Enter a number. Have the computer display the number and the square root of the number rounded to the nearest tenth.

33. **Problem:** Enter a number. Have the computer display the number and the square root of the number rounded to the nearest thousandth.

34. **Problem:** Enter a number. Give the user the option to have the square root of the number rounded to the

a. nearest tenth,
b. nearest hundredth, or
c. nearest thousandth.

Section 6.2 CIRCUMFERENCE AND AREA (MAKING PROGRAMS MORE FRIENDLY)

Take a circular dish and measure the distance around the dish as well as the distance across its center. We will do the same with our clock at the right.

Our clock has a distance around, called the *circumference*, of 17.3 cm, and a distance across, called the *diameter*, of 5.5 cm.

Divide the circumference by the diameter. We get:

$$\frac{\text{circumference}}{\text{diameter}} = \frac{17.3 \text{ cm}}{5.5 \text{ cm}} = 3.145$$

which is very close to what you should get with your measurements.

Over the years, people dividing these measurements noticed that they all came very close to a particular value, depending upon the accuracy of measurement. This particular value is approximately

3.14159265358979323846 . . . .

We use the Greek letter π (pi) to represent this value.

In the Bible, $\pi = 3$. (I Kings 7:23 gives a description of the pool in the courtyard of Solomon's Temple.) In more recent years, mathematicians have calculated the value of π to thousands of places by using computers.

We have $\frac{C}{D} = \pi$, from which we can get $C = \pi D$. Since the diameter of a circle is equal to two radii ($D = 2 \cdot R$), we have

$C = \pi D$

$C = \pi(2 \cdot R)$ or

$C = 2\pi R$

$\pi \approx$ 3.14159265358979323846264338327950288419716939937510
58209749445923078164062862089986280348253421170679
82148086513282306647093844609550582231725359408128
48111745028410270193852110555964462294895493038196
44288109756659334461284756482337867831652712019091
45648566923460348610454326648213393607260249141273
72458700660631558817488152092096282925409171536436
78925903600113305305488204665213841469519415116094
33057270365759591953092186117381932611793105118548
07446237996274956735188575272489122793818301194912
98336733624406566430860213949463952247371907021798
60943702770539217176293176752384674818467669405132
00056812714526356082778577134275778960917363717872
14684409012249534301465495853710507922796892589235
42019956112129021960864034418159813629774771309960
51870721134999999837297804995105973173281609631859
50244594553469083026425223082533446850352619311881
71010003137838752886587533208381420617177669147303
59825349042875546873115956286388235378759375195778
18577805321712268066130019278766111959092164201989
38095257201065485863278865936153381827968230301952
03530185296899577362259941389124972177528347913151
55748572424541506959508295331168617278558890750983
81754637464939319255060400927701671139009848824012
85836160356370766010471018194295559619894676783744
94482553797747268471040475346462080466842590694912
93313677028989152104752162056966024058038150193511
25338243003558764024749647326391419927260426992279
67823547816360093417216412199245863150302861829745
55706749838505494588586926995690927210797509302955
32116534498720275596023648066549911988183479775356
63698074265425278625518184175746728909777727938000
81647060016145249192173217214772350141441973568548
16136115735255213347574184946843852332390739414333
45477624168625189835694855620992192221842725502542
56887671790494601653466804988627232791786085784383
82796797668145410095388378636095068006422512520511
73929848960841284886269456042419652850222106611863
06744278622039194945047123713786960956364371917287
46776465757396241389086583264599581339047802759009
94657640789512694683983525957098258...

Example 1 Find the circumference, to the nearest tenth, of an automobile tire of diameter 60 cm. Use $\pi = 3.14$.

Solution: Since we are given the length of the diameter, we can use the formula $C = \pi D$.

$C = \pi D$

$C = (3.14)(60)$

$C = 188.4$ cm

Example 2 Find the circumference, to the nearest tenth, of the circle with radius 5 in. Use $\pi = 3.14$.

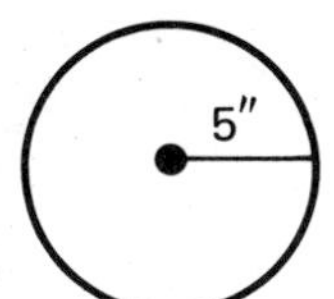

Solution: Since we are given the length of the radius, we use the formula $C = 2\pi R$.

$C = 2\pi R$

$C = 2(3.14)(5)$

$C = (6.28)(5)$

$C = 31.4$ in.

Class Exercise 1 Find the circumference, to the nearest hundredth, of each circle. Use $\pi = 3.1416$.

1. 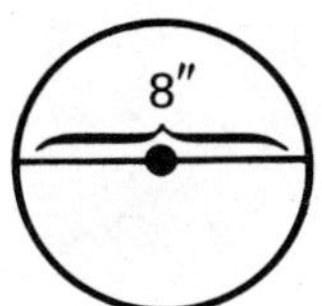

2. 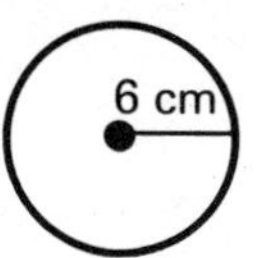

3. 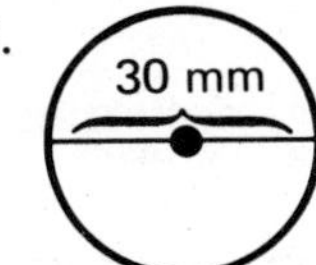

4. 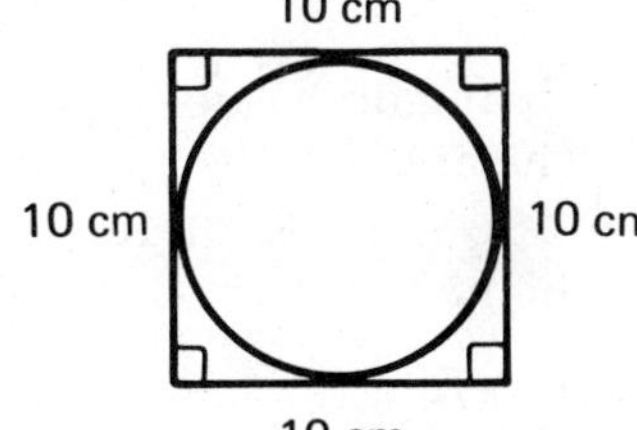

The area of a circle can be estimated by counting the number of square regions inside the circle.

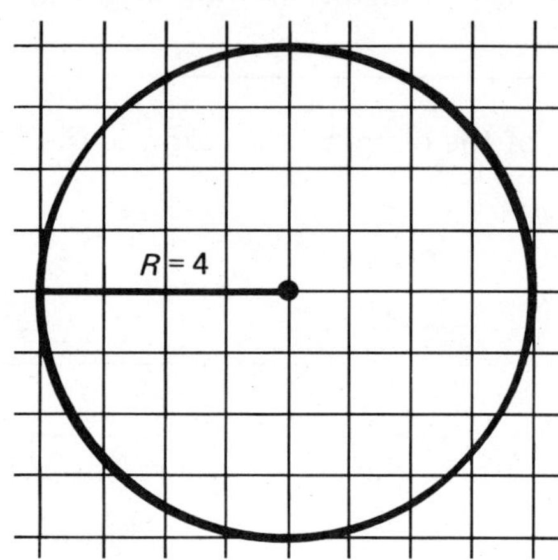

Number of complete square regions in circle:	32
Guess at number of partial square regions:	18
Total estimated area of the circle:	50 sq. units

Let us develop a formula for finding the area of a circle more accurately.
Consider a circle divided into 24 equal parts.

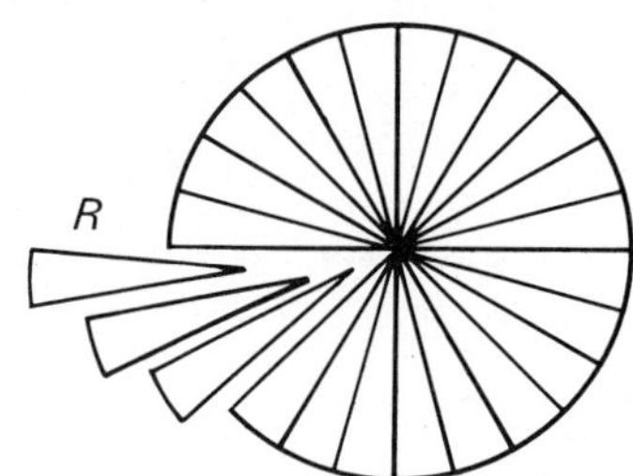

We can cut apart the circle and rearrange the 24 parts as shown below.

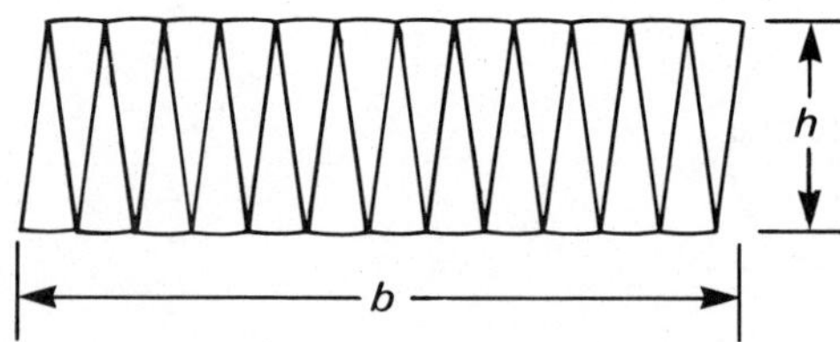

This figure is an "almost parallelogram" because the bases are not straight line segments. The area of this "parallelogram" is close to $b \cdot h$.

So the area of our original circle is $b \cdot h$, but $b = \frac{1}{2}(\text{circumference}) = \frac{1}{2}(2\pi R) = \pi R$, and the height, h, is R. Therefore, the area $= \pi R \cdot R$ or

$$A = \pi R^2$$

Example 3 Find, to the nearest tenth, the area of a circle of radius 4 units. Use $\pi = 3.14$.

Solution: We use the formula $A = \pi R^2$.

$A = \pi R^2$

$A = 3.14(4)^2$

$A = 3.14(16)$

$A = 50.24$

$A = 50.2$ sq. units (rounded to the nearest tenth)

Compare this answer to our approximation by counting the squares. We did a pretty good job of counting!

Class Exercise 2 Find the area of the following circles. Use $\pi = 3.14$ and round the answers to the nearest tenth.

1. 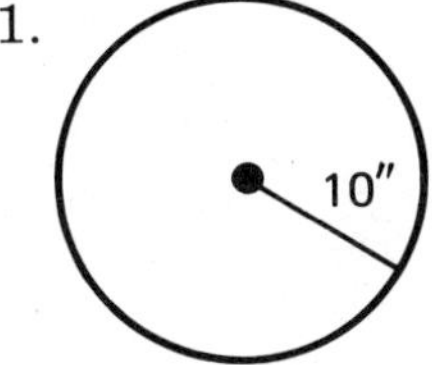

2. 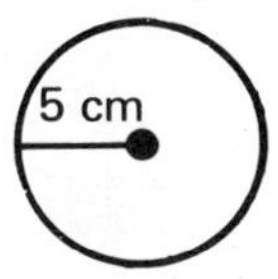

3. 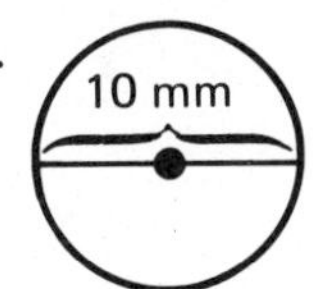

4. 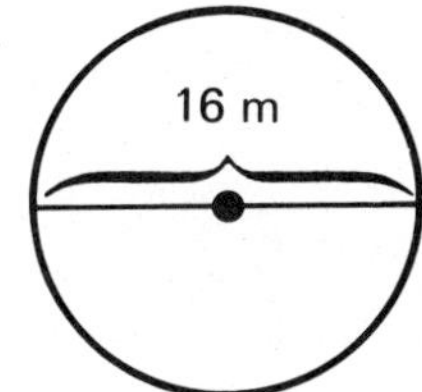

Now let us use the computer to solve the following problem.

Problem Enter the length of a radius. Find the area of the circle, rounded to the nearest hundredth. (Use $\pi = 3.14159$.)

Solution:

a. *Analysis:* To find the area of a circle we use the formula $A=\pi R^2$.

b. *Plan:*

1. Put in a value for R.
2. Is $R \leqslant 0$?
   a) If yes, then go to 1.
   b) If no, continue.
3. Calculate: $A=\pi R^2$.
4. Round A to the nearest hundredth, AR.
5. Print AR (A rounded).
6. Go to 1.

c. *Flowchart*

START
1
INPUT
R
IS
R ≤ 0
?
YES
NO
PRINT
RADIUS MUST
BE POSITIVE
1
CALCULATE
A = πR²
ROUND A TO
THE NEAREST
HUNDREDTH
(AR)
PRINT
AR
(A ROUNDED)
1

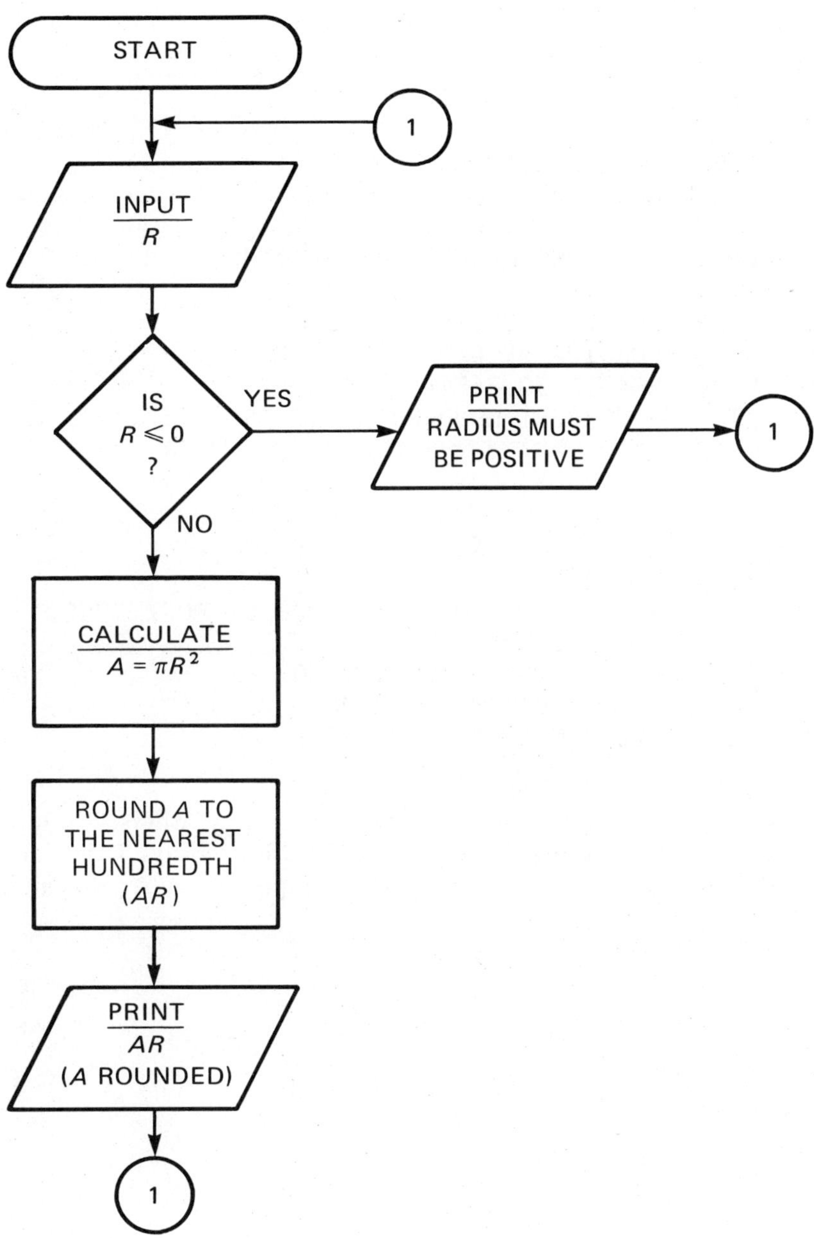

d. *Planned Output*

```
HELLO, MY NAME IS COMPUTER.

TYPE IN YOUR NAME ? AL

HELLO AL.  WE ARE GOING TO
FIND AREAS OF CIRCLES.

AL, ENTER THE LENGTH OF A RADIUS.
? 4
THANKS AL. I WILL NOW CALCULATE.
FOR A RADIUS OF 4
THE AREA IS 50.26544
TO THE NEAREST HUNDREDTH, IT IS 50.27

TYPE IN YES TO CONTINUE ? YES

AL, ENTER THE LENGTH OF A RADIUS.
? -4
THANKS AL.  BUT NO THANKS.  ONLY POSITIVES.

TYPE IN YES TO CONTINUE ? NO
THANKS FOR WORKING WITH ME.
AL, HAVE A NICE DAY. BYE-BYE!
```

Discussion of the Planned Output:

■ When writing programs that will be used by others, it is a practice of computer programmers to make the programs more friendly. This is accomplished by setting up a dialogue between the user and the computer. See if you can do this in your next program.

e. *The Program*

```
  10   PRINT "HELLO, MY NAME IS COMPUTER."
  20   PRINT
  30   INPUT "TYPE IN YOUR NAME"; A$
  40   PRINT
  50   PRINT "HELLOƀ"; A$; ".ƀƀWE ARE GOING TO"
  60   PRINT "FIND AREAS OF CIRCLES."
  70   PRINT
  80   PRINT A$; ",ƀENTER THE LENGTH OF A RADIUS."
  90   INPUT R
  100  IF R<=0 THEN GOTO 240
  110  PRINT "THANKSƀ"; A$; ".ƀƀI WILL NOW CALCULATE."
  120  PRINT "FOR A RADIUS OF"; R
  130  A=3.14159*R*R
  140  PRINT "THE AREA IS"; A
■ 150  N=A+.005
■ 160  AR=INT(100*N)/100
  170  PRINT "TO THE NEAREST HUNDREDTH, IT IS"; AR
  180  PRINT
  190  INPUT "TYPE IN YES TO CONTINUE"; X$
  200  IF X$="YES" THEN GOTO 70
  210  PRINT "THANKS FOR WORKING WITH ME."
  220  PRINT A$; ",ƀHAVE A NICE DAY.  BYE-BYE!"
  230  GOTO 260
  240  PRINT "THANKSƀ"; A$; ".ƀƀBUT NO THANKS. ONLY POSITIVES."
  250  GOTO 180
  260  END
```

Discussion of the Program:

- Lines 150 and 160 round the area to the nearest hundredth.

The next exercises will give you practice finding the circumference and area of circles.

Exercise 6.2 Find the circumference of each circle. Use $\pi = 3.1416$ and round the circumference to the nearest hundredth.

1. 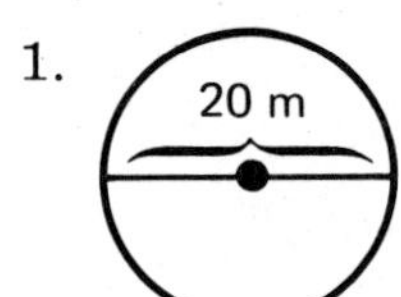

2. 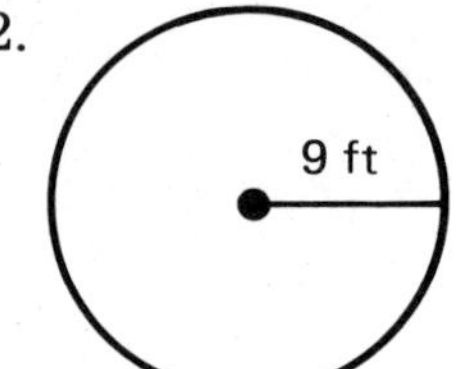

3. 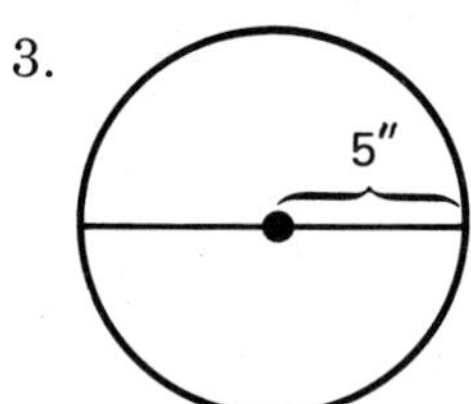

4. 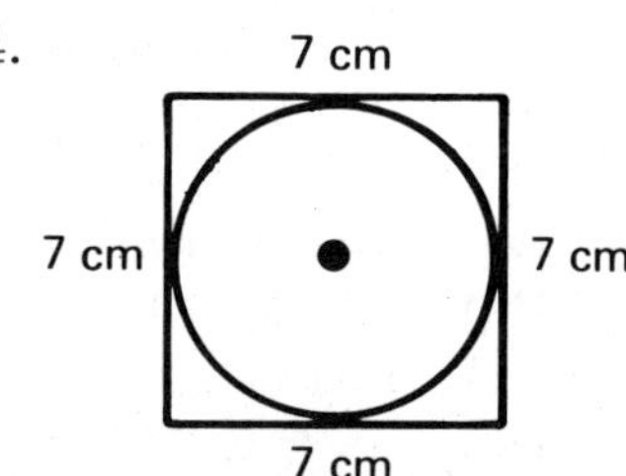

Find the area of each circle that follows. Use $\pi = 3.14$ and round the answers to the nearest tenth.

5. 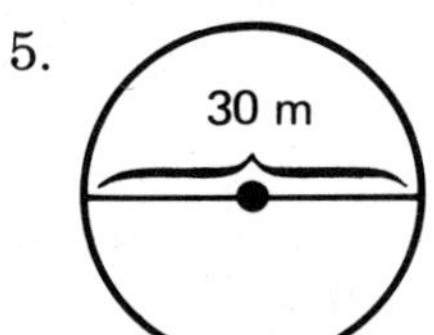

6. 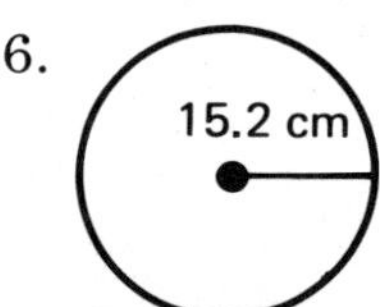

7. 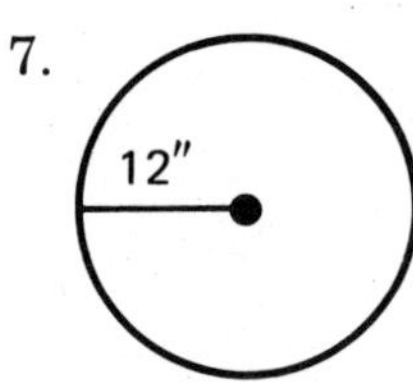

8. 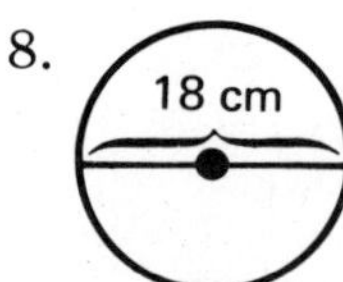

Revise the program that was written in this section to produce the following outputs.

9.

```
AREA OF A CIRCLE


ENTER THE LENGTH OF A RADIUS ? 4
THE AREA IS 50.26544
TO THE NEAREST HUNDREDTH: 50.27


ENTER THE LENGTH OF A RADIUS ? -4
ONLY POSITIVE NUMBERS


ENTER THE LENGTH OF A RADIUS?
```

10.

```
FINDING AREA OF A CIRCLE



FOR A RADIUS OF 4
AREA IS 50.26544
ROUNDED TO THE NEAREST HUNDREDTH: 50.27



FOR A RADIUS OF -4
WE DON'T CALCULATE
RADIUS MUST BE POSITIVE



OUT OF DATA
```

■ See if you can eliminate the OUT OF DATA message.

For the given problem, flowchart and planned output, code a program to solve the problem.

11. **Problem:** Enter the inside radius, $R1$, and the outside radius, $R2$, of a metal washer. Find the area of the metal washer to the nearest hundredth. (We are looking for the shaded area. See figure at left. Use $\pi = 3.1416$.)

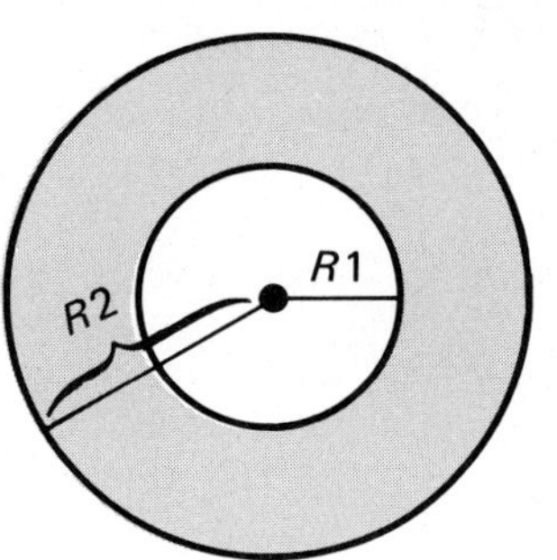

Analysis: The shaded area = area outside circle − area of inside circle.

Flowchart

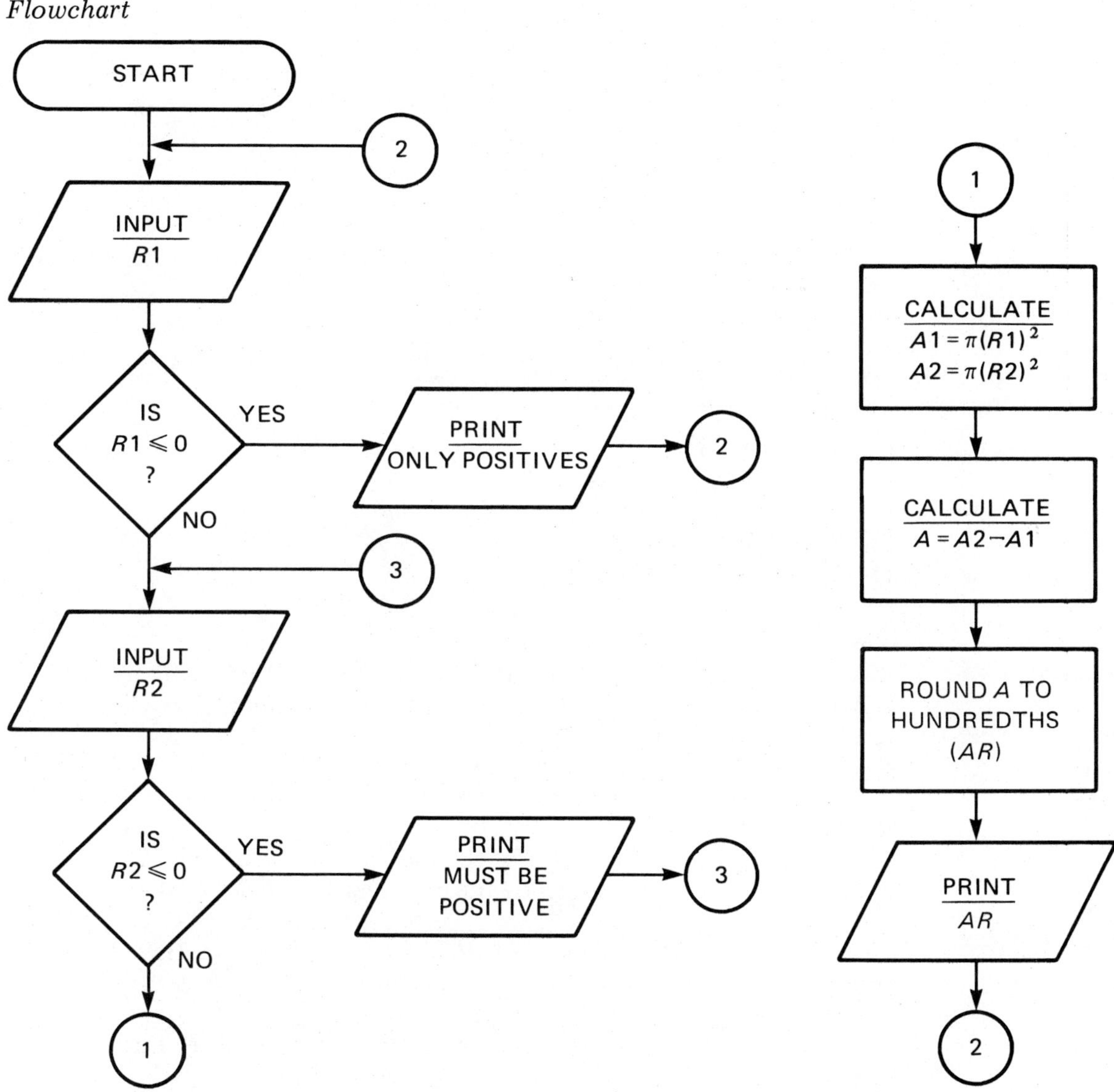

(As an extra feature, check to see if the outside radius is greater than the inside radius.)

Planned Output

```
AREA OF A METAL WASHER


ENTER INSIDE RADIUS ? 5
ENTER OUTSIDE RADIUS ? 11
AREA INSIDE = 78.54
AREA OUTSIDE = 380.1336
AREA OF METAL WASHER = 301.5936
ROUNDED TO HUNDREDTHS = 301.59


ENTER INSIDE RADIUS ? -5
ONLY POSITIVES
ENTER INSIDE RADIUS ? 8
ENTER OUTSIDE RADIUS ? -4
RADIUS MUST BE POSITIVE
ENTER OUTSIDE RADIUS ?
```

For each of the following problems:
a. Write an analysis which includes the development of an algorithm.
b. Write a plan for a computer solution to the problem.
c. Develop a flowchart.
d. Design a planned output.
e. Code a program to solve the problem and RUN it on the computer.

12. **Problem:** Enter the length of the radius of a circle. Have the computer find the circumference of the circle to the nearest tenth. (Use $\pi = 3.14$.)

13. **Problem:** Enter the length of the diameter of a circle. Have the computer find the area of the circle to the nearest tenth. (Use $\pi = 3.14$.)

14. **Problem:** Enter the area of a circle. Have the computer find the length of the diameter to the nearest hundredth. (Use $\pi = 3.1416$.)

Section 6.3 IS IT A TRIANGLE? (LOGICAL OPERATORS AND, OR)

Take three pencils where the combined length of two of them is shorter than the third pencil. Now try to form a triangle with these pencils (Figure 1). You should soon realize that it is impossible!

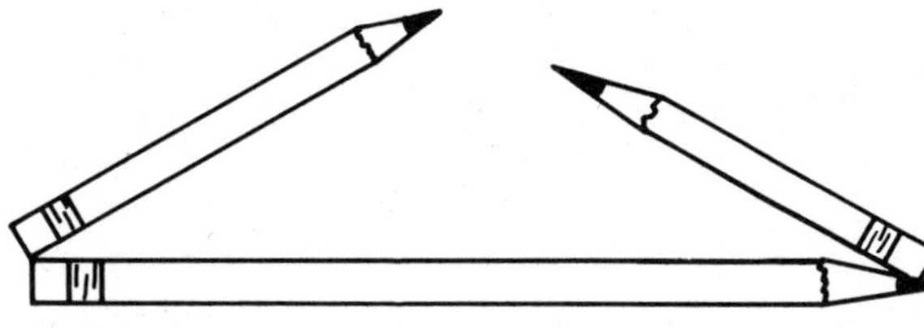

Figure 1

Would you be able to form a triangle if the combined lengths of the two shorter pencils equals the length of the third pencil (Figure 2)?

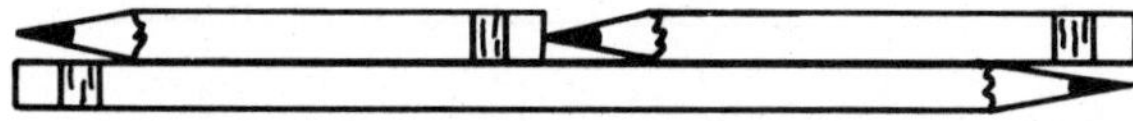

Figure 2

You probably noticed that as soon as you lift the two smaller pencils to make a triangle, they do not touch. Therefore, a triangle cannot be formed.

Now take three pencils where the combined lengths of the two shorter pencils is greater than the length of the third pencil (Figure 3). We get a triangle!

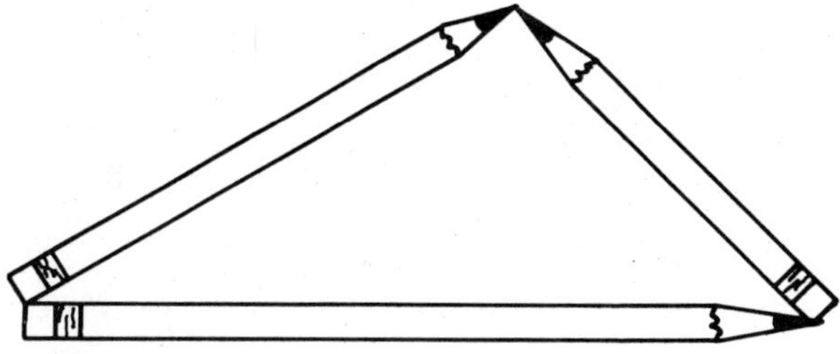

Figure 3

We can summarize the results of our experiment as follows.

> For a triangle, the sum of the lengths of any two sides must be greater than the length of the third side.

For $\triangle ABC$ that follows, we can say:

$a + b > c$

$a + c > b$

$b + c > a.$

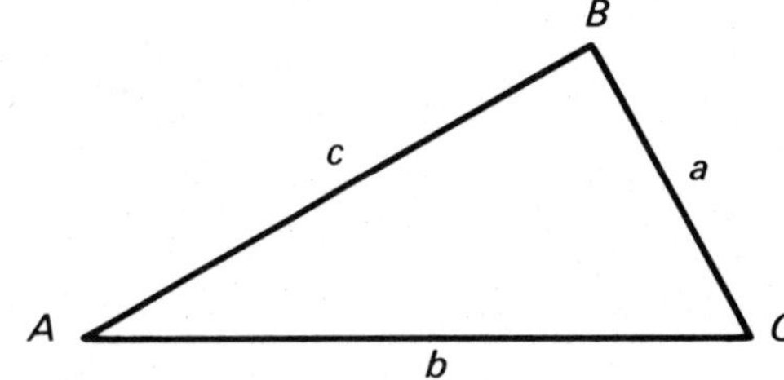

Example 1 Which of the following can represent the lengths of the sides of a triangle?

1. $\{5, 6, 7\}$ 2. $\{3, 1, 2\}$ 3. $\{5, 8, 14\}$

Solution: We shall inspect each set separately.

1. Since $5 + 6 > 7$,

$5 + 7 > 6$, and

$6 + 7 > 5$,

the set $\{5, 6, 7\}$ can represent the lengths of the sides of a triangle.

2. Although $3 + 1 > 2$ and

$3 + 2 > 1$,

$1+2$ is not greater than ($\not>$) 3. Therefore, $\{3, 1, 2\}$ *cannot* represent the lengths of the sides of a triangle. All *three* inequalities must be true in order to be able to form a triangle.

3. Since $5 + 8 \not> 14$, the set of lengths $\{5, 8, 14\}$ *cannot* be used for form a triangle.

Class Exercise 1 Which of the following can represent the lengths of the sides of a triangle?

1. $\{6, 8, 7\}$ 2. $\{6, 11, 5\}$ 3. $\{6, 4, 2\}$ 4. $\{4, 6, 8\}$

Now we turn to the computer to solve the following problem.

Problem

Enter three numbers. Determine if these numbers can represent the lengths of the sides of a triangle.

Solution:

a. *Analysis:* To form $\triangle CDE$, we learned that

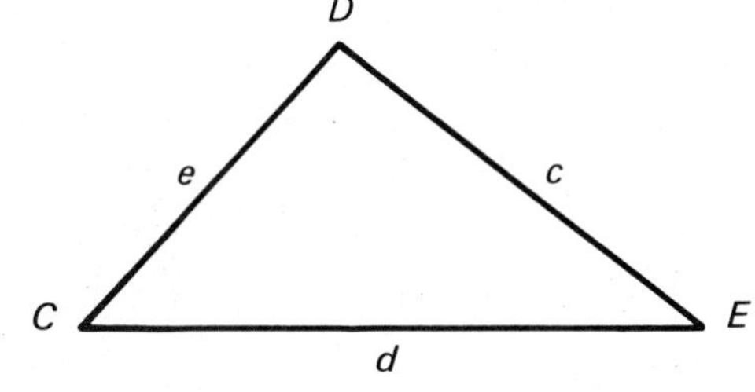

$c + d > e,$

$c + e > d,$ and

$e + d > c.$

In programming a computer it is sometimes easier to approach the solution to a problem from the opposite point of view. That is, instead of examining the above inequalities we might choose to see if

$c + d \leqslant e$

$c + e \leqslant d,$ and

$d + e \leqslant c.$

If any of the above is *true*, the lengths *c*, *d* and *e* *cannot* form the sides of a triangle. We will use this approach in our plan.

b. *Plan:*

1. Put in values for *C*, *D*, *E*.
2. Is *C*, *D* or $E \leqslant 0$?
   a) If yes, print "sides must be positive." Go to 1.
   b) If no, continue.
3. Is $C + D \leqslant E$?
   a) If yes, print "not a triangle." Go to 1.
   b) If no, continue.
4. Is $C + E \leqslant D$?
   a) If yes, print "not a triangle." Go to 1.
   b) If no, continue.
5. Is $D + E \leqslant C$?
   a) If yes, print "not a triangle." Go to 1.
   b) If no, continue.
6. Print "It's a triangle." Go to 1.

c. *Flowchart*

START
2
INPUT
C, D, E
IS
C ≤ 0
?
YES
PRINT
SIDES MUST BE
POSITIVE
2
NO
3
IS
D ≤ 0
?
YES
3
NO
IS
E ≤ 0
?
YES
3
NO
1

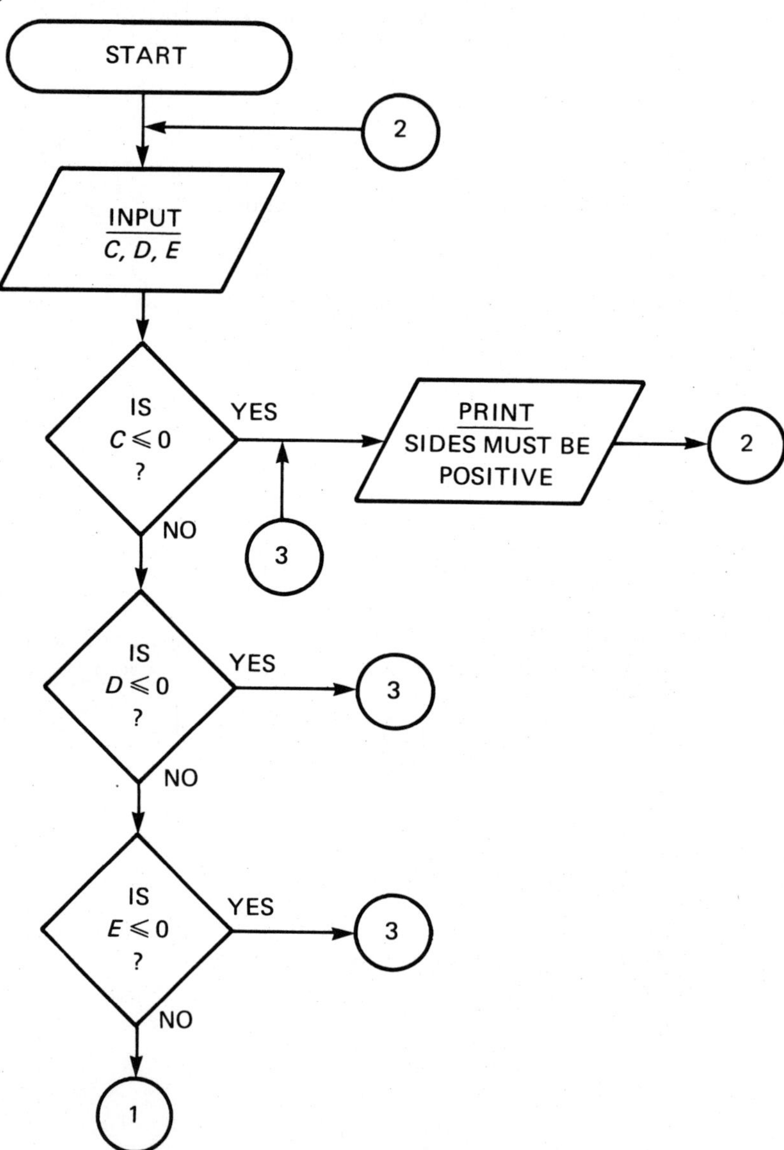

c. *Flowchart* (cont'd)

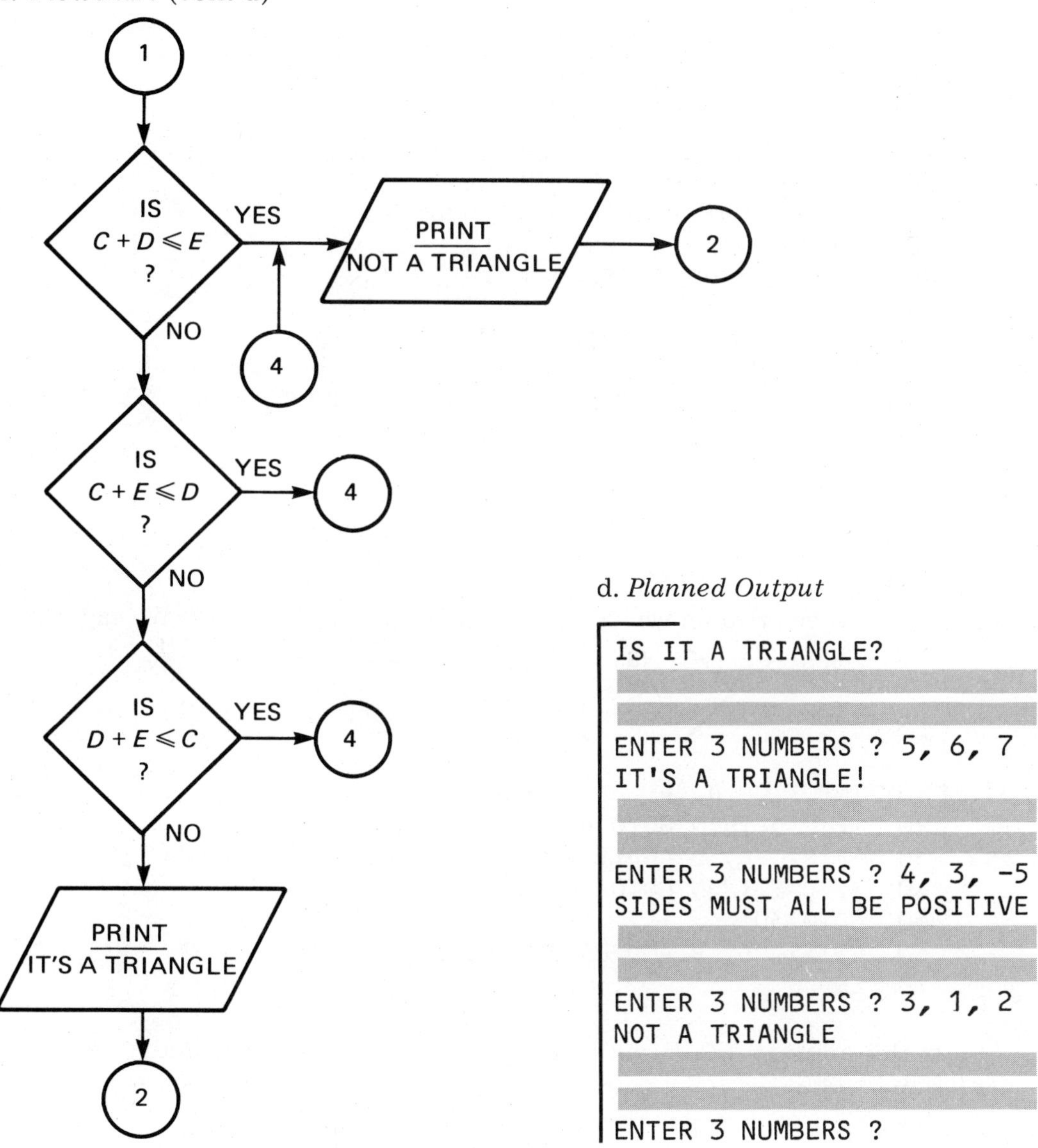

d. *Planned Output*

```
IS IT A TRIANGLE?


ENTER 3 NUMBERS ? 5, 6, 7
IT'S A TRIANGLE!


ENTER 3 NUMBERS ? 4, 3, -5
SIDES MUST ALL BE POSITIVE


ENTER 3 NUMBERS ? 3, 1, 2
NOT A TRIANGLE


ENTER 3 NUMBERS ?
```

e. *The Program*

```
10   PRINT "IS IT A TRIANGLE?"
20   PRINT : PRINT
30   INPUT "ENTER 3 NUMBERS"; C, D, E
40   IF C<=0 THEN GOTO 120
50   IF D<=0 THEN GOTO 120
60   IF E<=0 THEN GOTO 120
70   IF C+D<=E THEN GOTO 140
80   IF C+E<=D THEN GOTO 140
90   IF D+E<=C THEN GOTO 140
100  PRINT "IT'S A TRIANGLE!"
110  GOTO 20
120  PRINT "SIDES MUST ALL BE POSITIVE"
130  GOTO 20
140  PRINT "NOT A TRIANGLE"
150  GOTO 20
160  END
```

In the above program we had to use six IF-THEN statements. We can combine some of these statements using *logical operators.* Logical operators AND and OR may be used to combine IF-THEN statements.

For example, examine the following program.

```
10  INPUT C, D, E
20  IF C=D THEN GOTO 40
30  PRINT "C, D AND E ARE NOT EQUAL": GOTO 70
40  IF C=E THEN GOTO 60
50  GOTO 30
60  PRINT "C, D AND E ARE EQUAL"
70  END
```

In this program if *C*, *D and E* are equal then line 60 is executed. Otherwise, line 30 is executed.

Using the logical operator AND, the program may be rewritten as follows:

```
10  INPUT C, D, E
20  IF C=D AND C=E THEN GOTO 40
30  PRINT "C, D AND E ARE NOT EQUAL": GOTO 50
40  PRINT "C, D AND E ARE EQUAL"
50  END
```

Discussion of the Program:

- Line 20 uses the logical operator AND. The statement

20 IF $C=D$ AND $C=E$ THEN GOTO 40

passes control to line 40 only when *both* $C=D$ and $C=E$ are true. Otherwise, control goes to the next line in the program.

The next program uses the logical operator OR. What do you think will be the output?

```
10  READ B, C
20  IF B=4 OR C=9 THEN GOTO 50
30  PRINT "NEITHER STATEMENT IS TRUE"
40  GOTO 70
50  PRINT "ONE OR BOTH STATEMENTS ARE TRUE"
60  DATA 8, 9
70  END
```

Discussion of the Program:

- Line 20 uses the logical operator OR. The statement

20 IF $B=4$ OR $C=9$ THEN GOTO 50

passes control to line 50 when *either* statement ($B=4$, $C=9$) or *both* statements are true. If neither statement is true, control goes to the next line in the program.

Have you figured out the output for this program?

The output will be

ONE OR BOTH STATEMENTS ARE TRUE

since $C=9$.

Class Exercise 2 Show the output for each program.

1.
```
10  READ A, B
20  IF A=2 AND B=7 THEN 60
30  DATA 2, 5
40  PRINT "NOT TRUE"
50  GOTO 70
60  PRINT "TRUE"
70  END
```

2.
```
10  READ A, B
20  IF A=2 OR B=7 THEN 60
30  DATA 2, 5
40  PRINT "NOT TRUE"
50  GOTO 70
60  PRINT "TRUE"
70  END
```

3.†
```
10  READ A, B, C
20  DATA 5, 6, 6
30  IF A=B AND B=C THEN 60
40  PRINT "NOT ALL EQUAL"
50  GOTO 70
60  PRINT "ALL EQUAL"
70  END
```

4.†
```
10  READ A, B, C
20  DATA 5, 6, 6
30  IF A=B OR B=C THEN 60
40  PRINT "NOT ALL EQUAL"
50  GOTO 70
60  PRINT "ALL EQUAL"
70  END
```

Exercise 6.3 Which of the following can represent the lengths of the sides of a triangle?

1. $\{6, 9, 3\}$ 2. $\{6, 9, 12\}$ 3. $\{5, 6, 7\}$ 4. $\{7, 16, 8\}$

Show the output for each of the following programs.

5.
```
10  READ Q, R
20  IF Q=6 AND R>11 THEN 50
30  PRINT "WRONG"
40  GOTO 70
50  PRINT "CORRECT"
60  DATA 6, 11
70  END
```

6.
```
10  READ Q, R
20  IF Q=6 OR R>11 THEN 50
30  PRINT "WRONG"
40  GOTO 70
50  PRINT "CORRECT"
60  DATA 6, 11
70  END
```

†Is the statement in the output true for the data in the program?

7.
```
10  READ X, Y
20  IF X=5 AND Y<X THEN 80
30  FOR C=1 TO X
40  PRINT C;
50  NEXT C
60  DATA 5, 5
70  GOTO 90
80  PRINT "NO CALCULATION"
90  END
```

8.
```
10  READ X, Y
20  IF X=5 OR Y<X THEN 80
30  FOR C=1 TO X
40  PRINT C;
50  NEXT C
60  DATA 5, 5
70  GOTO 90
80  PRINT "NO CALCULATION"
90  END
```

Revise the program that was written in this section to produce the following outputs.

9.

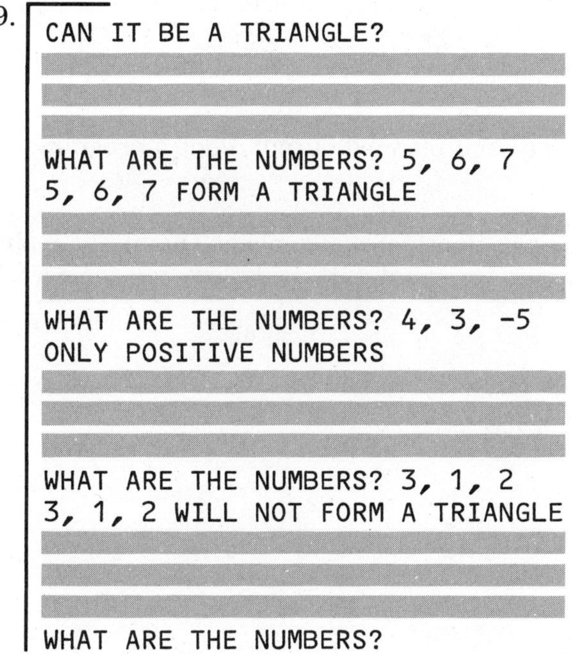

```
CAN IT BE A TRIANGLE?



WHAT ARE THE NUMBERS? 5, 6, 7
5, 6, 7 FORM A TRIANGLE



WHAT ARE THE NUMBERS? 4, 3, -5
ONLY POSITIVE NUMBERS



WHAT ARE THE NUMBERS? 3, 1, 2
3, 1, 2 WILL NOT FORM A TRIANGLE



WHAT ARE THE NUMBERS?
```

10.

```
IS IT A TRIANGLE?


FOR 5, 6, 7 WE HAVE A TRIANGLE


FOR 4, 3, -5 WE DO NOT HAVE A TRIANGLE
ONLY POSITIVE NUMBERS FOR SIDES


FOR 3, 1, 2 WE DO NOT HAVE A TRIANGLE


OUT OF DATA
```

■ See if you can eliminate the OUT OF DATA message.

For the given problem, flowchart and planned output, code a program to solve the problem.

11. **Problem:** Enter the lengths of the sides of a triangle. Determine if the triangle is equilateral, isosceles, or scalene. (Assume that the sides entered form a triangle.)

Flowchart

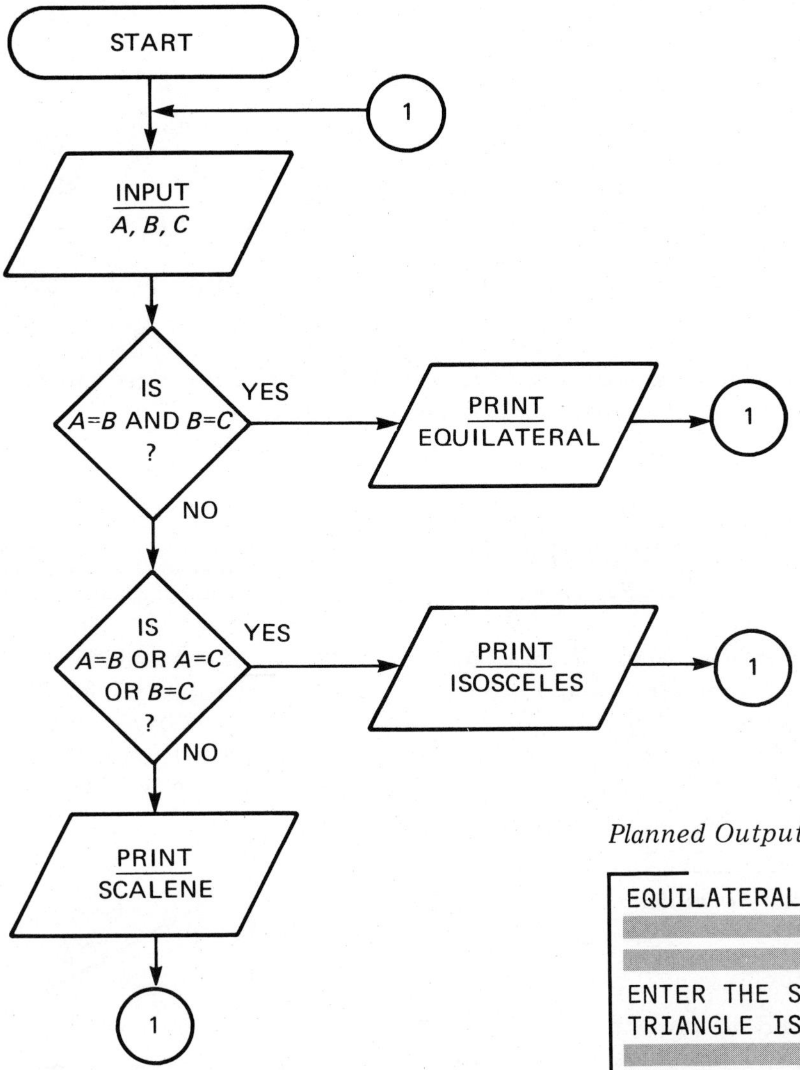

Planned Output

```
EQUILATERAL, ISOSCELES OR SCALENE


ENTER THE SIDES ? 6, 7, 8
TRIANGLE IS SCALENE


ENTER THE SIDES ? 6, 6, 6
TRIANGLE IS EQUILATERAL


ENTER THE SIDES ? 6, 6, 2
TRIANGLE IS ISOSCELES


ENTER THE SIDES ?
```

For each of the following problems:
a. Write an analysis which includes the development of an algorithm.
b. Write a plan for a computer solution to the problem.
c. Develop a flowchart.
d. Design a planned output.
e. Code a program to solve the problem and RUN it on the computer.

12. **Problem:** Enter three numbers. Have the computer determine if the numbers can represent the lengths of the sides of a triangle. Use logical operators.

13. **Problem:** Enter three numbers. Have the computer display the largest number. Use logical operators.

14. **Problem:** Enter the measures of *two* angles of a triangle. Have the computer determine whether the triangle is equilateral, isosceles, or scalene.

Section 6.4 PYTHAGORAS' FORMULA

The right triangle is a very special triangle in the study of geometry. In about 540 B.C., the Greek mathematician Pythagoras developed a formula for the sides of a right triangle which is still one of the most important formulas in mathematics.

> *Pythagoras' Formula:* For right triangle ABC,
>
> $c^2 = a^2 + b^2$,
>
> where a and b represent the lengths of the legs and c represents the length of the hypotenuse.

This can be stated as:

In any right triangle, the square of the length of the hypotenuse (longest side) is equal to the sum of the squares of the lengths of the two legs (shorter sides).

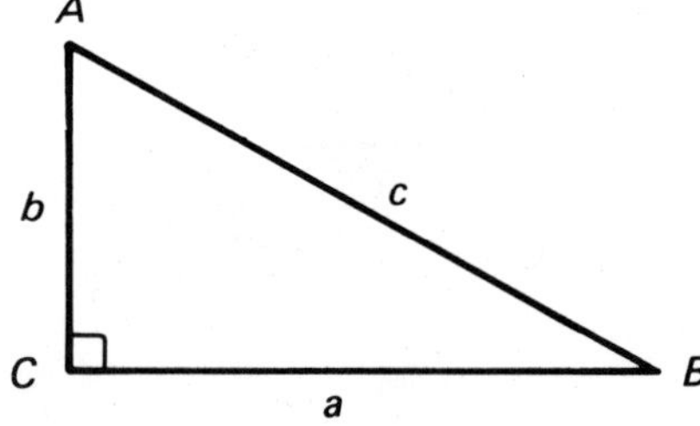

Example 1 Find the length of the hypotenuse of the right triangle below.

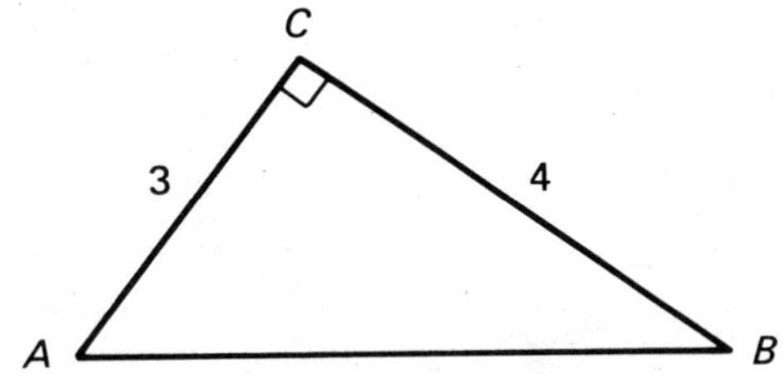

Solution: In this example we are looking for the length of the hypotenuse. Using Pythagoras' formula, we get:

$c^2 = a^2 + b^2$

$c^2 = 4^2 + 3^2$

$c^2 = 16 + 9$

$c^2 = 25$

$c = \sqrt{25} = 5$

Example 2 Find the value of x in the right triangle below.

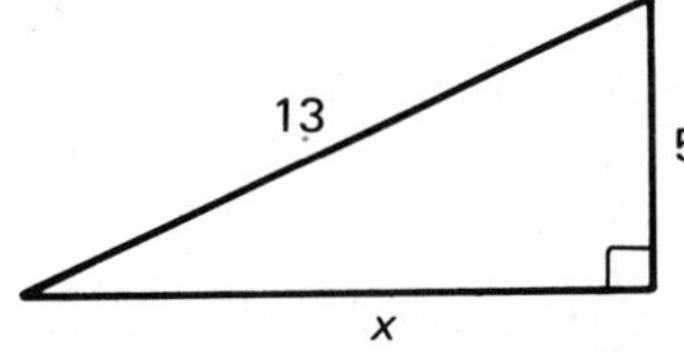

Solution: In this example we are looking for the length of one of the legs. Using Pythagoras' formula, we get:

$$
\begin{aligned}
c^2 &= a^2 + b^2 \\
13^2 &= x^2 + 5^2 \\
169 &= x^2 + 25 \\
-25 &= \quad\;\; -25 \\
\hline
144 &= x^2 \\
\sqrt{144} &= x \\
12 &= x
\end{aligned}
$$

Remember the hypotenuse is always the longest side of the right triangle. (Its length is always the value "c" in the formula.) The hypotenuse is opposite the right angle.

Class Exercise 1 Find the missing side length for each of the following triangles.

1.

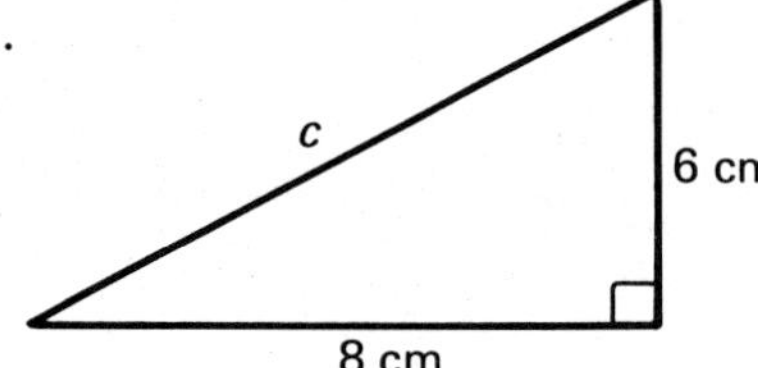

2.

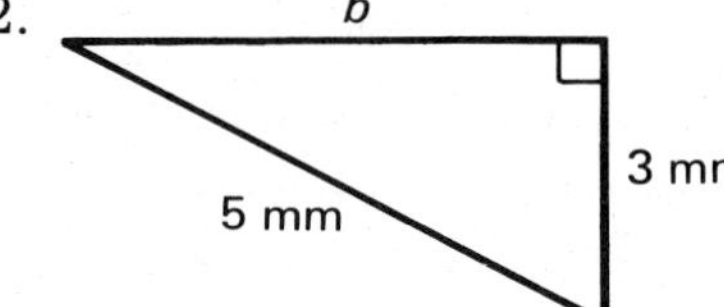

3.

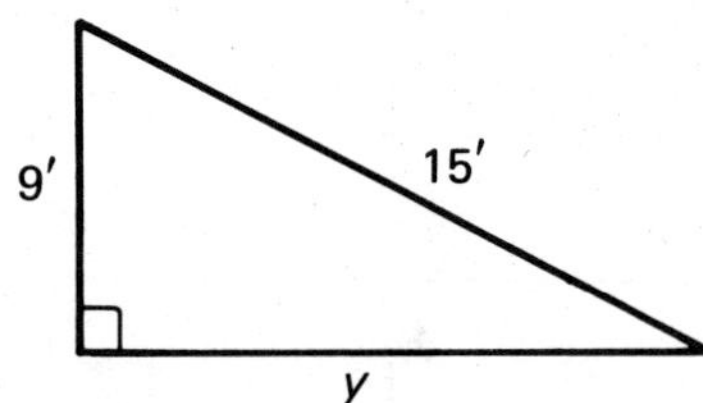

4.

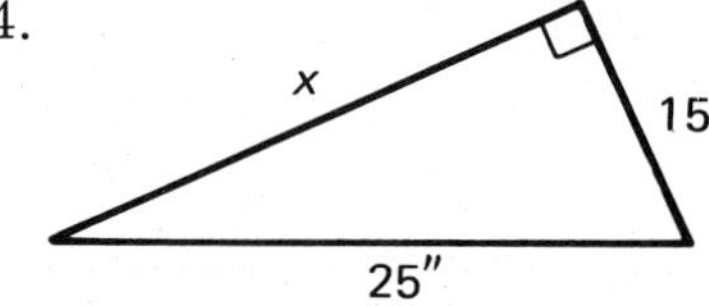

Now let's put the computer to work to solve the following problem.

Problem

Enter the lengths of the legs of a right triangle. Have the computer find the length of the hypotenuse to the nearest thousandth.

Solution:

a. *Analysis:* To find the length of the hypotenuse we will use the formula:

$c = \sqrt{a^2 + b^2}$

Include checks to make sure that the lengths of the legs are positive.

b. *Plan:*

1. Enter values for the lengths of the legs, A, B.
2. Is A or $B \leq 0$?
   a) If yes, print "only positive values." Go to 1.
   b) If no, continue.
3. Calculate: $C = \sqrt{A^2 + B^2}$.
4. Round C to the nearest thousandth (CR).
5. Print CR (C rounded).
6. Go to 1.

c. *Flowchart*

START
1
INPUT
A, B
IS
A OR B ≤ 0
?
YES
PRINT
ONLY POSITIVE
VALUES
1
NO
CALCULATE
C = √(A² + B²)
ROUND C TO
THE NEAREST
THOUSANDTH (CR)
PRINT
CR
1

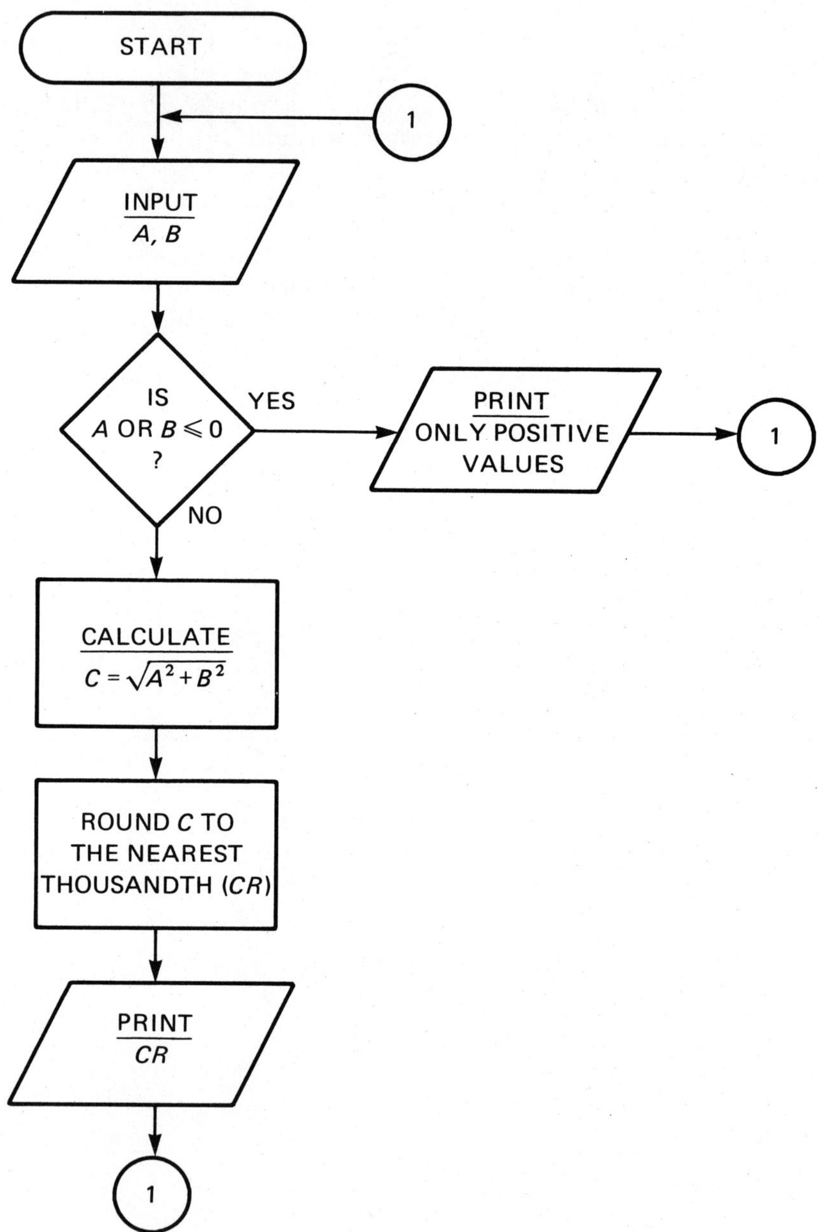

d. *Planned Output*

```
PYTHAGORAS' FORMULA


ENTER THE LENGTH OF THE LEGS ? 5, 7
THE LENGTH OF THE HYPOTENUSE IS 8.602


ENTER THE LENGTH OF THE LEGS ? 6, -2
ONLY POSITIVE VALUES PLEASE


ENTER THE LENGTH OF THE LEGS ?
```

e. *The Program*

```
10   PRINT "PYTHAGORAS' FORMULA"
20   PRINT : PRINT
30   INPUT "ENTER THE LENGTH OF THE LEGS"; A, B
40   IF A<=0 OR B<=0 THEN GOTO 100
50   C=SQR(A↑2+B↑2)
60   R=C+.0005
70   CR=INT(1000*R)/1000
80   PRINT "THE LENGTH OF THE HYPOTENUSE IS"; CR
90   GOTO 20
100  PRINT "ONLY POSITIVE VALUES PLEASE"
110  GOTO 20
120  END
```

The next set of exercises will give you practice working with Pythagoras' formula.

Exercise 6.4 Find x in each of the following (round to the nearest tenth).

1.

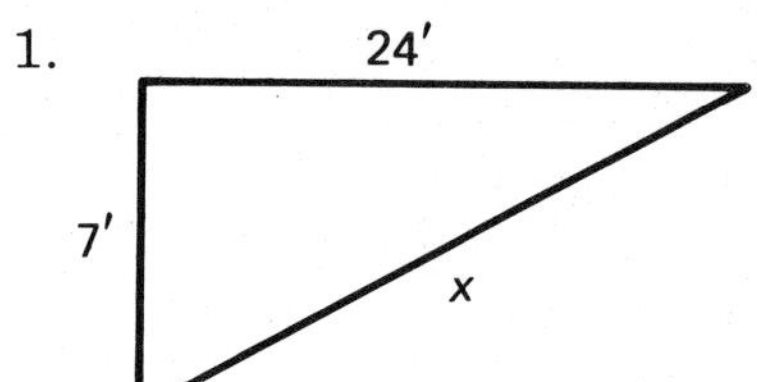

2.

x
5 cm
7 cm

3.

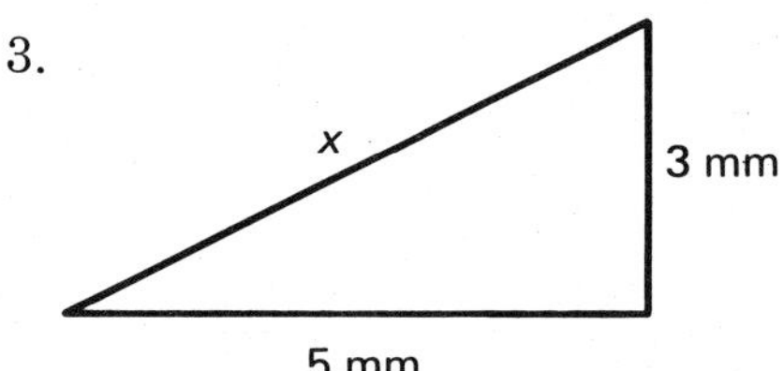

4.

12″
x
5″

Revise the program that was written in this section to produce the following outputs.

5.

```
FINDING THE HYPOTENUSE


WHAT ARE THE LEGS? 5, 7
FOR LEGS OF 5 AND 7
THE HYPOTENUSE IS 8.602


WHAT ARE THE LEGS? 6, -2
ONLY POSITIVE VALUES PLEASE


WHAT ARE THE LEGS?
```

6.

```
WHAT IS THE HYPOTENUSE?



FOR LEGS OF 5 AND 7
THE HYPOTENUSE IS 8.602



FOR LEGS OF 6 AND -2
WE DON'T CALCULATE. ONLY POSITIVES.



THE END
```

For the given problem, flowchart and planned output, code a program to solve the problem.

7. **Problem:** Enter the length of a hypotenuse and a leg. Have the computer display the length of the other leg rounded to the nearest hundredth.

Flowchart

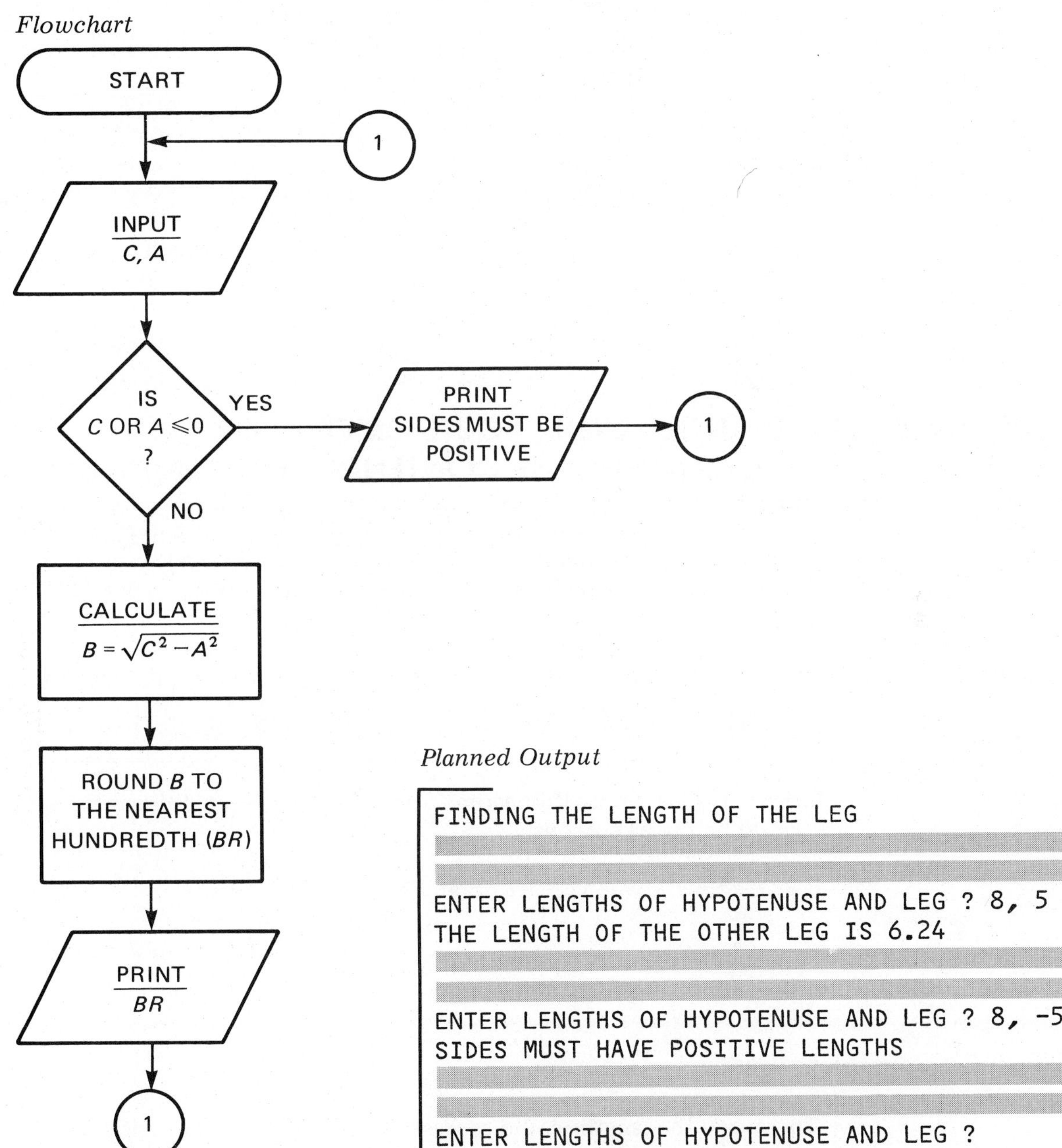

For each of the following problems:
a. Write an analysis which includes the development of an algorithm.
b. Write a plan for a computer solution to the problem.
c. Develop a flowchart.
d. Design a planned output.
e. Code a program to solve the problem and RUN it on the computer.

8. **Problem:** Enter the length of the legs of a right triangle. Have the computer display the length of the hypotenuse rounded to the nearest hundredth.

9. **Problem:** Enter two numbers; one to represent the length of the hypotenuse of a right triangle, the other the length of a leg. (The numbers may be entered in any order.) Have the computer display the length of the other leg rounded to the nearest tenth.

10. **Problem:** Enter either the length of two legs or the lengths of a hypotenuse and leg (the user's choice). Have the computer display the length of the remaining side rounded to the nearest hundredth.

Section 6.5 IS A TRIANGLE RIGHT, ACUTE OR OBTUSE? (INTERCHANGING THE CONTENTS OF TWO MEMORY CELLS)

We can also use Pythagoras' formula in reverse. That is, we can use it to tell if a given triangle is a right triangle by showing that the lengths of the sides "fit" into the formula.

Example 1 Which sets have lengths that could represent the sides of right triangles?

1. $\{3, 4, 5\}$ 2. $\{6, 5, 4\}$ 3. $\{5, 13, 12\}$

Solution: We just apply Pythagoras' formula to each set. Remember, only the longest side could be the hypotenuse!

1. Is $c^2 = a^2 + b^2$ when $c = 5$, $a = 3$ and $b = 4$?

$$5^2 \stackrel{?}{=} 3^2 + 4^2$$

$$25 \stackrel{?}{=} 9 + 16$$

$$25 = 25$$

Therefore, the lengths could represent the sides of a right triangle.

2. Is $c^2 = a^2 + b^2$ when $c=6$, $a=5$ and $b=4$?

$6^2 \stackrel{?}{=} 5^2 + 4^2$

$36 \stackrel{?}{=} 25 + 16$

$36 \neq 41$

Therefore, the lengths could not represent the sides of a right triangle.

3. Is $c^2 = a^2 + b^2$ when $c=13$, $a=5$ and $b=12$?

$13^2 \stackrel{?}{=} 5^2 + 12^2$

$169 \stackrel{?}{=} 25 + 144$

$169 = 169$

Therefore, the lengths could represent the sides of a right triangle.

The above can be summarized in this rule:

To show a triangle is a right triangle: The square of the length of the longest side must equal the sum of the squares of the lengths of the two shorter sides.

Class Exercise 1 Which sets have lengths that could represent the sides of a right triangle?

1. $\{7, 24, 25\}$ 2. $\{8, 9, 10\}$ 3. $\{15, 9, 12\}$ 4. $\{15, 8, 17\}$

Suppose we find that a triangle is not a right triangle. How can we determine if it is acute or obtuse? Pythagoras' formula can be extended to help us determine if a non-right triangle is acute or obtuse.

In any $\triangle ABC$, where c is the longest side, we already know that if

$c^2 = a^2 + b^2$

it's a *right* triangle. It is also true that:

a. if $c^2 < a^2 + b^2$, the triangle is *acute*.
b. if $c^2 > a^2 + b^2$, the triangle is *obtuse*.

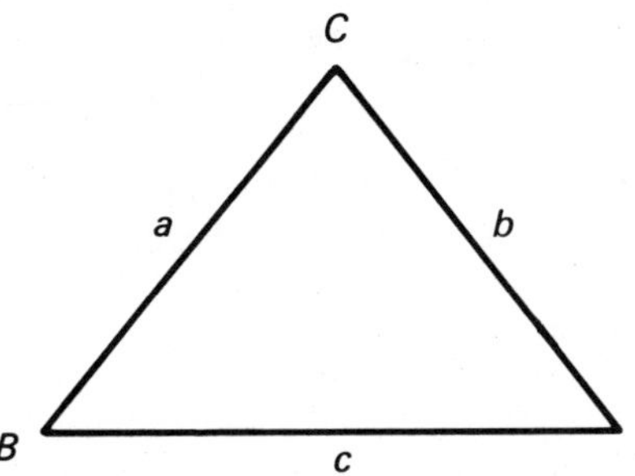

Example 2 Tell the type of triangle each of the following sets of lengths can form.

1. $\{6, 8, 11\}$ 2. $\{5, 6, 7\}$ 3. $\{12, 20, 16\}$

Solution: Using c as the largest number, we will examine the relationship between c^2 and $a^2 + b^2$.

1. $c^2 \ ? \ a^2 + b^2$ when $c = 11$, $a = 6$ and $b = 8$.

$11^2 \ ? \ 6^2 + 8^2$

$121 \ ? \ 36 + 64$

$121 > 100$

So the triangle is obtuse.

2. $c^2 \ ? \ a^2 + b^2$ when $c = 7$, $a = 5$ and $b = 6$.

$7^2 \ ? \ 5^2 + 6^2$

$49 \ ? \ 25 + 36$

$49 < 61$

So the triangle is acute.

3. $c^2 \ ? \ a^2 + b^2$ when $c = 20$, $a = 12$ and $b = 16$.

$20^2 \ ? \ 12^2 + 16^2$

$400 \ ? \ 144 + 256$

$400 = 400$

So the triangle is right.

Class Exercise 2 For each of the following, state whether the measures can represent the side lengths of a right, acute or obtuse triangle.

1. $\{8, 9, 10\}$ 2. $\{10, 24, 26\}$ 3. $\{8, 10, 14\}$ 4. $\{2, 3, 4\}$

We will now develop a computer solution to the following problem.

Problem

Enter the lengths of three sides of a triangle. Determine whether the triangle is acute, right, or obtuse.

Solution:

a. *Analysis:* To classify a triangle by angles, using the lengths of the sides, we use the following relationships:

If $C^2 < A^2 + B^2$, the triangle is acute.

If $C^2 = A^2 + B^2$, the triangle is right.

If $C^2 > A^2 + B^2$, the triangle is obtuse.

To use these relationships, "C" must always be the length of the longest side. We must, therefore, begin the program by finding the largest of the three numbers and placing it into memory cell C.

In this problem we will assume that the numbers entered represent the lengths of the sides of a triangle.

b. *Plan:*

1. Put in values for A, B, C.
2. Is $C \geqslant A$ and $C \geqslant B$?
   a) If yes, then C is the largest number (or $C=A=B$). Go to 5.
   b) If no, continue.
3. Is $B \geqslant C$ and $B \geqslant A$?
   a) If yes, B is the largest number. Interchange the contents of B and C (now C is the largest number). Go to 5.
   b) If no, continue.
4. Getting to this line means that A is the largest number. Interchange the contents of A and C (now C is the largest number). Continue.
5. Is $C^2 > A^2 + B^2$?
   a) If yes, triangle is obtuse. Go to 1.
   b) If no, continue.
6. Is $C^2 = A^2 + B^2$?
   a) If yes, triangle is right. Go to 1.
   b) If no, continue.
7. Getting to this line means that the triangle is not obtuse and not right. Therefore, the triangle is acute. Go to 1.

c. *Flowchart*

START
2
INPUT
A, B, C
IS
C≥A AND
C≥B
?
YES
1
NO
IS
B≥C AND
B≥A
?
YES
INTERCHANGE
THE CONTENTS
OF B AND C
1
NO
INTERCHANGE
THE CONTENTS
OF A AND C
1

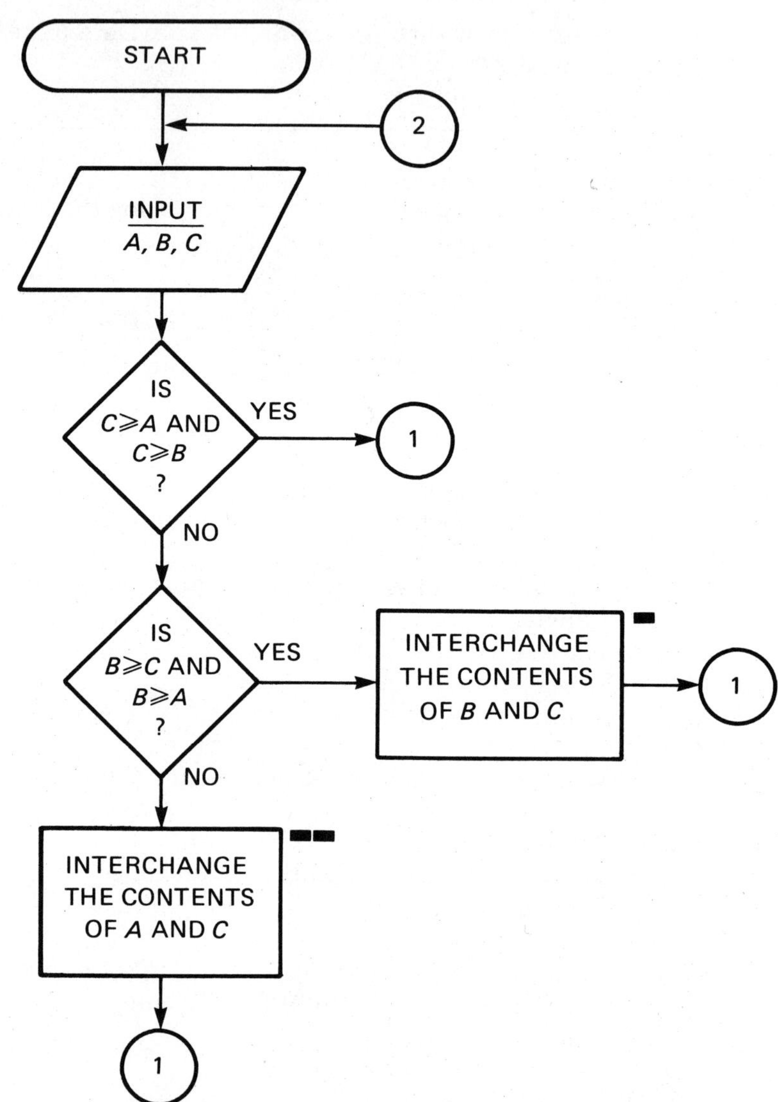

c. *Flowchart* (cont'd)

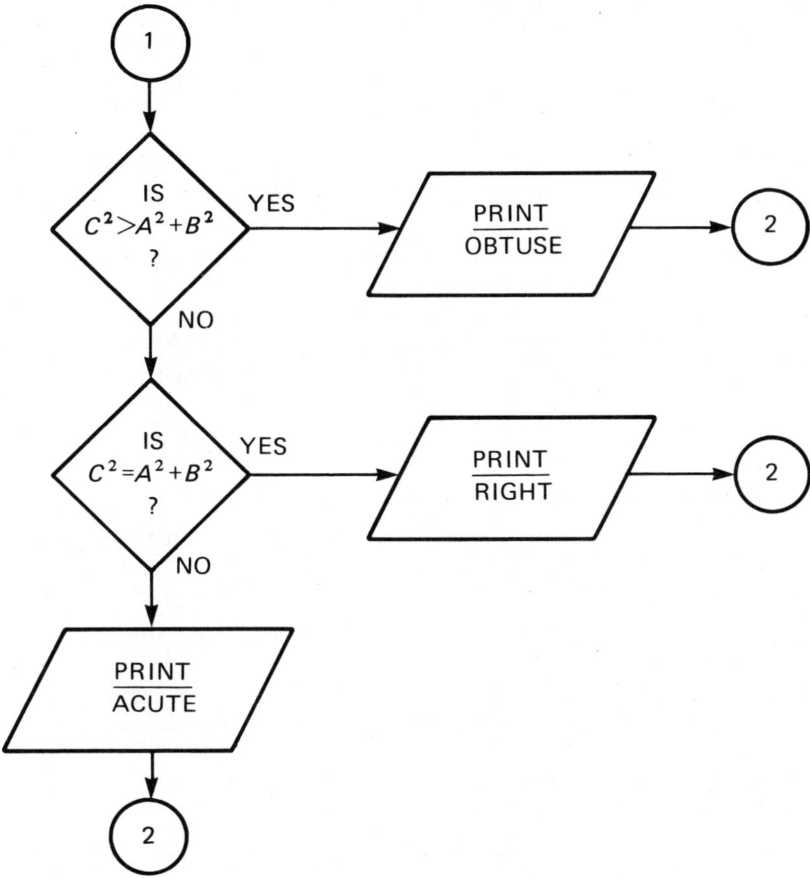

Discussion of the Flowchart:

■ Before coding our program, we must develop instructions for interchanging memory cells. To understand how you interchange memory cells, see if you can first solve this puzzle.

Puzzle: One morning at the kitchen table, John discovered that his mother put his orange juice in his milk glass and his milk in his orange juice glass. How can John interchange the milk and juice?

How can this problem be solved?

Solution: We introduce another glass to act as a temporary storage:

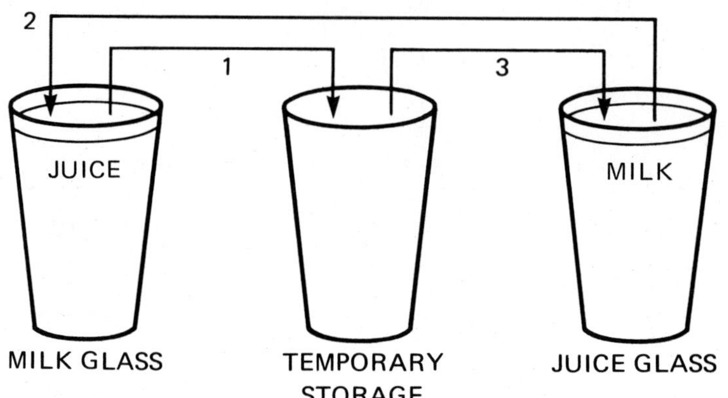

1. Pour the *J*uice into the *T*emporary Storage glass. (T becomes J.)
2. Pour the *M*ilk into the glass that had the *J*uice. (J becomes M.)
3. Pour what is in the *T*emporary Storage glass into the glass that had the *M*ilk. (M becomes T.)

Interchanging two memory cells is like interchanging the milk and the juice by introducing a temporary storage.

Look at the chart below and see how the temporary memory cell is used when interchanging the contents of memory cells B and C.

	C	B	T
We begin with the following in C and B:	3	4	
$T = C$	3	4	3
$C = B$	4	4	3
$B = T$	4	3	3

Using the instructions $T=C$, $C=B$, $B=T$, we interchanged the contents of C and B.

■■ How can we interchange the contents of A and C? Again, we introduce T as a temporary storage cell. Our algorithm becomes:

$T = A$

$A = C$

$C = T$

d. *Planned Output*

```
IS THE TRIANGLE ACUTE, RIGHT OR OBTUSE?


ENTER THE LENGTHS OF THE SIDES ? 7, 5, 6
THE TRIANGLE IS ACUTE


ENTER THE LENGTHS OF THE SIDES ? 6, 11, 8
THE TRIANGLE IS OBTUSE


ENTER THE LENGTHS OF THE SIDES ? 16, 12, 20
THE TRIANGLE IS RIGHT


ENTER THE LENGTHS OF THE SIDES ?
```

e. *The Program*

```
   10   PRINT "IS THE TRIANGLE ACUTE, RIGHT OR OBTUSE?"
   20   PRINT : PRINT
   30   INPUT "ENTER THE LENGTHS OF THE SIDES"; A, B, C
   40   IF C>=A AND C>=B THEN GOTO 90
   50   IF B>=C AND B>=A THEN GOTO 130
■  60   T=A
■  70   A=C
■  80   C=T
   90   IF C*C>A*A+B*B THEN GOTO 170
   100  IF C*C=A*A+B*B THEN GOTO 190
   110  PRINT "THE TRIANGLE IS ACUTE"
   120  GOTO 20
■■ 130  T=C
■■ 140  C=B
■■ 150  B=T
   160  GOTO 90
   170  PRINT "THE TRIANGLE IS OBTUSE"
   180  GOTO 20
   190  PRINT "THE TRIANGLE IS RIGHT"
   200  GOTO 20
   210  END
```

Discussion of the Program:

■ Lines 60-80 interchange the contents of memory cells *A* and *C*.

■■ Lines 130-150 interchange the contents of memory cells *B* and *C*.

Exercise 6.5

Which sets have lengths that could represent the sides of a right triangle?

1. $\{2.5, 6, 6.5\}$ 2. $\{5, 4, 6\}$ 3. $\{1.5, 2, 2.5\}$ 4. $\{10, 11, 12\}$

For each of the following, state whether the measures can represent the side lengths of a right, acute, or obtuse triangle.

5. $\{11, 5, 7\}$ 6. $\{2, 15, 14\}$ 7. $\{15, 25, 20\}$ 8. $\{5, 6, 9\}$

Revise the program that was written in this section to produce the following outputs.

9.

```
ACUTE, RIGHT OR OBTUSE?



WHAT ARE THE SIDES? 7, 5, 6
FOR 7, 5, 6 TRIANGLE IS ACUTE



WHAT ARE THE SIDES? 6, 11, 8
FOR 6, 11, 8 TRIANGLE IS OBTUSE



WHAT ARE THE SIDES? 16, 12, 20
FOR 16, 12, 20 TRIANGLE IS RIGHT



WHAT ARE THE SIDES?
```

10.

```
CLASSIFYING TRIANGLES BY ANGLES


FOR 7, 5, 6 ACUTE


FOR 6, 11, 8 OBTUSE


FOR 16, 12, 20 RIGHT


THIS ENDS THE DATA
```

For the given problem, flowchart and planned output, code a program to solve the problem.

11. **Problem:** Have the computer read the lengths of three sides of a triangle. Have the computer determine if the triangle is an obtuse scalene triangle. Have the computer read the four sets of data which follow:

3, 6, 8
3, 4, 5
4, 5, 6
5, 6, 10

Flowchart

START
FOR I = 1 TO 4
READ
A, B, C
IS
C ≥ A AND
C ≥ B?
YES
2
NO
IS
B ≥ C AND
B ≥ A?
YES
INTERCHANGE
THE CONTENTS
OF B AND C
2
NO
1
1
INTERCHANGE
THE CONTENTS
OF A AND C
2
IS
C² ≤ A² + B²
?
YES
PRINT
NOT AN OBTUSE
SCALENE
TRIANGLE
3
NO
4
IS
A = B
?
YES
4
NO
PRINT
TRIANGLE IS
OBTUSE
SCALENE
3
NEXT I
END

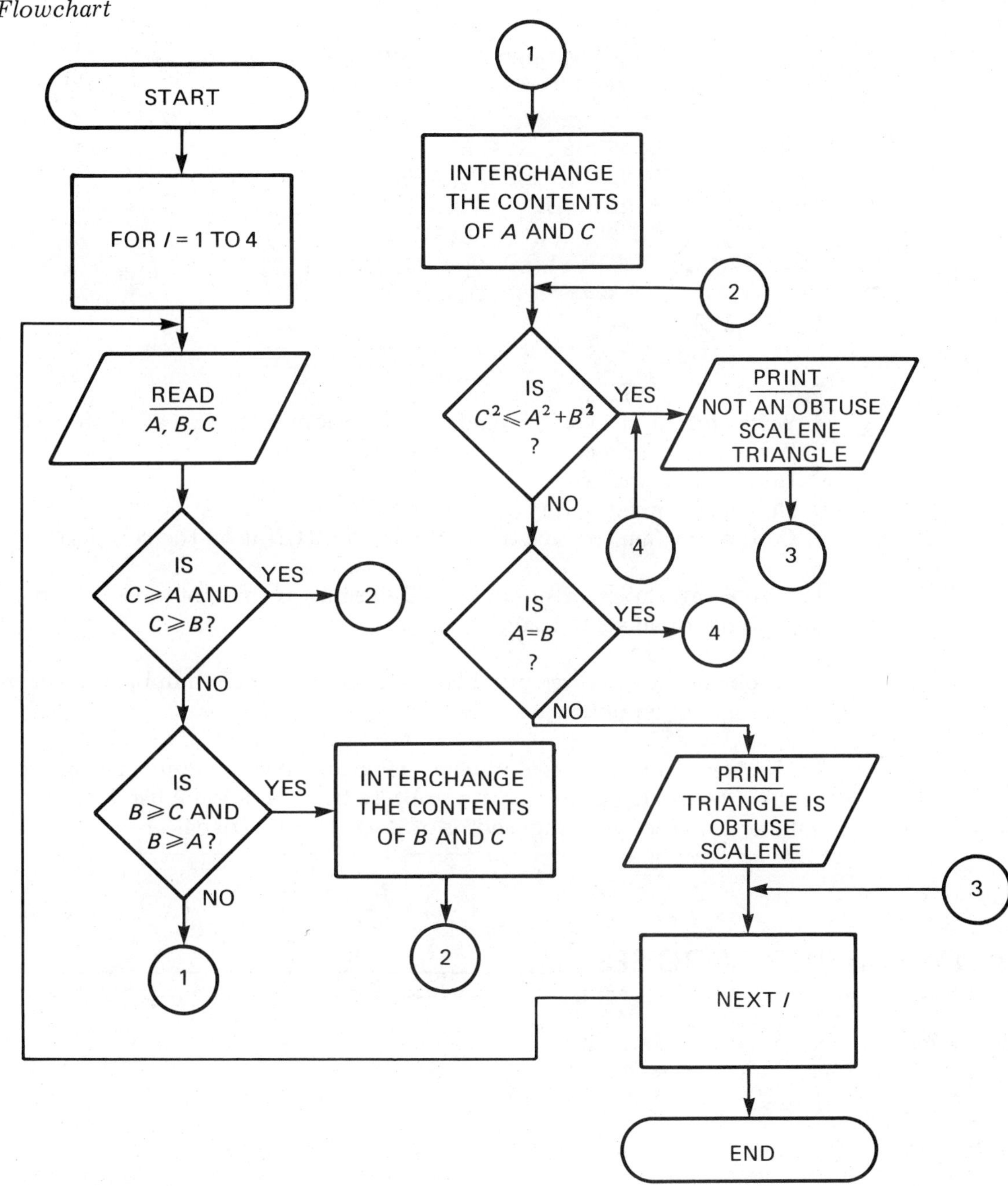

Planned Output

```
IS IT AN OBTUSE SCALENE TRIANGLE?


FOR 3, 6, 8 OBTUSE SCALENE
FOR 3, 4, 5 NOT OBTUSE SCALENE
FOR 4, 5, 6 NOT OBTUSE SCALENE
FOR 5, 6, 10 OBTUSE SCALENE


THAT'S ALL!
```

For each of the following problems:
a. Write an analysis which includes the development of an algorithm.
b. Write a plan for a computer solution to the problem.
c. Develop a flowchart.
d. Design a planned output.
e. Code a program to solve the problem and RUN it on the computer.

12. **Problem:** Enter three numbers. Determine if these numbers can represent the lengths of the sides of an acute triangle.

13. **Problem:** Enter three numbers. Determine if these numbers can represent the lengths of the sides of an obtuse triangle.

14. **Problem:** Enter three numbers. Determine if these numbers can represent the lengths of the sides of a(n) acute, right, or obtuse triangle. (You must first determine if the numbers can represent the lengths of the sides of a triangle.)

END OF CHAPTER EXERCISES

Section 6.1

Find:

1. $\sqrt{100}$ 2. $\sqrt{121}$ 3. $\sqrt{64}$ 4. $\sqrt{196}$

Find the value of each of the following to the nearest tenth.

5. $\sqrt{90}$ 6. $\sqrt{45}$ 7. $\sqrt{34}$ 8. $\sqrt{140}$

Find the value of each of the following rounded to the nearest hundredth.

9. $\sqrt{42}$ 10. $\sqrt{3}$ 11. $\sqrt{6}$ 12. $\sqrt{98}$

Find the positive value of X rounded to the nearest hundredth (only when necessary).

13. $X^2 = 121$ 14. $X^2 = 1$ 15. $X^2 = 13$ 16. $X^2 = 76$

Find the value assigned to G after executing the instructions.

17.
```
10 H=50
20 G=SQR(2*H)
```

18.
```
30 B=INT(64.31)
40 G=SQR(B)
```

19.
```
50 R=60
60 G=INT(SQR(R))
```

20.
```
60 T=30
70 G=INT(SQR(T))
```

Use the algorithm to round the number N to the nearest hundredth. The *Discussion of the Flowchart* in Section 6.1 can serve as a model solution.

Algorithm: $R = N + .005$

$NR = \text{INT}(100 * R)/100$

NR ← N rounded to the nearest hundredth

21. 8.047 22. 6.424 23. 9.3645 24. 45.7253

Use the algorithm to round the number N to the nearest tenth.

Algorithm: $R = N + .05$

$NR = \text{INT}(10 * R)/10$

NR ← N rounded to the nearest tenth

25. 16.64 26. 18.86 27. 7.056 28. 12.304

For the given problem, flowchart and planned output, code a program to solve the problem.

29. **Problem:** Enter the area of a square. Have the computer find the length of a side and the perimeter, both rounded to the nearest hundredth.

Flowchart

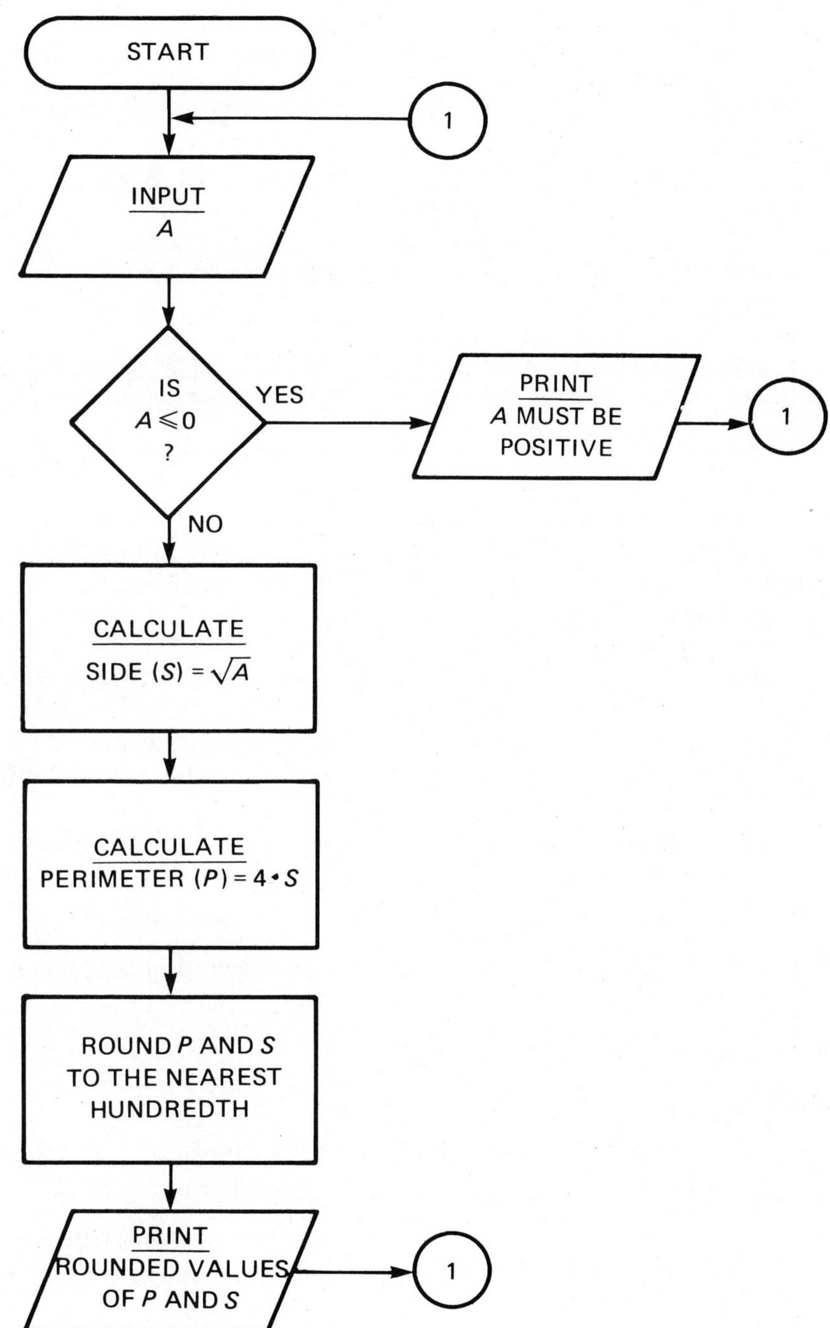
START
1
INPUT
A
IS
A ≤ 0
?
YES
PRINT
A MUST BE
POSITIVE
1
NO
CALCULATE
SIDE (S) = √A
CALCULATE
PERIMETER (P) = 4 • S
ROUND P AND S
TO THE NEAREST
HUNDREDTH
PRINT
ROUNDED VALUES
OF P AND S
1

Planned Output

```
FINDING THE LENGTH OF THE SIDE AND THE PERIMETER

ENTER THE AREA ? 75
THE LENGTH OF A SIDE OF THE SQUARE IS 8.66
THE PERIMETER IS 34.64

ENTER THE AREA ? -75
THE AREA MUST BE POSITIVE

ENTER THE AREA ?
```

For each of the following problems:
a. Write an analysis which includes the development of an algorithm.
b. Write a plan for a computer solution to the problem.
c. Develop a flowchart.
d. Design a planned output.
e. Code a program to solve the problem and RUN it on the computer.

30. **Problem:** Enter a number. Have the computer display the number, the square and then the square root of the number rounded to the nearest tenth.

31. **Problem:** Have the computer display a chart of the squares and square roots of all the whole numbers from 40 to 50. The square roots should be rounded to the nearest hundredth.

32. **Problem:** To find the distance between two points in a coordinate plane, we use the formula:

$$d = \sqrt{(x_2 - x_1)^2 + (y_2 - y_1)^2}$$

where d = distance,

(x_1, y_1) = coordinates of the first point,

(x_2, y_2) = coordinates of the second point.

Enter the coordinates of two points. Have the computer find the distance between the points rounded to the nearest hundredth.

Section 6.2 Find the circumference of each circle. Use $\pi = 3.1416$ and round the circumference to the nearest hundredth.

33.

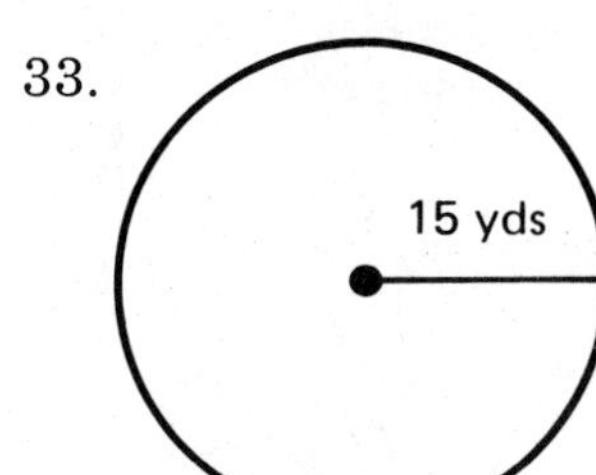

34.

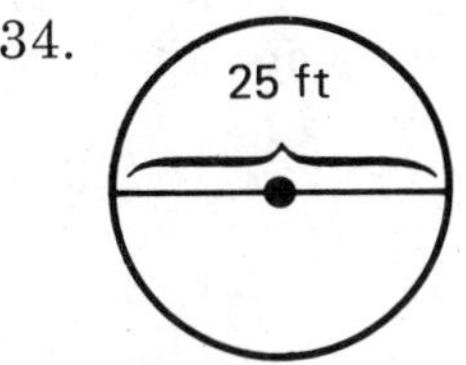

35.

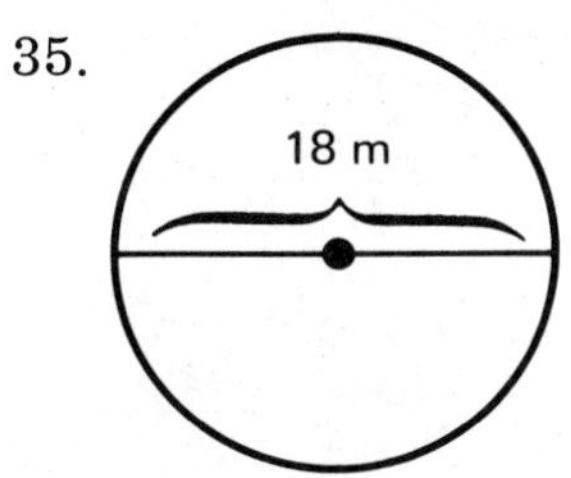

36.

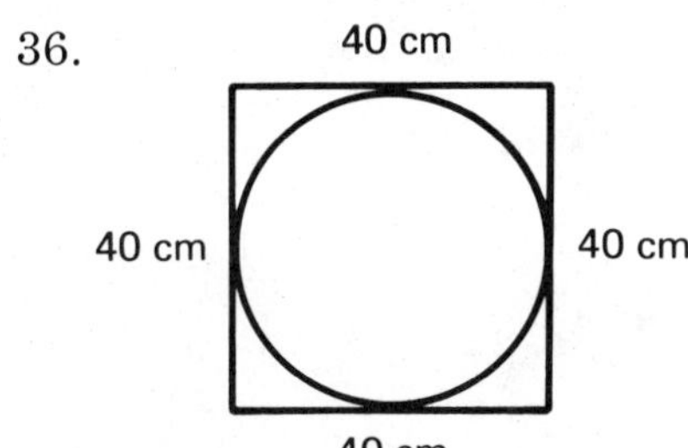

Find the area of each of the following circles. Use $\pi = 3.14$ and round the answers to the nearest tenth.

37.

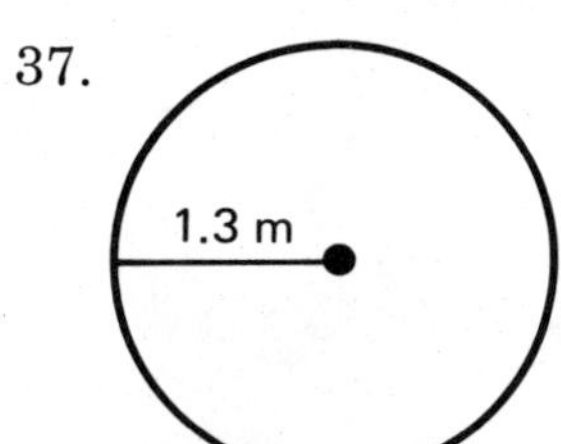

38.

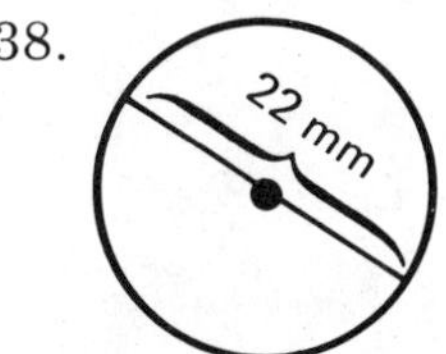

39.

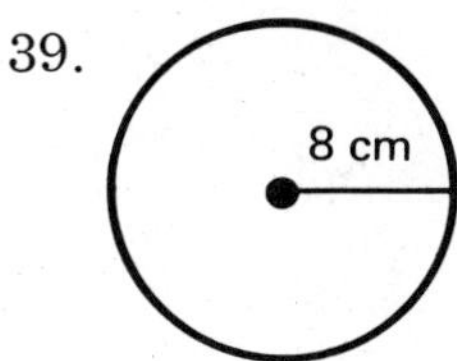

40.

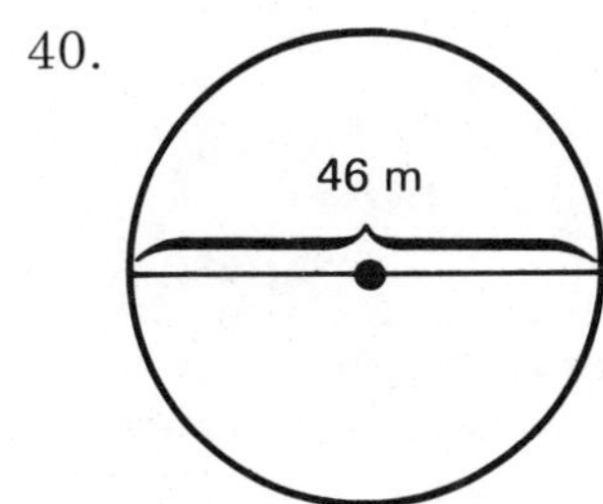

For the given problem, flowchart and planned output, code a program to solve the problem.

41. **Problem:** A circle is inscribed in a square. Enter the length of a side of the square. Find the shaded area to the nearest hundredth. (Use $\pi = 3.1416$.)

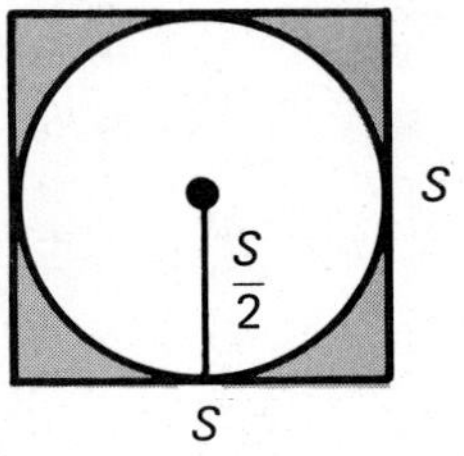

S

$\frac{S}{2}$

S

Flowchart

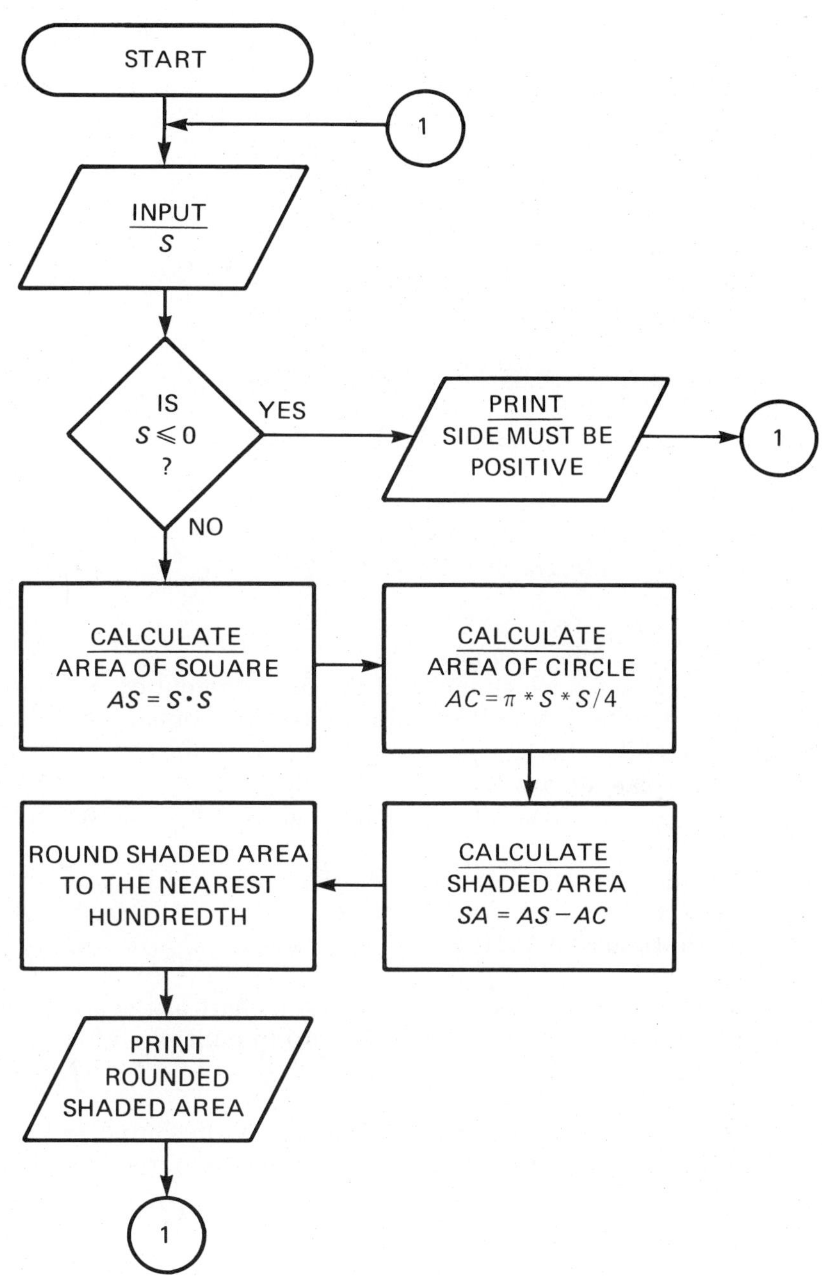
START
1
INPUT
S
IS
S ⩽ 0
?
YES
PRINT
SIDE MUST BE
POSITIVE
1
NO
CALCULATE
AREA OF SQUARE
AS = S • S
CALCULATE
AREA OF CIRCLE
AC = π * S * S/4
ROUND SHADED AREA
TO THE NEAREST
HUNDREDTH
CALCULATE
SHADED AREA
SA = AS − AC
PRINT
ROUNDED
SHADED AREA
1

Planned Output

```
FINDING THE SHADED AREA



ENTER THE LENGTH OF THE SIDE ? 6
AREA OF SQUARE IS 36
AREA OF CIRCLE IS 28.2744
SHADED AREA IS 7.73



ENTER THE LENGTH OF THE SIDE ? -6
ONLY POSITIVE NUMBERS ACCEPTED



ENTER THE LENGTH OF THE SIDE ?
```

For each of the following problems:
a. Write an analysis which includes the development of an algorithm.
b. Write a plan for a computer solution to the problem.
c. Develop a flowchart.
d. Design a planned output.
e. Code a program to solve the problem and RUN it on the computer.

42. **Problem:** Have the computer display a chart of the areas of circles with radii of lengths from 10 to 50 in multiples of 5 (10, 15, 20, . . . , 50). Round the area to the nearest hundredth. (Use $\pi = 3.1416$.)

43. **Problem:** Have the computer display a chart of the areas and circumferences of circles with radii of lengths from 20 to 40 in multiples of two. Round the areas and circumferences to the nearest thousandth. (Use $\pi = 3.14159$.)

44. **Problem:** Enter the area of a circle. Have the computer find the lengths of the radius, diameter and the circumference to the nearest hundredth. (Use $\pi = 3.14159$.)

Section 6.3

Which of the following can represent the lengths of the sides of a triangle?

45. $\{3, 4, 5\}$ 46. $\{1, 2, 3\}$ 47. $\{13, 6, 7\}$ 48. $\{13, 7, 7\}$

Show the output for each of the following programs.

49.
```
10  READ C, D
20  IF C=17 OR D<9 THEN 50
30  PRINT "FALSE"
40  GOTO 70
50  PRINT "TRUE"
60  DATA 17, 10
70  END
```

50.
```
10  READ C, D
20  IF C=17 AND D<9 THEN 50
30  PRINT "FALSE"
40  GOTO 70
50  PRINT "TRUE"
60  DATA 17, 10
70  END
```

51.
```
10  X=6 : Y=12
20  IF X=6 OR Y=8 THEN 50
30  PRINT "INCORRECT"
40  GOTO 60
50  PRINT "CORRECT"
60  END
```

52.
```
10  X=6 : Y=12
20  IF X=6 AND Y=12 THEN 50
30  PRINT "INCORRECT"
40  GOTO 60
50  PRINT "CORRECT"
60  END
```

For the given problem, flowchart and planned output, code a program to solve the problem.

53. **Problem:** Enter integer lengths of two sides of a triangle. Have the computer display all the possible integer lengths of the third side.

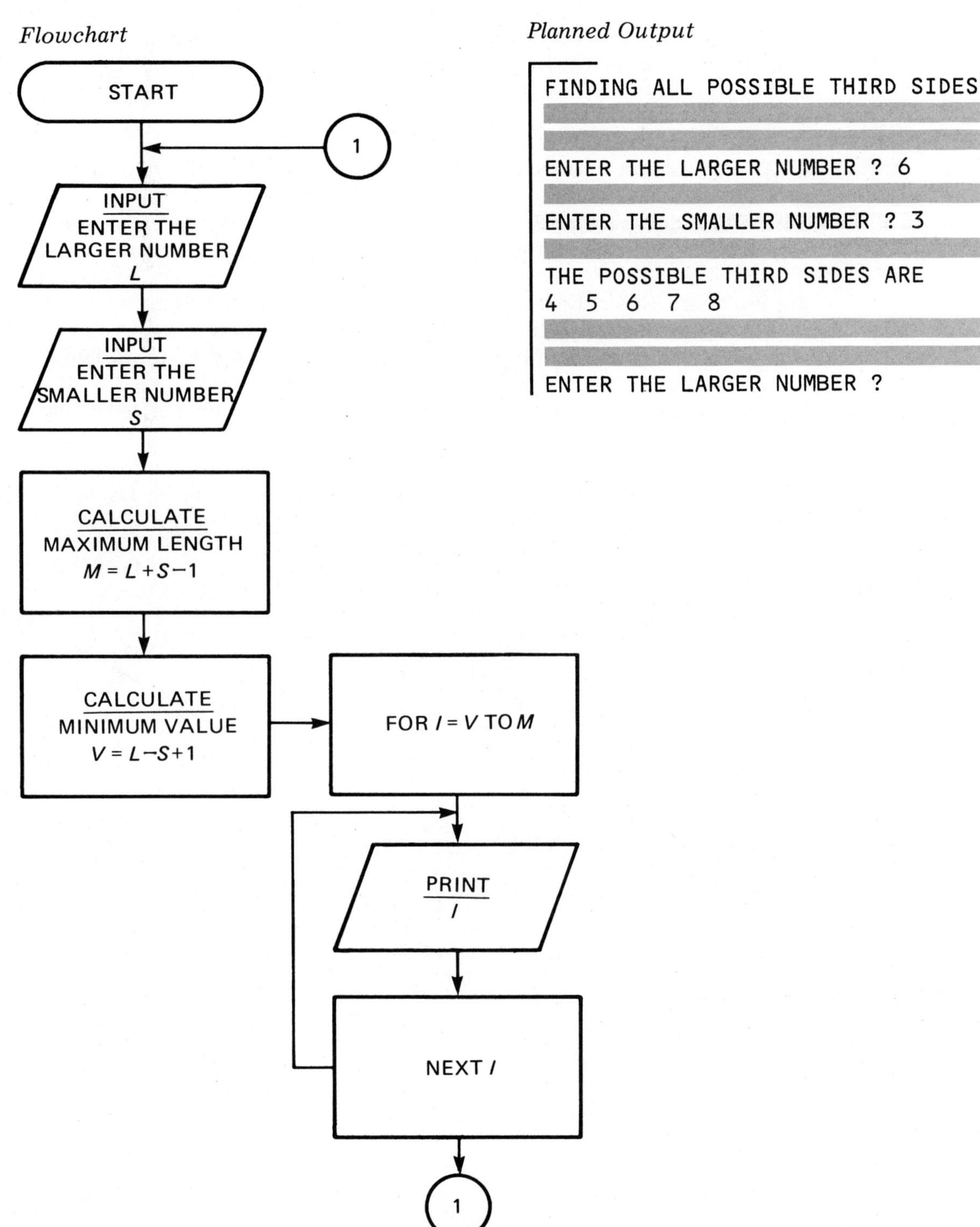
Flowchart
START
1
INPUT
ENTER THE
LARGER NUMBER
L
INPUT
ENTER THE
SMALLER NUMBER
S
CALCULATE
MAXIMUM LENGTH
M = L + S − 1
CALCULATE
MINIMUM VALUE
V = L − S + 1
FOR I = V TO M
PRINT
I
NEXT I
1
Planned Output
FINDING ALL POSSIBLE THIRD SIDES
ENTER THE LARGER NUMBER ? 6
ENTER THE SMALLER NUMBER ? 3
THE POSSIBLE THIRD SIDES ARE
4 5 6 7 8
ENTER THE LARGER NUMBER ?

For each of the following problems:
a. Write an analysis which includes the development of an algorithm.
b. Write a plan for a computer solution to the problem.
c. Develop a flowchart.
d. Design a planned output.
e. Code a program to solve the problem and RUN it on the computer.

54. **Problem:** Enter three numbers. Have the computer find the perimeter of the triangle formed with sides of lengths equal to the three numbers. (Be sure that you have a triangle.)

55. **Problem:** Enter three numbers. Have the computer determine if the three numbers represent the lengths of the sides of an equilateral, isosceles or scalene triangle. (Do not assume that the three numbers form the lengths of the sides of a triangle.)

56. **Problem:** Enter three names. Have the computer alphabetize the names.

Section 6.4

Find the missing side length for each of the following triangles. (Round to the nearest tenth where necessary.)

57.

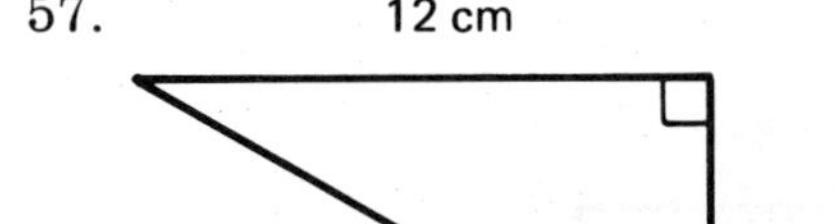

58.

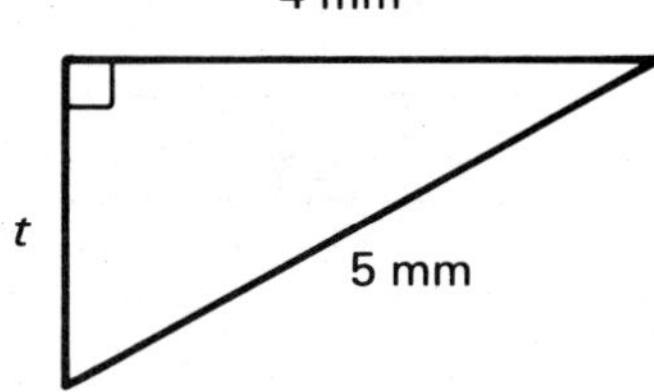

59.

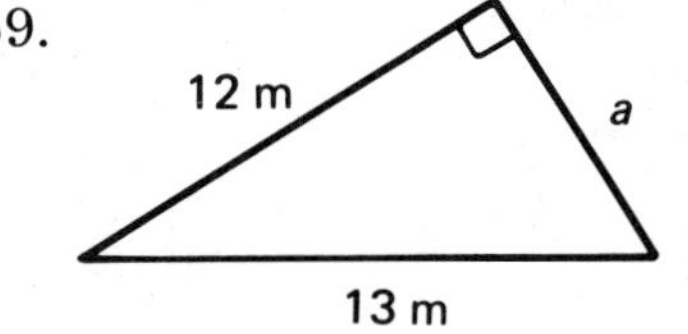

60.

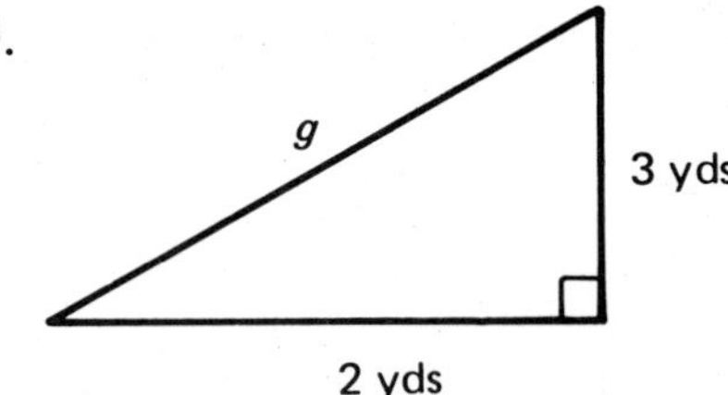

For the given problem, flowchart and planned output, code a program to solve the problem.

61. **Problem:** A triangle inscribed in a semi-circle (half a circle) is a right triangle. Notice also that half the length of the hypotenuse is the length of the radius of the circumscribed circle (the circle that touches all vertices of the triangle).

Enter the lengths of the legs of a right triangle. Find the area of the circle circumscribed around the right triangle. (Use $\pi = 3.1416$ and round to the nearest hundredth.)

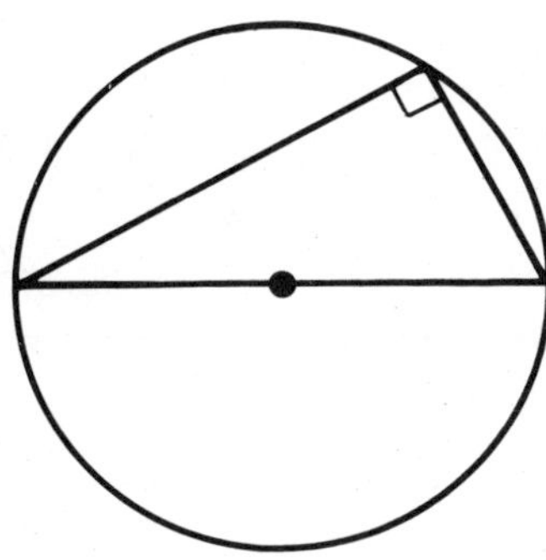

Planned Output

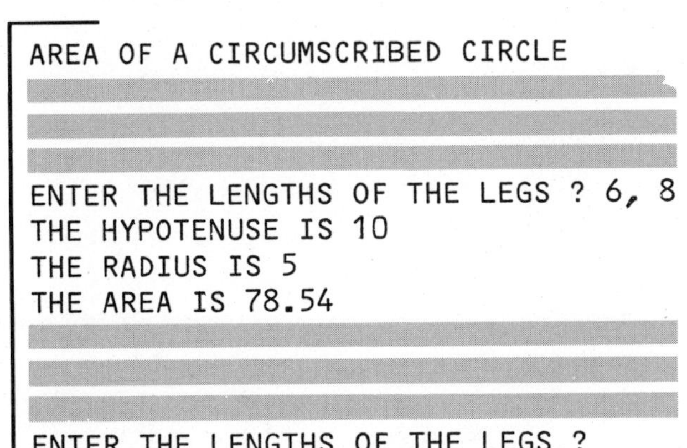

```
AREA OF A CIRCUMSCRIBED CIRCLE



ENTER THE LENGTHS OF THE LEGS ? 6, 8
THE HYPOTENUSE IS 10
THE RADIUS IS 5
THE AREA IS 78.54



ENTER THE LENGTHS OF THE LEGS ?
```

Flowchart

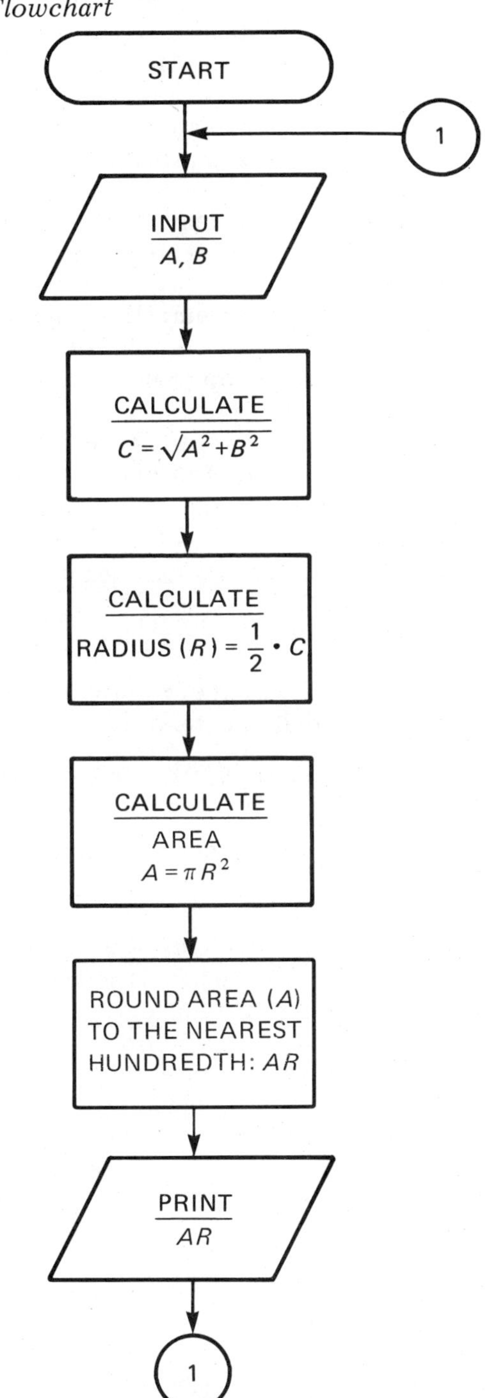

For each of the following problems:
a. Write an analysis which includes the development of an algorithm.
b. Write a plan for a computer solution to the problem.
c. Develop a flowchart.
d. Design a planned output.
e. Code a program to solve the problem and RUN it on the computer.

62. **Problem:** Enter the lengths of the two legs of a right triangle. Have the computer display the area and the hypotenuse rounded to the nearest hundredth.

63. **Problem:** Have the computer read the lengths of seven pairs of legs of a right triangle. Have the computer print a chart listing the legs, hypotenuses, perimeters, and areas. All values are to be rounded to the nearest tenth.

64. **Problem:** Enter the length of the hypotenuse of a right isosceles triangle. Have the computer display the length of the legs rounded to the nearest hundredth.

Section 6.5

Which sets have lengths that could represent the sides of a right triangle?

65. {16, 30, 34}   66. {7, 8, 9}   67. {6, 8, 10}   68. {10, 12, 14}

For each of the following, state whether the measures can represent the side lengths of a(n) right, acute, or obtuse triangle.

69. {6, 7, 9}   70. {24, 10, 26}   71. {8, 7, 12}   72. {4, 5, 6}

For the given problem, flowchart and planned output, code a program to solve the problem.

73. **Problem:** Enter the lengths of three sides of a triangle. Have the computer determine if the triangle is an acute isosceles triangle. (Assume that the lengths of the three sides form a triangle.)

Flowchart

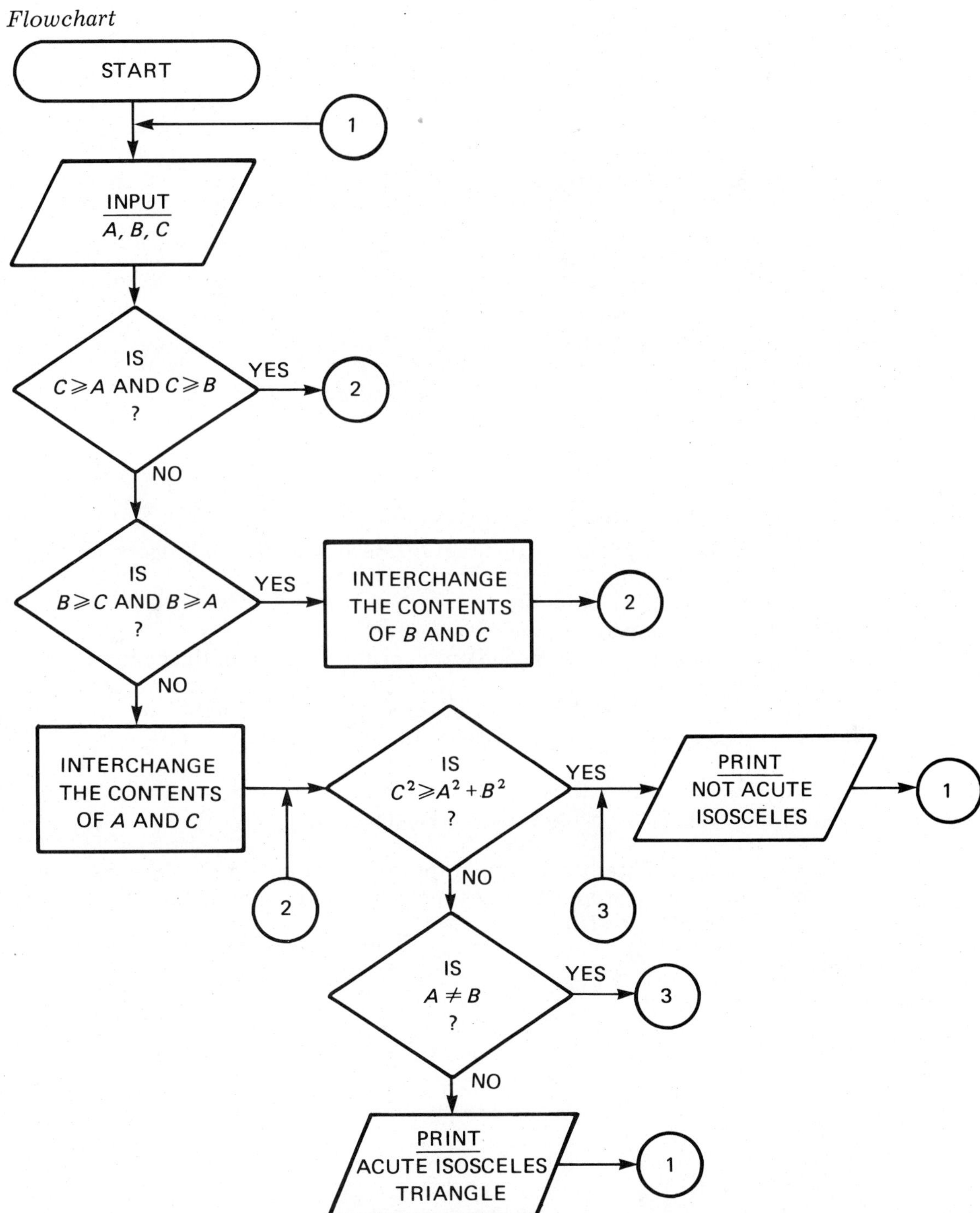
START
1
INPUT
A, B, C
IS
C≥A AND C≥B
?
YES
2
NO
IS
B≥C AND B≥A
?
YES
INTERCHANGE
THE CONTENTS
OF B AND C
2
NO
INTERCHANGE
THE CONTENTS
OF A AND C
IS
C²≥A²+B²
?
YES
PRINT
NOT ACUTE
ISOSCELES
1
2
NO
3
IS
A≠B
?
YES
3
NO
PRINT
ACUTE ISOSCELES
TRIANGLE
1

Planned Output

```
IS IT AN ACUTE ISOSCELES TRIANGLE?



ENTER THE LENGTHS OF 3 SIDES ? 4, 4, 5
IT'S AN ACUTE ISOSCELES TRIANGLE



ENTER THE LENGTHS OF 3 SIDES ? 3, 4, 5
IT'S NOT AN ACUTE ISOSCELES TRIANGLE



ENTER THE LENGTHS OF 3 SIDES ?
```

For each of the following problems:
a. Write an analysis which includes the development of an algorithm.
b. Write a plan for a computer solution to the problem.
c. Develop a flowchart.
d. Design a planned output.
e. Code a program to solve the problem and RUN it on the computer.

74. **Problem:** Enter the lengths of three sides of a triangle. Have the computer determine if the triangle is an obtuse isosceles triangle.

75. **Problem:** Enter the lengths of three sides of a triangle. Have the computer determine if the triangle is an acute scalene triangle.

76. **Problem:** Enter the length of three sides of a triangle. Have the computer classify the triangle according to *both* angles and sides. (For example, the triangle may be acute isosceles, obtuse scalene, etc.)

Chapter 7

Advanced Computer Programming Concepts

Section 7.1 SEARCHING FOR ITEMS IN A LIST (SUBSCRIPTED VARIABLE)

Warehouses use computers to keep a record (inventory) of the items that they have in stock. This often requires the storage of information in many memory cells. This section will give you an idea of how this may be accomplished.

Storing information in many memory cells can be done more efficiently if we use a *subscripted variable.*

> A *subscripted variable* is a variable written as $X(0)$, $X(1)$, $X(2)$, . . . , $X(I)$ and read X sub 0, X sub 1, etc.

The subscript of the variable always appears in parentheses after the variable. This type of variable is sometimes referred to as an array.

Let us see how we use this type of variable in a program.

Program #1

```
■ 10  READ X(0), X(1), X(2), X(3), X(4), X(5), X(6), X(7)
■ 20  PRINT X(0); X(1); X(2); X(3); X(4); X(5); X(6); X(7)
  30  DATA 9, 8, 7, 6, 5, 4, 3, 2
  40  END
  RUN
  9  8  7  6  5  4  3  2
```

Discussion of Program #1:

■ The variables $X(0)$, $X(1)$, . . . , $X(7)$ are treated here as if they were memory cells such as $X0$, $X1$, . . . , $X7$ or A, B, C, etc.

Then why do we ever need subscripted variables?

Examine the next program.

Program #2

```
  10  FOR I=0 TO 7
■ 20  READ X(I)
■ 30  PRINT X(I);
  40  NEXT I
  50  DATA 9, 8, 7, 6, 5, 4, 3, 2
  60  END
  RUN
  9 8 7 6 5 4 3 2
```

Discussion of Program #2:

■ In this program the subscripted variable $X(I)$ becomes eight separate memory cells named $X(0)$, $X(1)$, $X(2)$, $X(3)$, $X(4)$, $X(5)$, $X(6)$ and $X(7)$.

Which values are assigned to each memory cell?

$X(0) = 9$	$X(2) = 7$	$X(4) = 5$	$X(6) = 3$
$X(1) = 8$	$X(3) = 6$	$X(5) = 4$	$X(7) = 2$

When the computer encounters a subscripted variable $X(I)$, it automatically reserves 11 memory cells $X(0)$, $X(1)$, . . . , $X(10)$. If you wish to make the value of the subscript greater than ten, you must give the computer that information at the beginning of the program in a DIMension statement.

The next program will show how the DIMension statement is used.

Program #3

```
 ■ 10  DIM Y(20)
   20  FOR I=1 TO 12
■■ 30  READ Y(I)
   40  NEXT I
   50  FOR C=12 TO 1 STEP -1
■■ 60  PRINT Y(C);
   70  NEXT C
   80  DATA 0, 1, 2, 3, 4, 5, 6, 7, 8, 9, 10, 11
   90  END
   RUN
   11  10  9  8  7  6  5  4  3  2  1  0
```

Discussion of Program #3:

■ DIM $Y(20)$ tells the computer to reserve 21 memory cells called $Y(0)$, $Y(1)$, . . . , $Y(20)$. Note that you can reserve more memory cells than you actually use in a program; however, you cannot reserve fewer cells than you use.

▪▪ Are $Y(I)$ and $Y(C)$ the same memory cells? In this program the variables *are* the same. Remember that subscripted variables are not memory cells until the subscript becomes an integer. Since I and C both take on the values 1 to 12, $Y(I)$ and $Y(C)$ both become $Y(1)$, $Y(2), \ldots, Y(12)$ and thus are the same memory cells.

What values are assigned to $Y(1), Y(2), \ldots, Y(12)$?

$Y(1) = 0$	$Y(4) = 3$	$Y(7) = 6$	$Y(10) = 9$
$Y(2) = 1$	$Y(5) = 4$	$Y(8) = 7$	$Y(11) = 10$
$Y(3) = 2$	$Y(6) = 5$	$Y(9) = 8$	$Y(12) = 11$

Why does the output show the numbers in reverse order?

Since C takes on the values 12, 11, 10, . . . , 1 (from line 50), the contents of $Y(12)$, $Y(11), \ldots, Y(1)$ are displayed in that order. Therefore, we see the list in reverse order.

Test your understanding of the subscripted variable by working the problems in the next exercise.

Class Exercise 1 Show the output for each program.

1.
```
10  FOR X=1 TO 5
20  READ E(X)
30  PRINT E(X);
40  NEXT X
50  DATA 7, 8, 9, 10, 11
60  END
```

2.
```
10  FOR I=1 TO 6
20  READ F(I)
30  NEXT I
40  FOR I=1 TO 6
50  PRINT F(I);
60  NEXT I
70  DATA 9, 8, 7, 6, 5, 4
80  END
```

3.
```
10  FOR Y=1 TO 4
20  READ X(Y)
30  NEXT Y
40  FOR M=1 TO 4
50  PRINT X(M);
60  NEXT M
70  DATA 5, 6, 7, 8
80  END
```

4.
```
10  FOR A=1 TO 3
20  READ G(A)
30  NEXT A
40  FOR A=1 TO 3
50  PRINT A, G(A)
60  NEXT A
70  DATA 12, 13, 14
80  END
```

5.

```
10  FOR T=1 TO 4
20  READ A(T), B(T)
30  NEXT T
40  DATA 1, 2, 3, 4, 5, 6, 7, 8
50  FOR T=1 TO 4
60  PRINT A(T), B(T)
70  NEXT T
80  END
```

6.

```
10  FOR D=1 TO 5
20  READ X$(D)
30  NEXT D
40  DATA Q, R, S, T, U, V
50  FOR D=1 TO 5
60  PRINT X$(D), D
70  NEXT D
80  END
```

■ Subscripted string variable.

7.

```
10   FOR E=1 TO 4
20   READ Q(E), R(E)
30   NEXT E
40   DATA 7, 8, 9, 10, 11, 12, 13, 14
50   FOR E=1 TO 4
60   S(E)=Q(E)+R(E)
70   NEXT E
80   FOR E=1 TO 4
90   PRINT Q(E), R(E), S(E)
100  NEXT E
110  END
```

8.

```
10   FOR N=1 TO 4
20   READ X(N)
30   NEXT N
40   FOR N=1 TO 3
50   IF X(N)<=X(N+1) THEN 80
60   PRINT X(N)
70   GOTO 90
80   PRINT X(N+1)
90   NEXT N
100  DATA 13, 15, 26, 17
110  END
```

Now we will use a subscripted variable to solve a problem.

Problem

The ABC Toy Company has ten different toys in stock. Below is a list of each toy with its item number.

Toy	*Number*
1. Checkers	2001
2. Chess	2002
3. Pick-Up Sticks	3001
4. Crayons	4001
5. Bingo	5001
6. Playing Cards	6001
7. Marbles	6002
8. Dress-Up-Doll	7001
9. Baseball Glove	8001
10. Baseball Hat	8002

Write a program so that a customer can enter an item number and the computer will display whether that item is in stock.

Solution:

a. *Analysis:* To solve this problem, two main tasks must be performed.

1. A list of ten numbers must be read in and stored.
2. Each customer number that is entered must be compared to each item number in the list.

b. *Plan:*

1. Read in ten numbers and store them in $P(I)$ where I takes on values from 1 to 10.
2. Put in the customer's number, X.
3. Compare X to each $P(I)$.
   a) If $X=P(I)$ then the item is in stock. Go to 2.
   b) If $X \neq P(I)$ then the item is not in stock. Go to 2.

c. *Flowchart*

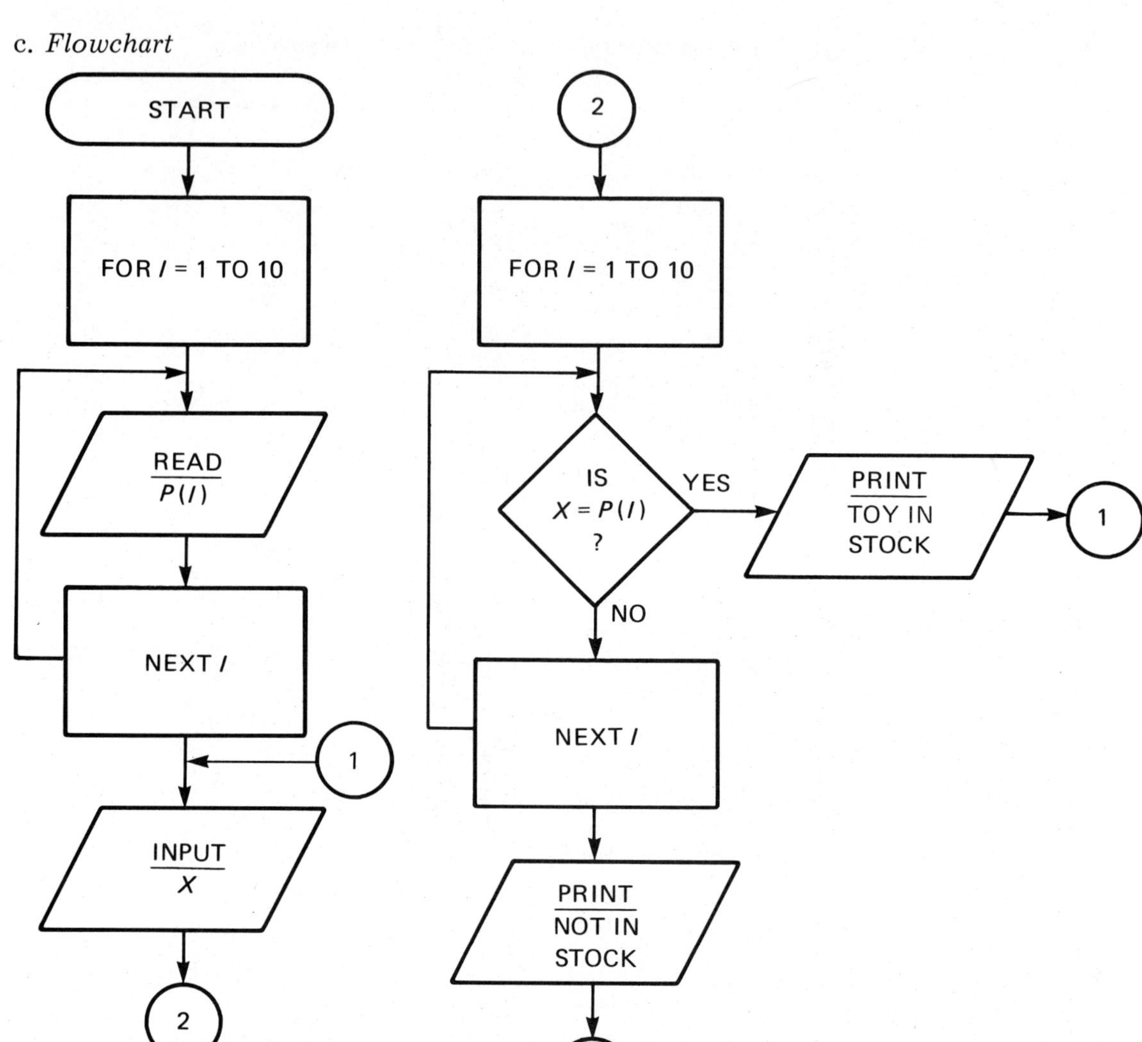
START
FOR I = 1 TO 10
READ
P(I)
NEXT I
1
INPUT
X
2
2
FOR I = 1 TO 10
IS
X = P(I)
?
YES
PRINT
TOY IN
STOCK
1
NO
NEXT I
PRINT
NOT IN
STOCK
1

d. *Planned Output*

```
IS THE TOY IN STOCK?
ENTER THE ITEM NUMBER ? 5001
TOY IS IN STOCK


ENTER THE ITEM NUMBER ? 5003
TOY IS NOT IN STOCK


ENTER THE ITEM NUMBER ?
```

e. *The Program*

```
10    PRINT "IS THE TOY IN STOCK?"
20    FOR I=1 TO 10
30    READ P(I)
40    NEXT I
50    INPUT "ENTER THE ITEM NUMBER"; X
60    FOR I=1 TO 10
70    IF X=P(I) THEN GOTO 100
80    NEXT I
90    GOTO 130
100   PRINT "TOY IS IN STOCK"
110   PRINT : PRINT
120   GOTO 50
130   PRINT "TOY IS NOT IN STOCK"
140   GOTO 110
150   DATA 2001, 2002, 3001, 4001, 5001, 6001, 6002, 7001, 8001, 8002
160   END
```

Exercise 7.1 Show the output for each program.

1.
```
10  FOR J=1 TO 6
20  READ I$(J)
30  PRINT I$(J);
40  NEXT J
50  DATA A, B, C, D, E, F
60  END
```

2.
```
10  FOR K=1 TO 8
20  READ A(K)
30  NEXT K
40  FOR K=2 TO 8 STEP 2
50  PRINT A(K)
60  NEXT K
70  DATA 20, 21, 22, 23, 24, 25, 26, 27
80  END
```

3.
```
10  DIM Y(15)
20  FOR I=1 TO 15
30  Y(I)=15-I
40  NEXT I
50  FOR I=3 TO 15 STEP 3
60  PRINT Y(I);
70  NEXT I
80  END
```

4.
```
10  FOR L=1 TO 4
20  READ B(L)
30  NEXT L
40  DATA 5, 6, 7, 8
50  FOR L=1 TO 4
60  PRINT B(L); L
70  NEXT L
80  END
```

5.
```
10  FOR M=1 TO 5
20  A(M)=2*M
30  NEXT M
40  FOR M=1 TO 5
50  PRINT M; A(M)
60  NEXT M
70  END
```

6.
```
10   FOR N=1 TO 6
20   A(N)=3
30   NEXT N
40   FOR N=1 TO 6
50   A(N)=A(N)+N
60   NEXT N
70   FOR N=1 TO 6
80   PRINT A(N);
90   NEXT N
100  END
```

7.

```
10    FOR P=1 TO 4
20    READ C(P), D(P)
30    NEXT P
40    DATA 4, 2, 6, 3, 8, 5, 10, 7
50    FOR P=1 TO 4
60    E(P)=C(P)-D(P)
70    NEXT P
80    FOR P=1 TO 4
90    PRINT C(P), D(P), E(P)
100   NEXT P
110   END
```

8.

```
10    FOR S=1 TO 3
20    READ Y(S)
30    NEXT S
40    FOR S=1 TO 2
50    IF Y(S)>=Y(S+1) THEN 70
60    Y(S+1)=Y(S)
70    NEXT S
80    DATA 7, 8, 9
90    FOR S=1 TO 3
100   PRINT Y(S);
110   NEXT S
120   END
```

Revise the program that was written in this section to produce the following outputs.

9.

```
SEARCHING FOR AN ITEM NUMBER


WHAT IS YOUR ITEM NUMBER? 5001
IT'S IN STOCK!


WHAT IS YOUR ITEM NUMBER? 5003
IT'S NOT IN STOCK!


WHAT IS YOUR ITEM NUMBER?
```

10.

```
CONDUCTING A SEARCH


PUT IN YOUR ITEM NUMBER ? 5001
5001 IS IN STOCK!


PUT IN YOUR ITEM NUMBER ? 5003
5003 IS NOT IN STOCK!


PUT IN YOUR ITEM NUMBER ?
```

For the following problem and flowchart, design a planned output and code a program to solve the problem.

11. **Problem:** Enter into an array quantities for seven items in a candy factory. A delivery of these candies was just made to the factory. Enter into a second array the numbers of each of these items that were just delivered. Print an inventory list that includes this new delivery.

Flowchart

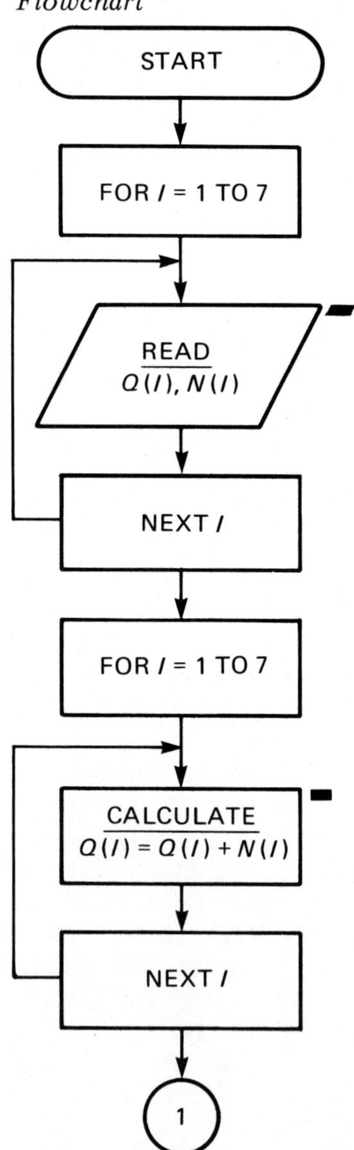

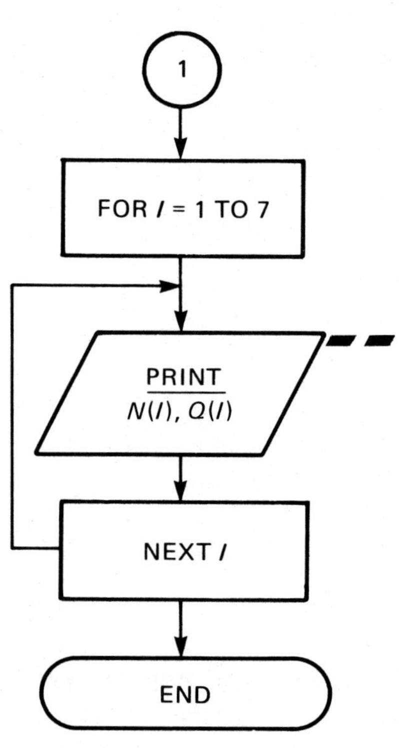

Discussion of the Flowchart:

▰ Read in the original quantities and the additional quantities from the new delivery for seven items.

▬ Add the quantity from the new delivery to the quantity already in stock.

▰ ▰ Print the new inventory.

For each of the following problems:
a. Write an analysis which includes the development of an algorithm.
b. Write a plan for a computer solution to the problem.
c. Develop a flowchart.
d. Design a planned output.
e. Code a program to solve the problem and RUN it on the computer.

12. **Problem:** Revise the program that was written in this section so that when an item number is found, the name of the toy is printed as well as the message.

13. **Problem:** Have the computer read ten numbers into an array. Have the computer display the sum of the numbers and then display each number.

14. **Problem:** Expand the program written for the problem developed in this section (page 389) as follows:

a. Read in each item number, the name of the item, the quantity of the item and the cost of the item.
b. Enter the customer's item number.
c. Is the item in stock?
  1. If *yes*:
    a) Print item number, name, quantity and cost of item.
    b) Let the user enter the number of items (quantity) he wishes to purchase.
    c) Print the cost of the purchase.
    d) Change the quantity of the item stored in the memory.
  2. If *no*, go to b.

Section 7.2 RANDOM NUMBERS

Computer games are usually developed using random numbers. A *random number* is a number where each digit has the same chance of being a 0, 1, 2, . . . , or a 9.

The following program was run on the Apple computer. Look at the output. It gives us (generates) five random numbers.

```
10  FOR I=1 TO 5
20  Y=RND(1)
30  PRINT Y
40  NEXT I
50  END
RUN
.973136996 \  Five
.103117626 |  random
.779343355 >  numbers
.551834438 |  between
.417714833 /  0 and 1
```

Discussion of the Output:

- RND(1)† generates a random number between 0 and 1.

With the use of random numbers, the computer can be programmed to give the effect (simulate) of throwing dice. Let us see how this can be accomplished.

Problem

Write a program to simulate (give the effect of) throwing a pair of dice. Determine if the sum of the two dice is equal to 7 or 11.

Solution:

a. *Analysis:* Our random function gives us nine digit decimals between 0 and 1. We want whole numbers from 1 to 6. Let us look at the smallest and largest random numbers that RND(1) produces to see how we can get numbers from 1 to 6.

Algorithm:		*Smallest* RND(1)	*Largest* RND(1)
		.000000001	.999999999
1. Multiply:	6 * RND(1)	.000000006	5.999999994
2. Use INT:	INT (6 * RND(1))	0	5

INT (6 * RND(1)) will give us random whole numbers from 0 to 5.

---

†On some computers you might have to use RND(0). Check your manual.

Then how do we get random numbers from 1 to 6?
We add 1:

INT (6 * RND (1)) + 1

b. *Plan:*

1. Simulate throwing a pair of dice.
   a) X = INT (6 * RND (1)) + 1.
   b) Y = INT (6 * RND (1)) + 1.
2. Sum (S) = $X + Y$.
3. Is S = 7?
   a) If yes, print "SUM IS 7". Then go to 5.
   b) If no, continue.
4. Is S = 11?
   a) If yes, print "SUM IS 11". Then go to 5.
   b) If no, continue.
5. Do you wish to continue? Answer yes or no.
   a) If yes, go to 1.
   b) If no, end.

c. *Flowchart*

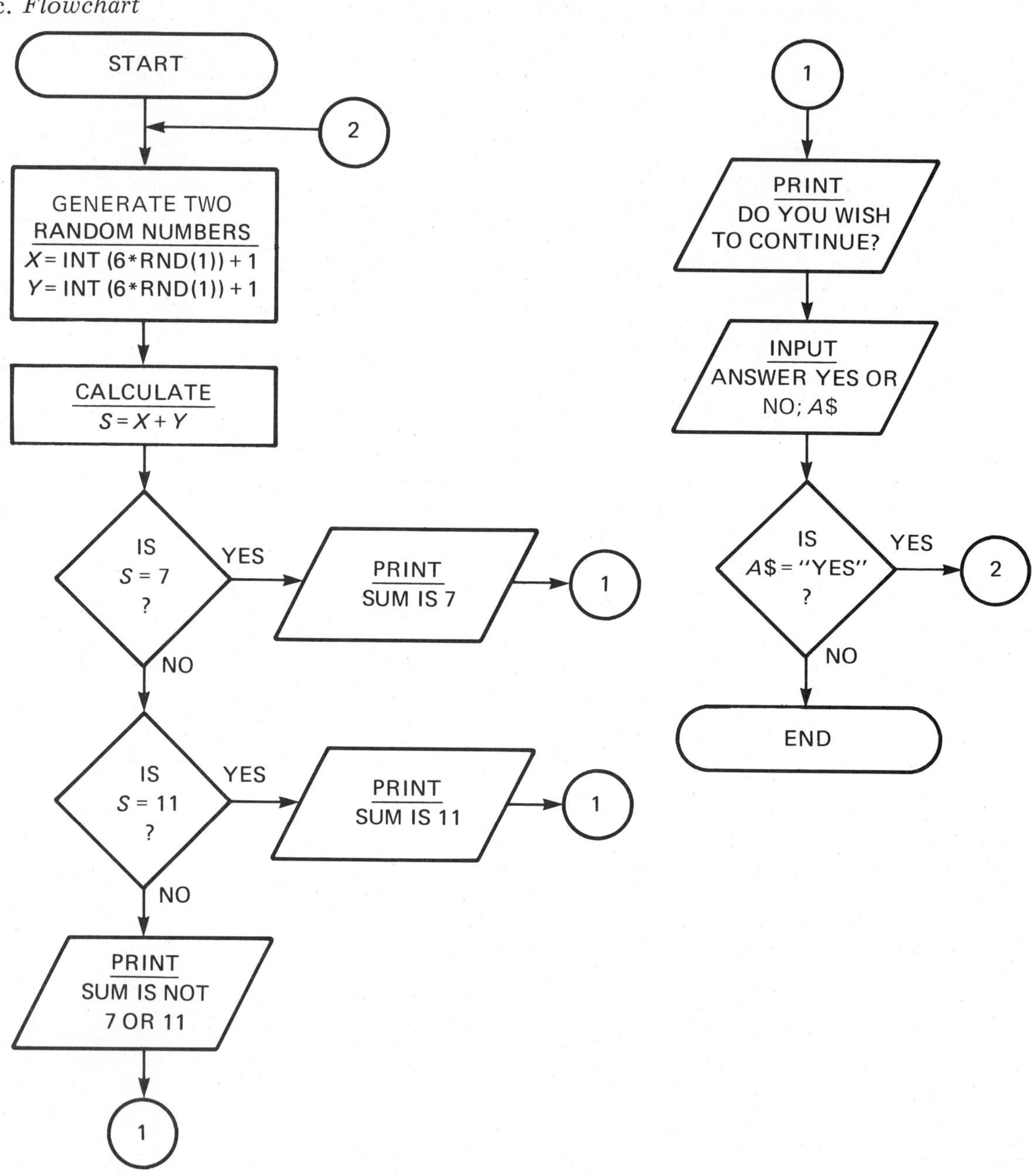
START
2
GENERATE TWO RANDOM NUMBERS
X = INT (6*RND(1)) + 1
Y = INT (6*RND(1)) + 1
CALCULATE
S = X + Y
IS S = 7 ?
YES
NO
PRINT SUM IS 7
1
IS S = 11 ?
YES
NO
PRINT SUM IS 11
1
PRINT SUM IS NOT 7 OR 11
1
1
PRINT DO YOU WISH TO CONTINUE?
INPUT ANSWER YES OR NO; A$
IS A$ = "YES" ?
YES
2
NO
END

d. *Planned Output*

```
SUMMING DICE


DIE  #1 IS 4
DIE  #2 IS 6
THE SUM IS NOT 7 OR 11


DO YOU WISH TO CONTINUE?
ANSWER YES OR NO ? YES


DIE  #1 IS 2
DIE  #2 IS 5
THE SUM IS 7


DO YOU WISH TO CONTINUE?
ANSWER YES OR NO ? NO


THE END
```

e. *The Program*

```
10   PRINT "SUMMING DICE"
20   PRINT : PRINT
30   X=INT(6*RND(1))+1
40   Y=INT(6*RND(1))+1
50   S=X+Y
60   PRINT "DIE  #1 IS"; X
70   PRINT "DIE  #2 IS"; Y
80   IF S=7 THEN GOTO 180
90   IF S=11 THEN GOTO 190
100  PRINT "SUM IS NOT 7 OR 11"
110  PRINT : PRINT
120  PRINT "DO YOU WISH TO CONTINUE?"
130  INPUT "ANSWER YES OR NO"; A$
140  IF A$="YES" THEN GOTO 20
150  PRINT : PRINT
160  PRINT "THE END"
170  GOTO 200
180  PRINT "THE SUM IS 7": GOTO 110
190  PRINT "THE SUM IS 11": GOTO 110
200  END
```

Exercise 7.2 Develop an algorithm to generate random numbers:

1. from 0 to 9
2. from 1 to 10
3. from 1 to 13
4. from 1 to 4

Revise the program in this section to produce the following outputs.

5.

```
IS IT 7 OR 11?



ROLLING!!!
YOUR ROLL IS A 4 AND 6
SUM IS NOT 7 OR 11


DO YOU WISH TO CONTINUE?
ANSWER YES OR NO ? YES


ROLLING!!!
YOUR ROLL IS A 2 AND 5
SUM IS 7


DO YOU WISH TO CONTINUE?
ANSWER YES OR NO ? NO


THANKS FOR PLAYING!
```

6.

```
ROLLING A 7 OR 11?



ROLLING!!! 4 AND 6
NOT 7 OR 11


DO YOU WISH TO CONTINUE?
ANSWER YES OR NO ? YES


ROLLING!!! 2 AND 5
YOU HIT A 7!


DO YOU WISH TO CONTINUE?
ANSWER YES OR NO ? NO


I HOPE YOU HAD FUN!!!!
```

For the following problem, flowchart and planned output, code a program to solve the problem.

7. **Problem:** Simulate the rolling of a pair of dice ten times. Have the computer display the numbers on the two dice, the sum on each pair and the number of times that the sum of the dice is 7.

Flowchart

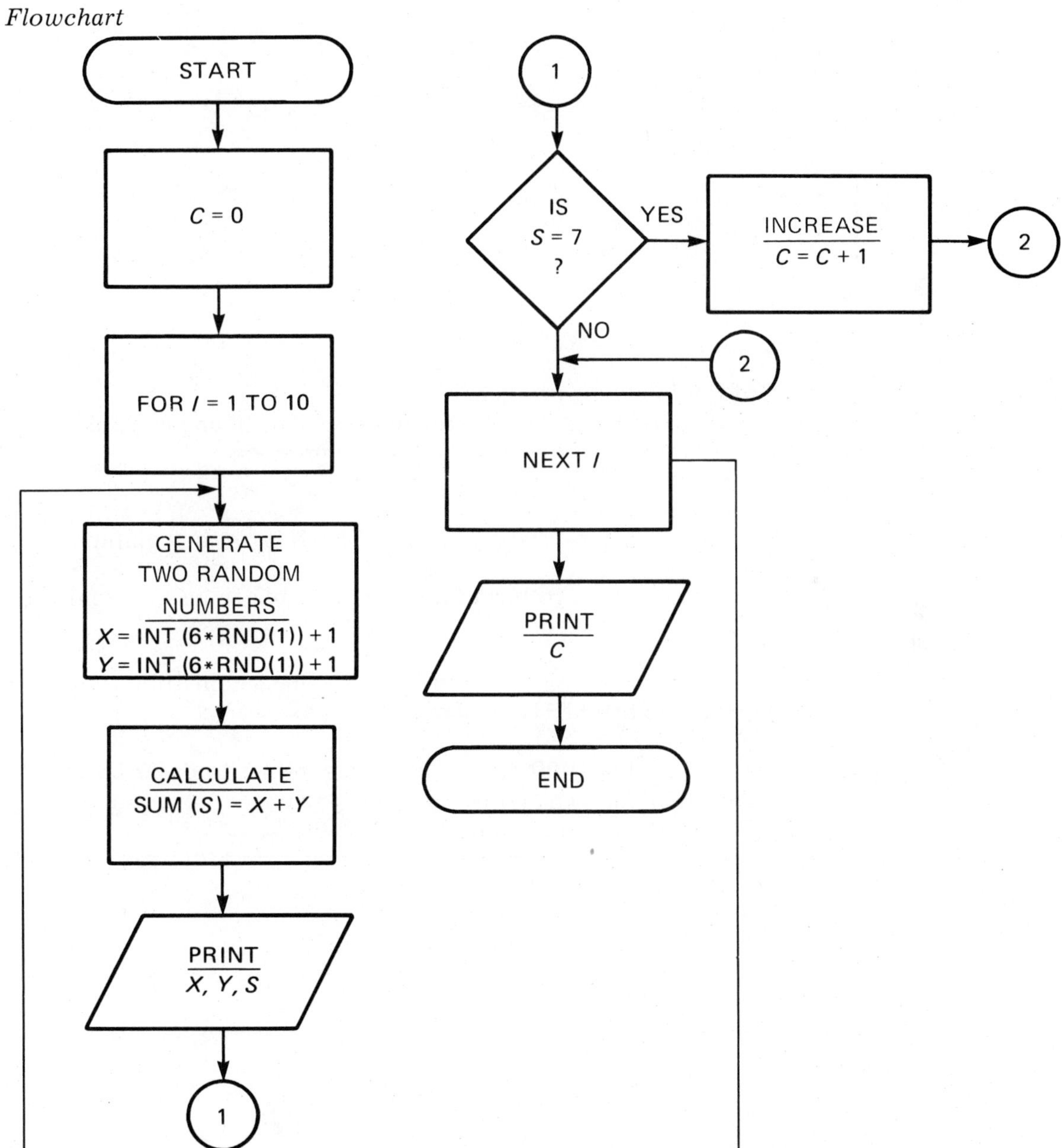

Planned Output

```
HOW MANY 7'S IN 10 ROLLS?


DIE #1      DIE #2      SUM
 3           4           7
 5           5          10
 2           2           4
 6           3           9
 1           6           7
 4           5           9
 6           4          10
 3           1           4
 2           6           8
 5           2           7
NUMBER OF 7'S ROLLED = 3
```

For each of the following problems:
a. Write an analysis which includes the development of an algorithm.
b. Write a plan for a computer solution to the problem.
c. Develop a flowchart.
d. Design a planned output.
e. Code a program to solve the problem and RUN it on the computer.

8. **Problem:** Simulate the rolling of a pair of dice ten times. Determine the number of times that a 7 or 11 is rolled.

9. **Problem:** Develop a program to simulate tossing a coin 1000 times. Record the number of "heads" that are tossed. (How many do you expect?)

10. **Problem:** Generate 1000 random numbers from 1 to 6. Determine the number of primes in the 1000 numbers. (How many do you expect?)
(*Hint:* 2, 3, and 5 are the primes from 1 to 6.)

Section 7.3 WORKING WITH STRINGS (MID$, LEN)

Computers can be programmed to get information from a message that was entered in a string variable. For example, if *A*$ = "PENNSYLVANIA", how could we determine the number of N's in *A*$?

Our plan would involve looking at each character of *A*$ and comparing it to the letter N. Each time a character is N, we must increase a count by one. To do this we need two new computer functions.

One function must determine the *LEN*gth of the string.

> *LEN* (string) returns the character length of a string variable.

When *A*$="PENNSYLVANIA" then LEN(*A*$) will return the numeric value 12.

If *D*$="WE LOVE MATH.", then LEN(*D*$) will return the numeric value 13. Note that blanks and punctuation marks within the quotes are counted as characters.

Class Exercise 1 For each string stored in *C*$, find LEN (*C*$):

1. *C*$="INDIANA"
2. *C*$="I LIKE ALGEBRA."
3. *C*$="JACK AND JILL"
4. *C*$="CROSS AT THE GREEN."

Another function must be able to examine each character of the string.

> *MID*$(*STRING*, *P*, *L*) returns a substring of length *L* beginning at position *P* of the STRING.

If *E*$="COMPUTERS ARE FUN", the MID$ function would work as follows:

MID$ (*E*$, 1, 1) = C

MID$ (*E*$, 2, 1) = O

MID$ (*E*$, 3, 1) = M

MID$ (*E*$, 11, 7) = ARE FUN

Example 1 If *F*$="I LIKE GEOMETRY", find the string assigned to *L*$ if *L*$=MID$(*F*$, 1, 8).

Solution: *F*$="I LIKE GEOMETRY", *L*$ begins at position number 1 and returns eight characters:

L$= MID$(*F*$, 1, 8) therefore

L$=I LIKE G

Class Exercise 2 If $F\$$="I LIKE GEOMETRY", find the string assigned to $L\$$ if:

1. $L\$$=MID$(F\$, 1, 1)
2. $L\$$=MID$(F\$, 2, 1)
3. $L\$$=MID$(F\$, 3, 1)
4. $L\$$=MID$(F\$, 3, 13)

Let us now use these two functions to solve the following problem.

Problem

Enter a message ($A\$$) of not more than 255† characters. Determine the number of times the letter N appears.

Solution:

a. *Analysis:* We will use LEN($A\$$) to determine the length of the string. MID$($A\$$, P, 1) will let us examine each character of $A\$$. The 1 in MID$($A\$$, P, 1) will always return a single character.

As P varies from 1 to LEN($A\$$) we will be able to look at every character one at a time.

For example, look at what MID$($A\$$, P, L) will return when $A\$$="NINE". L is held constant at 1, and P varies from 1 to LEN($A\$$). (*Note:* LEN($A\$$)=4.)

MID$($A\$$, 1, 1) = N
MID$($A\$$, 2, 1) = I
MID$($A\$$, 3, 1) = N
MID$($A\$$, 4, 1) = E

b. *Plan:*

1. Initialize a count: C=0.
2. Enter a string, $A\$$.
3. For P=1 TO LEN($A\$$).
4. Is MID$ ($A\$$, P, 1)="N"?
   a) If yes, C=C+1. Go to 5.
   b) If no, continue.
5. Next P.
6. Print C.
7. End.

†Some computers require a modification in the program to handle 255 characters. See your manual.

c. *Flowchart*

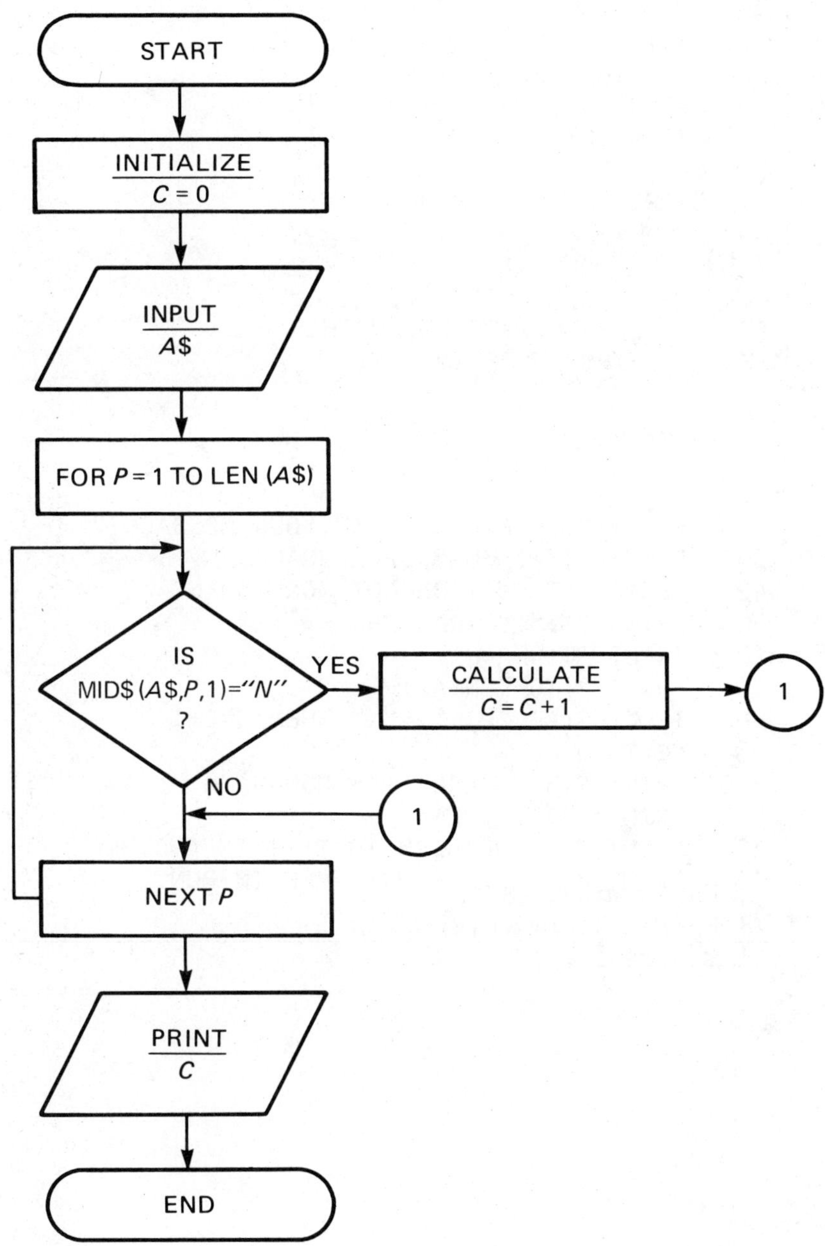
START
INITIALIZE
C = 0
INPUT
A$
FOR P = 1 TO LEN (A$)
IS
MID$ (A$,P,1)="N"
?
YES
CALCULATE
C = C + 1
1
NO
1
NEXT P
PRINT
C
END

d. *Planned Output*

```
HOW MANY N'S IN YOUR MESSAGE?
TYPE IN YOUR MESSAGE.
IT MUST BE NOT MORE THAN
255 CHARACTERS.
? NINE
THERE ARE 2 N'S IN NINE.



IF YOU WISH TO TYPE IN ANOTHER
MESSAGE, FIRST TYPE IN RUN.
```

e. *The Program*

```
10   C=0
20   PRINT "HOW MANY N'S IN YOUR MESSAGE?"
30   PRINT "TYPE IN YOUR MESSAGE."
40   PRINT "IT MUST BE NOT MORE THAN"
50   PRINT "255 CHARACTERS."
60   INPUT A$
70   FOR P=1 TO LEN(A$)
80   IF MID$(A$, P, 1)="N" THEN C=C+1
90   NEXT P
100  PRINT "THERE ARE"; C; "ƀN'S INƀ"; A$; "."
110  PRINT:PRINT:PRINT
120  PRINT "IF YOU WISH TO TYPE IN ANOTHER"
130  PRINT "MESSAGE, FIRST TYPE IN RUN."
140  END
```

Exercise 7.3 For each string stored in $Q\$$ find LEN ($Q\$$).

1. $Q\$$="CALIFORNIA"
2. $Q\$$="NEW ENGLAND"
3. $Q\$$="BE GOOD!"
4. $Q\$$="SHE IS BRIGHT."

If $R\$$="JEREMY IS SMART TO STUDY COMPUTER SCIENCE.", find the string assigned to $T\$$ if:

5. $T\$$=MID$ ($R\$$, 1, 8)
6. $T\$$=MID$ ($R\$$, 8, 4)
7. $T\$$=MID$ ($R\$$, 1, 15)
8. $T\$$=MID$ ($R\$$, 17, 26)

Revise the program in this section to produce the following outputs.

9.

```
LOOKING FOR N'S


TYPE IN YOUR MESSAGE ? NINE
2 N'S IN NINE


TYPE IN YOUR MESSAGE ?
```

10.

```
SEARCHING FOR THE LETTER N



TYPE IS YOUR MESSAGE ? NINE
YOUR MESSAGE IS - NINE
IT HAS 2 N'S


DO YOU WISH TO CONTINUE?
ANSWER YES OR NO ? YES


TYPE IN YOUR MESSAGE ?
```

For the following problem, flowchart and planned output, code a program to solve the problem.

11. **Problem:** Enter a message, $A\$$, of not more than 255 characters. Determine the number of vowels in the message.

Flowchart

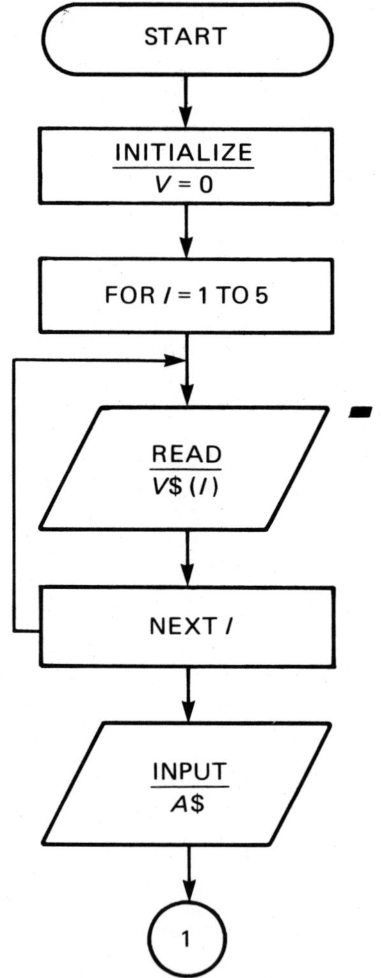

Discussion of the Flowchart:
▰ The vowels are read into $V\$(I)$ as follows: $V\$(1)$=A, $V\$(2)$=E, $V\$(3)$=I, $V\$(4)$=O and $V\$(5)$=U.
▬ Returns one character at a time from the message, $A\$$.
▬▬ Compares each character of $A\$$ to each vowel.

Flowchart (cont'd)

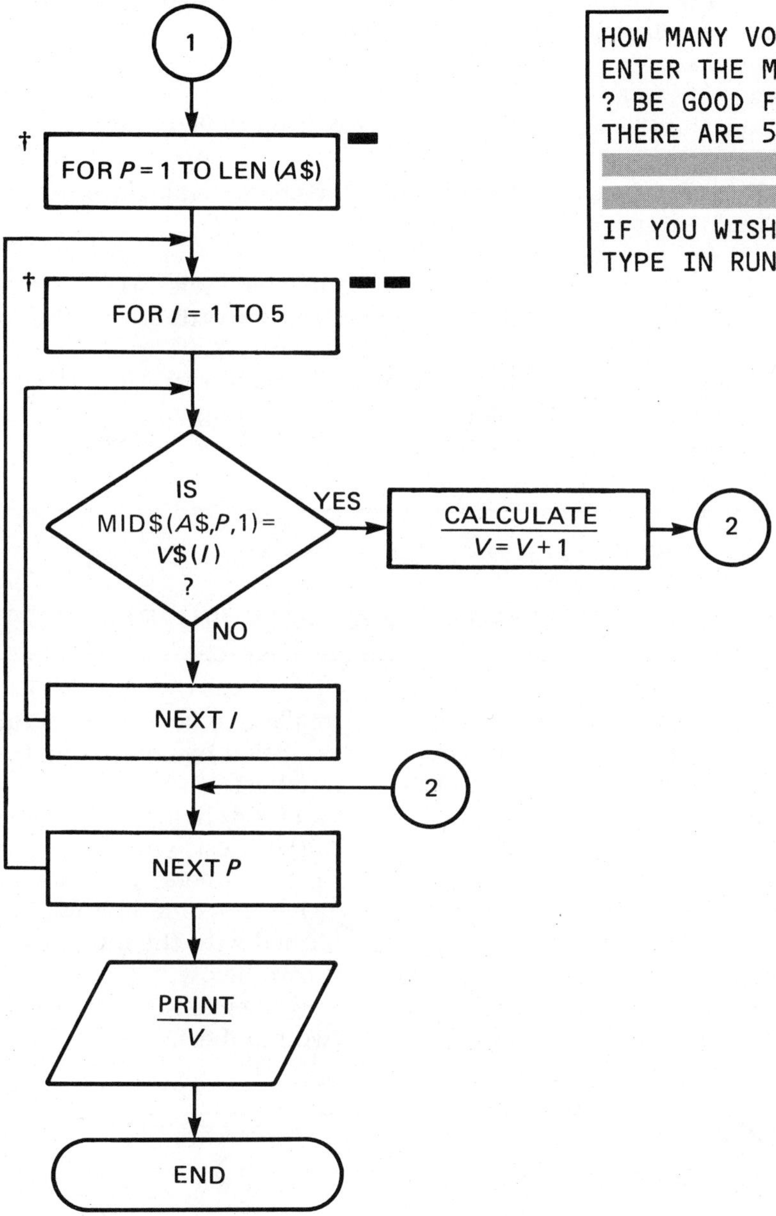

1

† FOR P = 1 TO LEN ($A$$)

† FOR I = 1 TO 5

IS MID$($A$$,P,1) = $V$$(I) ?

YES

CALCULATE
$V = V + 1$

2

NO

NEXT I

2

NEXT P

PRINT
V

END

Planned Output

```
HOW MANY VOWELS IN THE MESSAGE?
ENTER THE MESSAGE
? BE GOOD FRIENDS.
THERE ARE 5 VOWELS IN THE MESSAGE.


IF YOU WISH TO CONTINUE
TYPE IN RUN
```

†This construction is called a nested FOR-NEXT loop. For more information, see Section 3.3 of Elgarten, et. al, *Using Computers in Mathematics*, Addison-Wesley Publishing Company, Menlo Park, California, 1983.

For each of the following problems:
a. Write an analysis which includes the development of an algorithm.
b. Write a plan for a computer solution to the problem.
c. Develop a flowchart.
d. Design a planned output.
e. Code a program to solve the problem and RUN it on the computer.

12. **Problem:** Enter a message (not more than 255 characters). Have the computer determine the number of L's in the message.

13. **Problem:** Enter a sentence (not more than 255 characters). Have the computer determine the number of words in the sentence. (*Hint:* Words are separated by a blank.)

14. **Problem:** Enter a message (not more than 255 characters). Have the computer determine the number of words in the message. In this program the message may consist of several sentences. (*Hint:* Sentences are separated by a period, an exclamation mark, or a question mark followed by two blank spaces.)

Section 7.4 SORTING A LIST OF NUMBERS OR NAMES (BUBBLE SORT)

An important task that computers can be programmed to perform is sorting lists of numbers or names. Sorting lists of numbers means to put the numbers in either ascending order (smallest to largest) or descending order (largest to smallest). Sorting a list of names is called *alphabetizing* the list.

Efficient techniques for sorting, rearrange the original list while making comparisons. For example, to order numbers from smallest to largest (ascending order), you can begin by comparing the first two numbers. If the first is not the smallest, you interchange the numbers. If the first is the smallest, you compare the second and third numbers in the list, and so on.

Let us examine how this can be accomplished with the numbers stored in $X(I)$ as shown below.

We begin our comparison using the first two numbers.

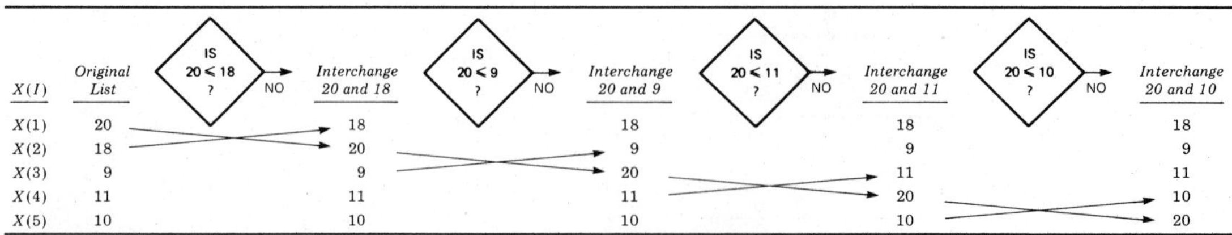

Note that the largest number fell to the bottom as the smaller numbers begin to float to the top like bubbles in water. That is why the technique is called the *bubble sort.*

We have just completed the first pass. Let us now make the second pass. This means that we begin again with the first two numbers, but we stop after we reach the number that is second from the bottom.

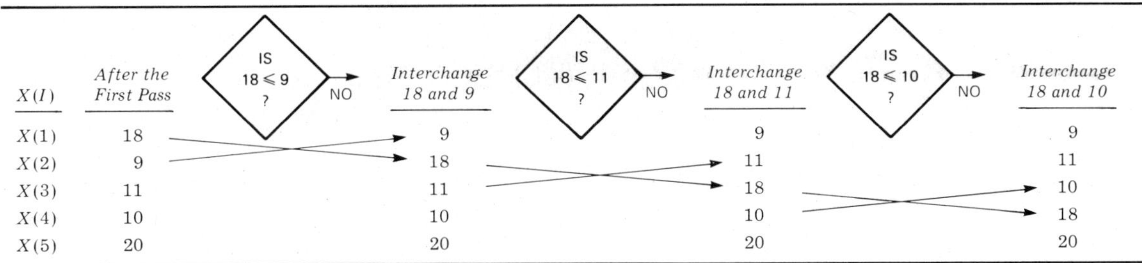

With the largest numbers at the bottom, we now make our third pass. Again, we start with the first two numbers. However, this time we stop after we reach the number that is third from the bottom.

IS $9 \leqslant 11$? → YES → List is Unchanged; IS $11 \leqslant 10$? → NO → Interchange 11 and 10

X(I)	*After the Second Pass*	*List is Unchanged*	*Interchange 11 and 10*
X(1)	9	9	9
X(2)	11	11	10
X(3)	10	10	11
X(4)	18	18	18
X(5)	20	20	20

With the three largest numbers at the bottom, we now make our fourth and final pass. Again, we start with the first two numbers. However, this time we stop after comparing the first and second (fourth from the bottom) numbers.

$X(I)$	*After the Third Pass*	IS $9 \leq 10$? → YES	*List is Unchanged*
$X(1)$	9		9
$X(2)$	10		10
$X(3)$	11		11
$X(4)$	18		18
$X(5)$	20		20

The numbers are now sorted from smallest to largest.

Class Exercise 1 Show how the bubble sort would place the following numbers in ascending order. Be sure to show the list for each comparison.

1. $X(1)=16$
   $X(2)=11$
   $X(3)=14$
   $X(4)=17$

2. $P(1)=9$
   $P(2)=8$
   $P(3)=7$
   $P(4)=6$
   $P(5)=5$

How can we use the bubble sort to alphabetize a list of names?

Recall from Section 5.6 that the computer interprets letters like numbers. That is $A<B<C<D \ldots <Z$.

For example, examine the following program.

```
■ 10  READ A$, B$
  20  IF A$<=B$ THEN 40
  30  PRINT B$, A$: GOTO 60
  40  PRINT A$, B$
  50  DATA JOHN, JOE
  60  END
  RUN
  JOE                    JOHN
```

Discussion of the Program:

■ In this program, $A\$$="JOHN" and $B\$$="JOE". Line 20 compares JOHN to JOE. The computer assigns a number to each letter of each name and then compares the numbers (the numbers assigned depend upon the computer used).

	J	O	H	N		J	O	E	ƀ
	↑	↑	↑	↑		↑	↑	↑	↑
ASCII† Number Codes:	74	79	72	78		74	79	69	32

Is 74797278⩽74796932? Since the answer is no, the computer goes to line 30 and prints the names in alphabetical order:

JOE JOHN ($B\$$, $A\$$)

We will now trace one pass of the bubble sort to see how this sorting technique alphabetizes a list of names.

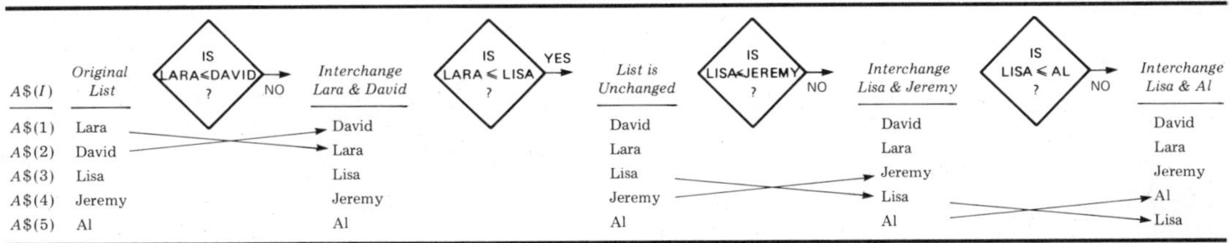

After the first pass, note that the name closest to Z in the alphabet fell to the bottom.

†American Standard Code Information Interchange

Class Exercise 2 Show how the bubble sort would alphabetize the following names. Be sure to show the lists for each comparison.

1. $B\$(1)$=RONALD
   $B\$(2)$=CAROL
   $B\$(3)$=ROSS
   $B\$(4)$=ROSE

2. $C\$(1)$=ARITHMETIC
   $C\$(2)$=COMPUTER
   $C\$(3)$=TRIGONOMETRY
   $C\$(4)$=ALGEBRA
   $C\$(5)$=GEOMETRY

Now let us have the computer use the bubble sort to solve the following problem.

Problem Have the computer alphabetize and display a list of five names using the bubble sort.

Solution:

a. *Analysis:* We review the procedure the computer uses to compare names (strings).
   1. The computer assigns a numerical value (called an ASCII code number) to each character in the string. These values are consecutive numbers resulting from the following relationship:

   $A<B<C<\ldots<Z.$

   2. The computer compares two strings as described on the previous page.

b. *Plan:*

1. Enter five names storing them in $N\$(I)$: $N\$(1)$, $N\$(2)$, . . . , $N\$(5)$.
2. Use $N\$(J)$ to represent a particular name in the list.
3. Use $N\$(J+1)$ to represent the name below $N\$(J)$ in the list. For five names we need four passes. We, therefore, keep track of the passes using a FOR-NEXT loop (as in 4).
4. FOR I=1 TO 4. We will also use a FOR-NEXT loop to move down the list (as in 5).
5. FOR J=1 TO 5−I.

   ↑ —— Since we do not use the last name in the list after each pass.

   a) Is $N\$(J) \leqslant N\$(J+1)$?
      1) If yes, go to b.
      2) If no, interchange the contents of $N\$(J)$ and $N\$(J+1)$. Continue.

   b) NEXT J.
6. NEXT I.
7. Print the list in alphabetical order.
8. End.

c. *Flowchart*

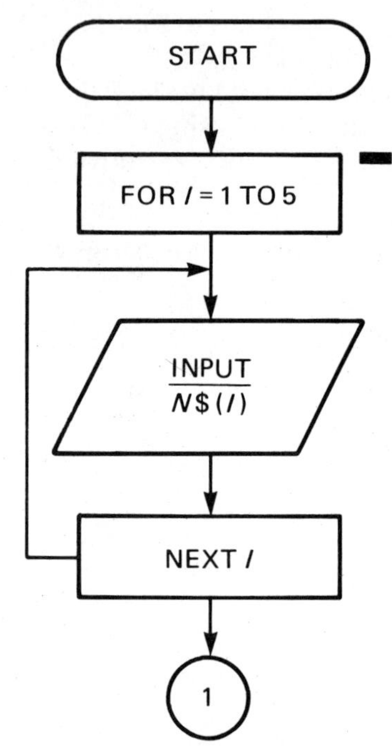
START
FOR I = 1 TO 5
INPUT
N$(I)
NEXT I
1

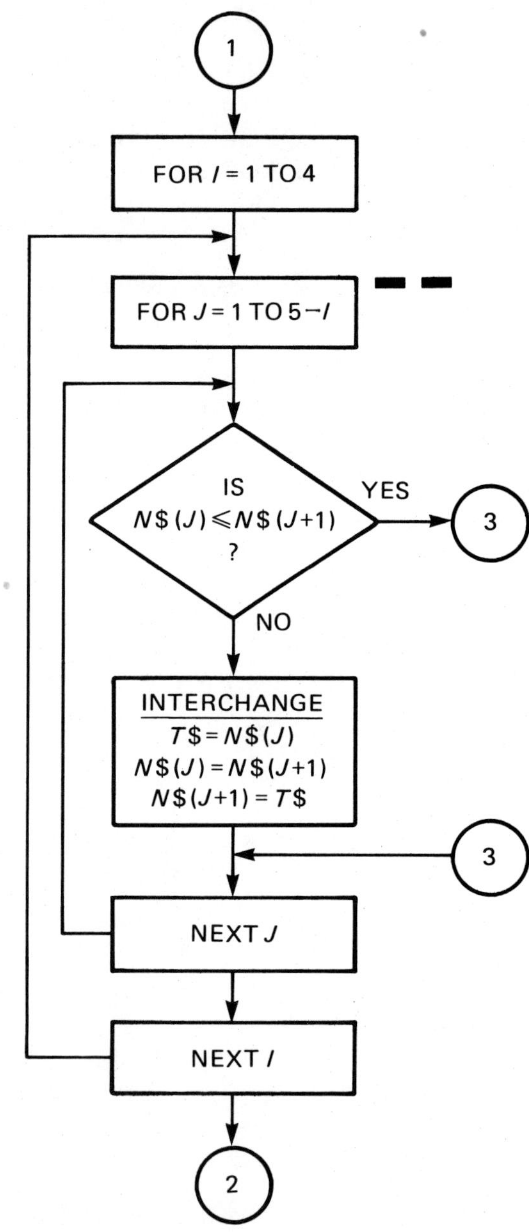
1
FOR I = 1 TO 4
FOR J = 1 TO 5−I
IS
N$(J) ≤ N$(J+1)
?
YES
3
NO
INTERCHANGE
T$ = N$(J)
N$(J) = N$(J+1)
N$(J+1) = T$
3
NEXT J
NEXT I
2

c. *Flowchart* (cont'd)

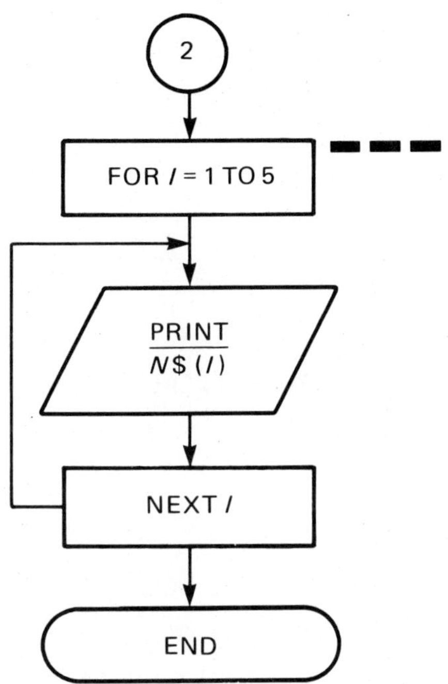

Discussion of the Flowchart:

■ The names are entered.

■■ The names are alphabetized. After each pass, the name closest to Z is moved to the bottom. The $5-I$ is used so that after each pass we do not use the last name in the list again.

■■■ The alphabetized names are displayed.

Notice that if a list was entered in alphabetical order, the computer would still make the same number of passes as if the names were scrambled. For a more efficient program, see Exercise 7.4, Problem 7.

d. *Planned Output*

```
ALPHABETIZING FIVE NAMES


ENTER THE NAMES
? JOE
? JIM
? JILL
? AL
? ADA

THE LIST IN ALPHABETICAL ORDER
ADA
AL
JILL
JIM
JOE
```

e. *The Program*

```
10   PRINT "ALPHABETIZING FIVE NAMES"
20   PRINT : PRINT
30   PRINT "ENTER THE NAMES"
40   FOR I=1 TO 5
50   INPUT N$(I)
60   NEXT I
70   PRINT
80   FOR I=1 TO 4
90   FOR J=1 TO 5-I
100  IF N$(J)<=N$(J+1) THEN GOTO 140
110  T$=N$(J)
120  N$(J)=N$(J+1)
130  N$(J+1)=T$
140  NEXT J
150  NEXT I
160  PRINT "THE LIST IN ALPHABETICAL ORDER"
170  FOR I=1 TO 5
180  PRINT N$(I)
190  NEXT I
200  END
```

Exercise 7.4

Show how the bubble sort would place each of the following lists of numbers in ascending order. Be sure to show the list for each comparison.

1. $Y(1)=9$
   $Y(2)=8$
   $Y(3)=6$
   $Y(4)=7$

2. $T(1)=13$
   $T(2)=19$
   $T(3)=14$
   $T(4)=12$
   $T(5)=10$

Show how the bubble sort would alphabetize each of the following lists of names. Be sure to show the list for each comparison.

3. $Y(1)$=SARA
   $Y(2)$=MARY
   $Y(3)$=AMY
   $Y(4)$=BILL

4. $T(1)$=JOE
   $T(2)$=JIM
   $T(3)$=JILL
   $T(4)$=AL
   $T(5)$=ADA

Revise the program that was written in this section to produce the following outputs.

5.

```
ALPHABETIZING USING THE BUBBLE SORT



WHAT ARE THE NAMES?
? JOE
? JIM
? JILL
? AL
? ADA


ALPHABETICAL LISTING
ADA
AL
JILL
JIM
JOE
```

6.

```
THE BUBBLE SORT


THE NAMES ARE
JOE
JIM
JILL
AL
ADA

IN ALPHABETICAL ORDER
ADA
AL
JILL
JIM
JOE
```

For the following problem and flowchart, design a planned output and code a program to solve the problem.

7. **Problem:** Our program (for the bubble sort) will make the same number of passes regardless of the order of the names. Write a program for the bubble sort so that if at the end of a pass no names are moved the alphabetized list is printed.

(*Hint:* We must set up a flag (signal) in the program. If the flag is zero after a pass, it will mean that no changes were made. This will indicate that the list is alphabetized. Truly a more efficient sort!)

Flowchart

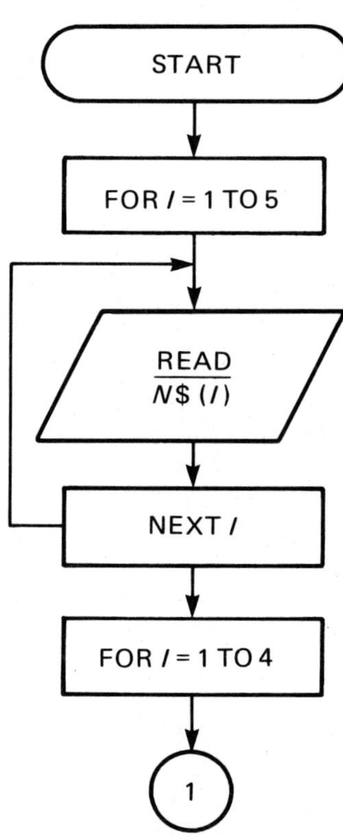

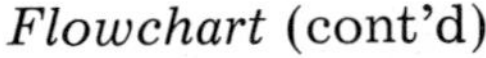

Flowchart (cont'd)

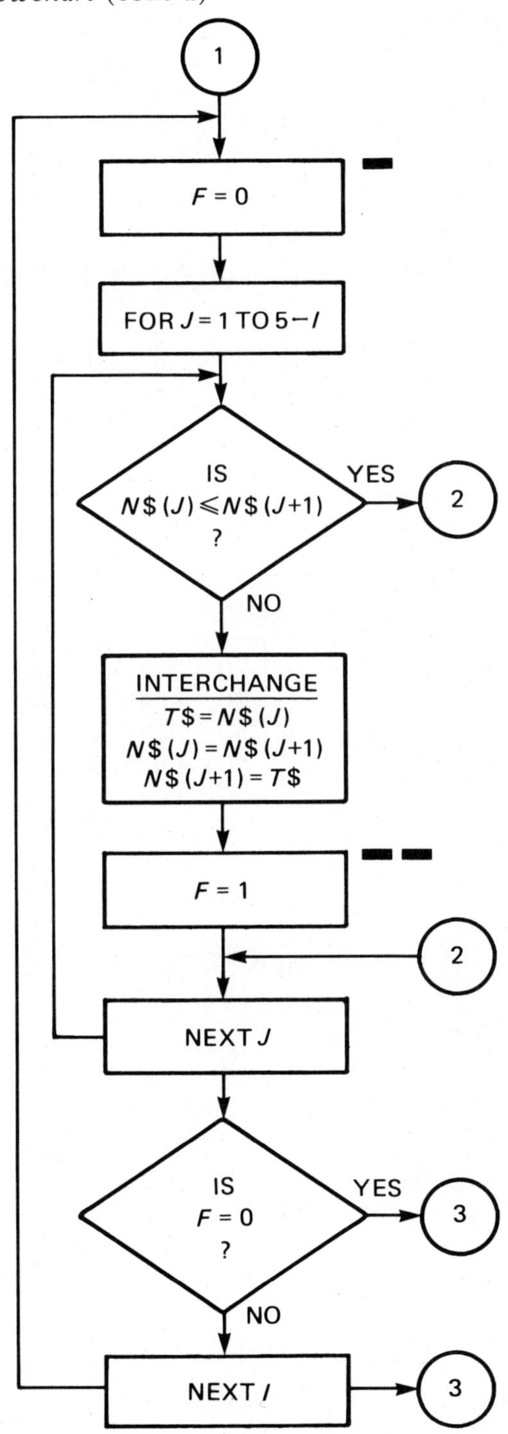

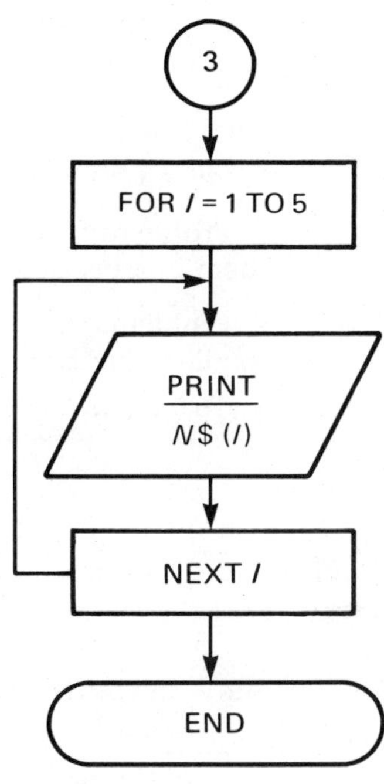

Discussion of the Flowchart:

■ Flag is set at 0 (zero).

■■ If an interchange takes place during a pass, the flag is assigned the value of 1. If no change is made, it remains 0.

For each of the following problems:
a. Write an analysis which includes the development of an algorithm.
b. Write a plan for a computer solution to the problem.
c. Develop a flowchart.
d. Design a planned output.
e. Code a program to solve the problem and RUN it on the computer.

8. **Problem:** Use the bubble sort to have the computer sort five numbers and display them in ascending order.

9. **Problem:** Enter ten numbers. Have the computer display the largest and then the smallest number.

10. **Problem:** Alphabetize a list of *N* names.

END OF CHAPTER EXERCISES

Section 7.1 Show the output for each program.

1.
```
10  FOR T=1 TO 6
20  READ E(T)
30  NEXT T
40  DATA 9, 8, 7, 6, 5, 4
50  FOR T=1 TO 6
60  PRINT E(T), T
70  NEXT T
80  END
```

2.
```
10  FOR V=1 TO 5
20  READ F$(V)
30  NEXT V
40  DATA G, H, I, J, K
50  FOR V=1 TO 5
60  PRINT V, F$(V)
70  NEXT V
80  END
```

3.
```
10  FOR W=1 TO 6
20  READ G(W)
30  NEXT W
40  FOR R=3 TO 6
50  PRINT G(R)
60  NEXT R
70  DATA 14, 15, 16, 17, 18, 19
80  END
```

4.
```
10  FOR Y=1 TO 7
20  H(Y)=7-Y
30  NEXT Y
40  FOR Y=1 TO 7
50  PRINT H(Y);
60  NEXT Y
70  END
```

5.

```
10  FOR Z=1 TO 6
20  A(Z)=3*Z
30  NEXT Z
40  FOR Z=1 TO 6 STEP 2
50  PRINT A(Z)
60  NEXT Z
70  END
```

6.

```
10   FOR B=1 TO 5
20   C(B)=1
30   NEXT B
40   FOR B=1 TO 5 STEP 2
50   C(B)=C(B)+2
60   NEXT B
70   FOR B=1 TO 5
80   PRINT C(B);
90   NEXT B
100  END
```

7.

```
10   C=0
20   FOR D=1 TO 6
30   READ E(D)
40   NEXT D
50   DATA 3, 4, 3, 3, 5, 3
60   FOR D=1 TO 6
70   IF E(D)=3 THEN C=C+1
80   NEXT D
90   PRINT C
100  END
```

8.

```
10   FOR F=1 TO 3
20   READ H(F), J(F)
30   NEXT F
40   DATA 9, 7, 4, 12, 5, 5
50   FOR F=1 TO 3
60   IF H(F)<=J(F) THEN 90
70   PRINT J(F)
80   GOTO 100
90   PRINT H(F)
100  NEXT F
110  END
```

For the following problem and flowchart, design a planned output and code a program to solve the problem.

9. **Problem:** Have the computer read ten numbers into an array. Have it find the average (mean) of the numbers. Have it display the average and then display the numbers in reverse order.

Flowchart

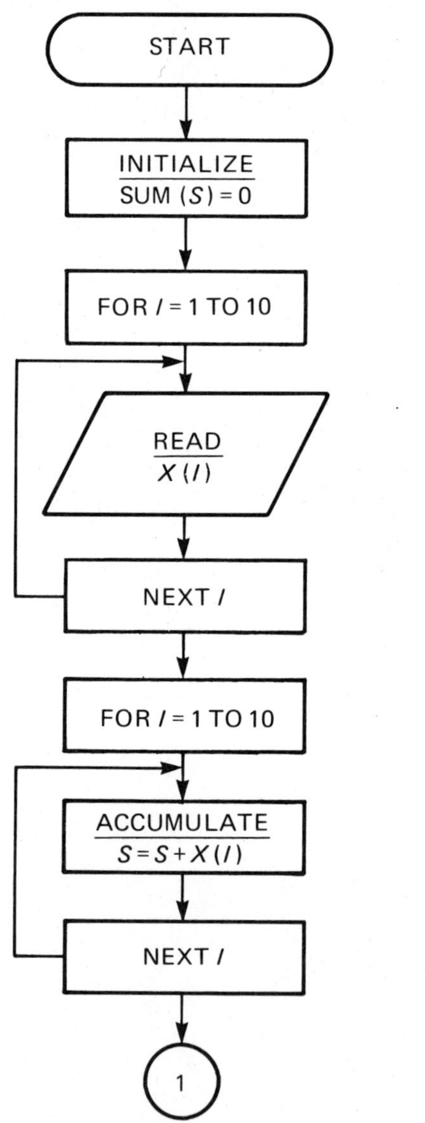

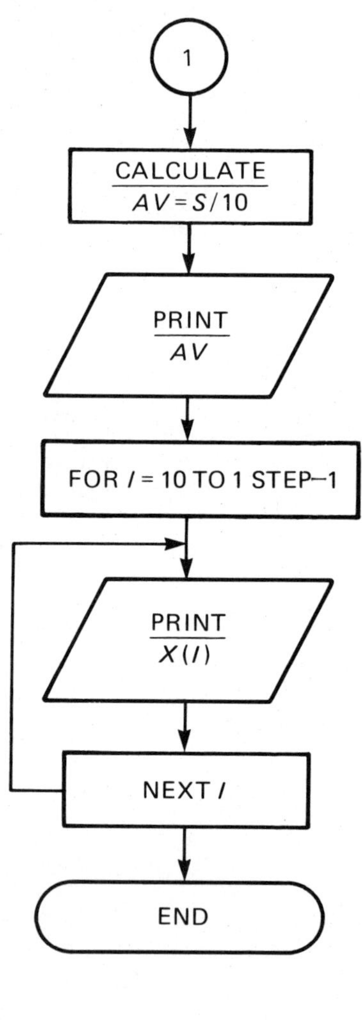

For each of the following problems:
a. Write an analysis which includes the development of an algorithm.
b. Write a plan for a computer solution to the problem.
c. Develop a flowchart.
d. Design a planned output.
e. Code a program to solve the problem and RUN it on the computer.

10. **Problem:** Have the computer read ten numbers into an array. Have it display the numbers and then display the numbers in reverse order.

11. **Problem:** Have the computer read ten numbers into an array. Have it print the sum of the second, fourth, sixth, eighth and tenth numbers. Then display a list of all the numbers.

12. **Problem:** Expand the inventory program in Exercise 7.1, Problem 14, by allowing the user the following options that are on the menu (list of options):

MENU

1. Give a listing of all items in stock.
2. Order an item. Print cost and update inventory.
3. New shipment. Update inventory.

The user will enter 1, 2 or 3. The computer will perform the task for the number entered.

Section 7.2

Develop an algorithm to generate random numbers:

13. from 0 to 20   14. from 1 to 20   15. from 0 to 100   16. from 1 to 100

For the given problem, flowchart and planned output, code a program to solve the problem.

17. **Problem:** Code a program to simulate dealing five cards from a deck.

Flowchart

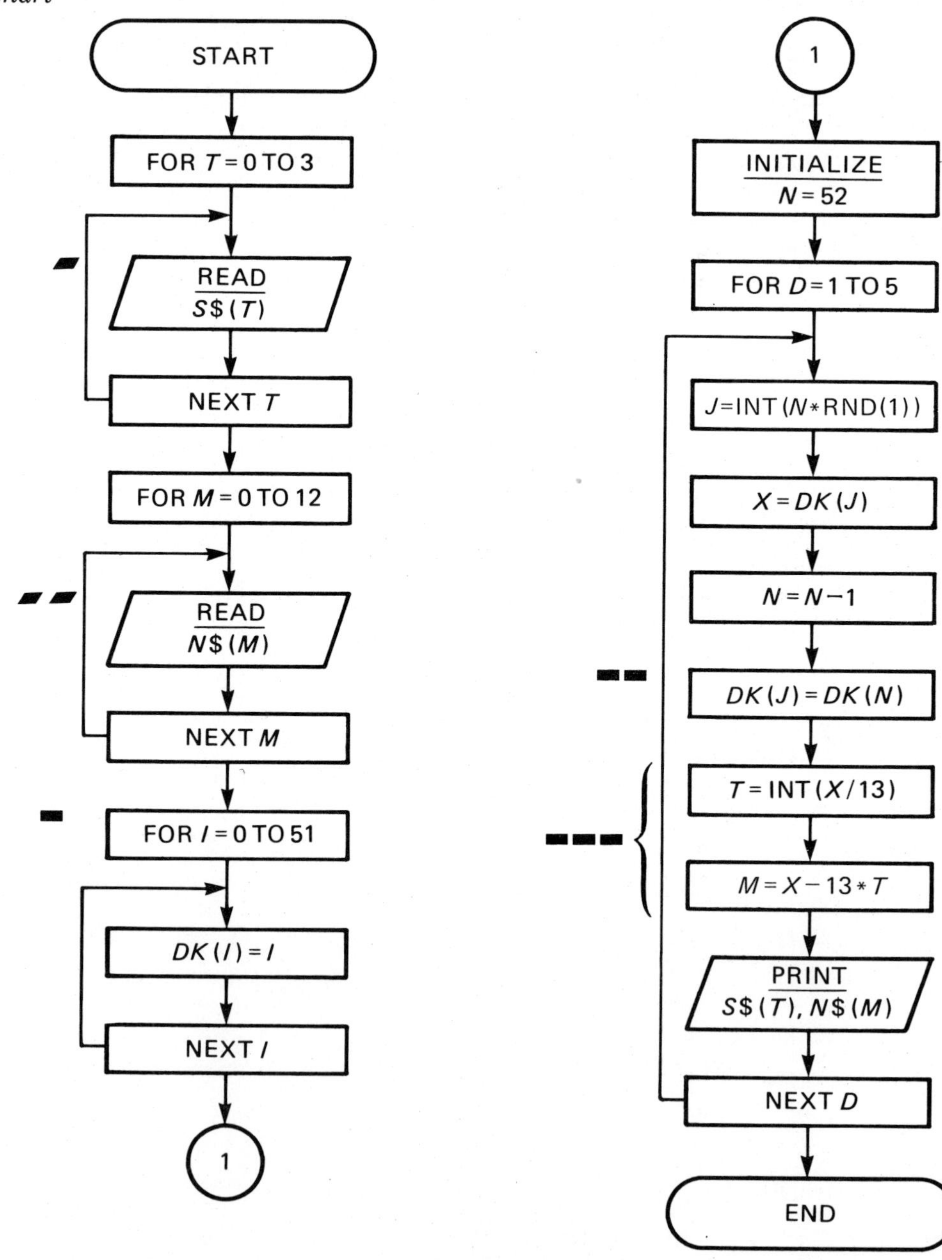

Discussion of the Flowchart:

▰ Reading the suits: hearts, diamonds, clubs and spades.

▰▰ Reading the value of the cards: ace, two, three, . . . , Jack, Queen, King.

▬ Assigning each of the 52 cards a value from 0 to 51.

▬▬ Each time a card is selected, it is replaced by the last card in the deck.

▬▬▬ Algorithm for getting the subscripts for the suit (T) and value (M).

Planned Output

```
DEALING FIVE CARDS



FOUR OF HEARTS
ACE OF SPADES
NINE OF DIAMONDS
JACK OF CLUBS
THREE OF SPADES
```

For each of the following problems:

a. Write an analysis which includes the development of an algorithm.
b. Write a plan for a computer solution to the problem.
c. Develop a flowchart.
d. Design a planned output.
e. Code a program to solve the problem and RUN it on the computer.

18. **Problem:** Develop a CAI program to give the user practice using the multiplication facts with the numbers 0 to 10. Generate ten random examples. After the tenth example, print the student's score (percent correct). Allow the user two chances to get the correct answer. After two incorrect guesses, print the answer for the user and continue.

19. **Problem:** Generate 10,000 random numbers from 0 to 100. Determine the number of primes in the 10,000 numbers. (How many do you expect?)

20. **Problem:** Develop a CAI program to give the user practice in dividing two numbers. All quotients must be integers. All divisors must be one digit. Dividends should not be greater than 270. (Use random numbers in this program.) Give the user ten examples. After the tenth example, print the percent correct. Allow the user two chances to get the correct answer. After two incorrect guesses, print the correct answer and continue.

Section 7.3

For each string stored in $R\$$, find LEN($R\$$):

21. $R\$$="I LIKE COMPUTERS"
22. $R\$$="I WORK HARD."
23. $R\$$="MATH IS GREAT FUN."
24. $R\$$="COMPUTERS ARE FUN"

If $C\$$="I LOVE DOING HOMEWORK.", find the string assigned to $F\$$ if:

25. $F\$$=MID\$($C\$$, 1, 22)
26. $F\$$=MID\$($C\$$, 14, 8)
27. $F\$$=MID\$($C\$$, 22, 1)
28. $F\$$=MID\$($C\$$, 8, 14)

For the given problem, flowchart and planned output, code a program to solve the problem.

29. **Problem:** Enter a message (not more than 255 characters). Have the computer display the first half of the message.

Flowchart

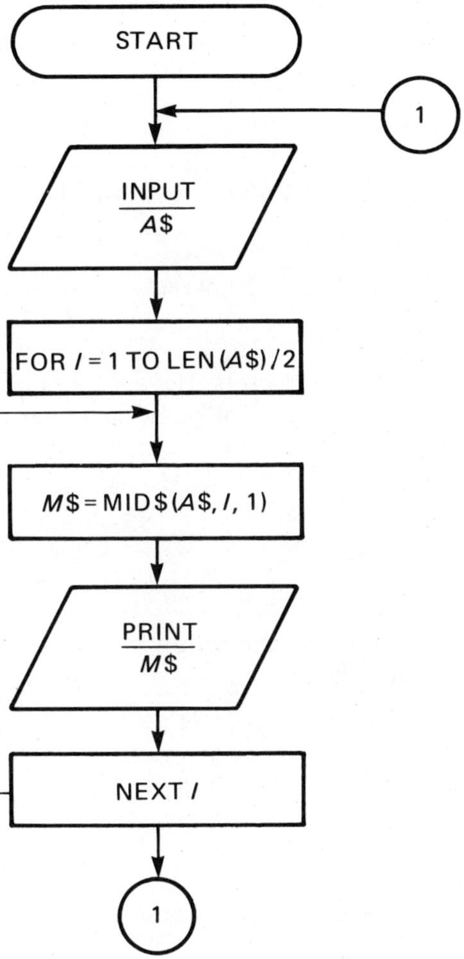

START

INPUT A$

FOR I = 1 TO LEN(A$)/2

M$ = MID$(A$, I, 1)

PRINT M$

NEXT I

1

Planned Output

```
PRINTING HALF THE MESSAGE


ENTER A MESSAGE
? MATHEMATICS IS FUN
HALF THE MESSAGE IS
MATHEMATI


ENTER A MESSAGE
?
```

For each of the following problems:
a. Write an analysis which includes the development of an algorithm.
b. Write a plan for a computer solution to the problem.
c. Develop a flowchart.
d. Design a planned output.
e. Code a program to solve the problem and RUN it on the computer.

30. **Problem:** Enter a message (not more than 255 characters). Have the computer determine the number of E's in the message.

31. **Problem:** Enter a message (not more than 255 characters). Have the computer display the second half of the message.

32. **Problem:** Enter a word of at least five characters. Have the computer display the first, third and fifth character. Tell the user the number of remaining characters. Have the user guess the word. After three guesses, have the computer display the word. As an extra feature, you may wish to have the user guess at each letter. After two incorrect guesses, give the correct one.
Do the above for three different words.

Section 7.4

Show how the bubble sort would place each of the following lists of numbers in ascending order. Be sure to show the list for each comparison.

33. *Y*(1)=12
Y(2)=11
Y(3)=13
Y(4)=15
Y(5)=6

34. *T*(1)=7
T(2)=5
T(3)=6
T(4)=8

Show how the bubble sort would alphabetize each of the following lists of words. Be sure to show the lists for each comparison.

35. *A*$(1)=LYSA
A$(2)=LYNN
A$(3)=LYNDA

36. *D*$(1)=THERE
D$(2)=AND
D$(3)=HE
D$(4)=LIVES

For the given problem, flowchart and planned output, code a program to solve the problem.

37. **Problem:** Enter a list of five numbers. Have the computer find the largest number. Do not use the bubble sort. (First use the values from the planned output to trace the flowchart and see how the largest value is found.)

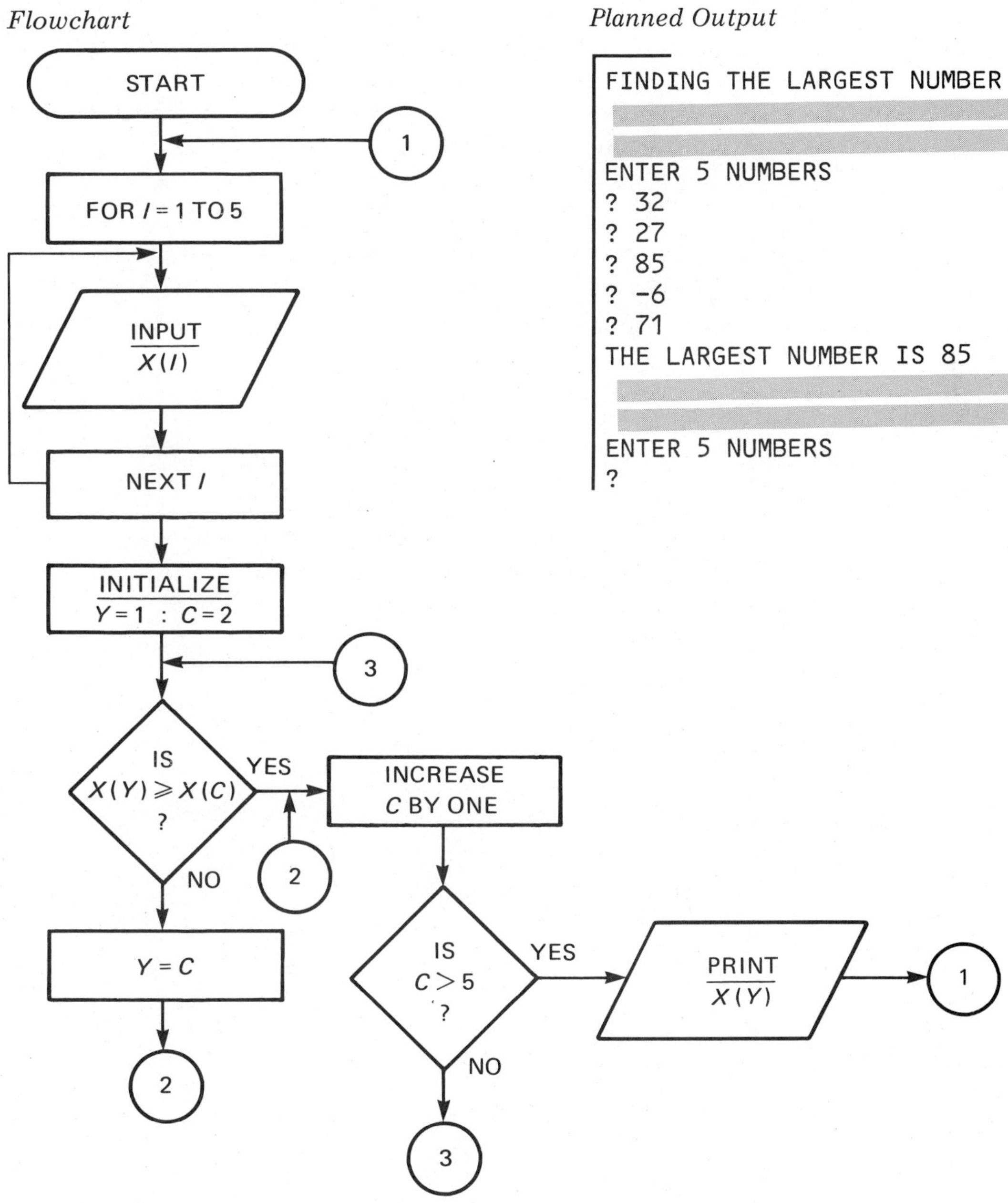

For each of the following problems:
a. Write an analysis which includes the development of an algorithm.
b. Write a plan for a computer solution to the problem.
c. Develop a flowchart.
d. Design a planned output.
e. Code a program to solve the problem and RUN it on the computer.

38. **Problem:** Use the bubble sort to have the computer alphabetize a list of five names. Have it display the list after each pass.

39. **Problem:** Use the sort in Problem 37 to have the computer find the largest in a list of *N* numbers.

40. **Problem:** The median is the middle number when the data are listed from largest to smallest (or smallest to largest). Enter an odd amount of numbers. Have the computer display the median.

How would this program be done with an even amount of numbers?

Appendix

Quick Reference of the BASIC Language

System and Utility Commands

1.	END	Ends execution of the program and returns to the command mode.
2.	LIST	Lists the entire program.
3.	NEW	Erases current program and initializes all variables to 0.
4.	RUN	Executes program beginning at the lowest line number.
5.	STOP	Halts execution of the program and indicates which line.

Statements: Input/Output

DATA 3, 26, 8	Holds the data that is to be assigned to the variables in a READ statement.
INPUT A	Computer awaits an entry from the keyboard and places it in memory cell A.
INPUT "A="; A	Computer prints A= and then awaits an entry from the keyboard which is placed in memory cell A.
PRINT X	Computer prints the contents of memory cell X.
PRINT TAB(5); X	Moves the cursor to the specified position 5 and prints the contents of memory cell X. On some computers the cursor moves to position 6.
READ A, B, C	Assigns values to the variables A, B, and C beginning with the first value in a DATA statement.

REM	Enables user to enter REMarks for documentation. The computer ignores this program line.

Flow of Control Commands

FOR I=1 TO 10 STEP 3	Executes all statements between the FOR statement and a corresponding NEXT. It begins with I=1 and increments I by 3. If no STEP is used, X is incremented by 1.
GOSUB 400	Transfers control to the subroutine beginning at line-number 400.
GOTO 50	Transfers control to line-number 50.
IF A=B THEN GOTO 60	If the A=B is true, control transfers to line 60. If the statement is false, control transfers to the next line. Omitting the word GOTO will not alter the command.
NEXT I	Defines bottom of FOR-NEXT loop.
RETURN	Ends the subroutine. Control passes to the statement following the most recent GOSUB.

Algebraic Operators

+	Addition	/	Division
−	Subtraction	↑ or ∧	Exponentiation
*	Multiplication	=	Assigns to the left side a value from the right side.

Relational and Logical Operators

=	Equal	<>	Not equal to
>	Greater than	<=	Less than or equal to
<	Less than	>=	Greater than or equal to
AND	Logical AND	OR	Logical OR

Strings and Arrays

DIM X (A, B)	Determines maximum subscripts for X starting with X (0, 0).
LEN (X$)	Returns the number of characters in X$.
MID (X$, P, L)	Returns a substring of X$ beginning with position P with length L.

Mathematics Functions

ABS (N)	Returns the absolute value of N.
ATN (N)	Returns arctangent of N in radians.
COS (R)	Returns cosine of R in radians.
INT (N)	Returns largest integer less than.
LOG (N)	Returns the natural logarithm of N.
RND (N)	Generates a pseudo-random number. See your manual for the specifics.
SGN (N)	Returns -1 if $N<0$, 0 if $N=0$, and 1 if $N>0$.
SIN (R)	Returns sine of R radians.
SQR (N)	Returns the positive square root of N.
TAN (R)	Returns the tangent of R radians.

Glossary

Accumulator a computer term referring to the summing of amounts.
Acute triangle a triangle with all acute angles (all angles less than 90°).
Algorithm step-by-step procedure to solve a problem.
Alphanumeric data characters and or numbers.
Angle formed by two rays with a common endpoint. The common endpoint is called the vertex.
Area the measurement of the inside of any polygon.
Arithmetic mean the average; the sum of a group of numbers divided by the number of addends in the group.
Arithmetic sequence a sequence with a common difference between terms.
Arithmetic series the sum of an arithmetic sequence.
Array a group of subscripted variables.
Ascending in order from smallest to largest.
ASCII code a numerical value assigned to each character used in BASIC.
Assignment statement an instruction that assigns a specific value from the right side of an equal sign to the variable on the left.

Base the type of grouping involved in a number system, e.g.,
25 = 2 eights and 5 ones,
233 = 2 twenty fives, 3 fives and 3 ones.
BASIC a computer language. The letters stand for *B*asic *A*ll-purpose *S*ymbolic *I*nstruction *C*ode.
Bubble sort a computer technique used for alphabetizing words or sorting numbers in ascending order.
Byte a term used to describe a part of the computer's memory.

C.A.I. stands for Computer Assisted Instruction. An interactional program between the user and the computer. The program is usually used to help the user practice a particular skill.
Celsius the metric unit used for measuring temperature.
Centimeter a metric unit of length equaling 1/100 of a meter.
Character any single figure that can be typed on the keyboard, including letters, numbers, and symbols. Examples: A, X, $, %, @, 1, 6, +, *
Circumference the distance around (perimeter of) a circle.
Circumscribed refers to the geometric figure whose sides are all tangent to the inscribed figure.
Coding translating a plan of action into a computer language.
Colon two dots arranged vertically and used in a program to carry more than one instruction on the same line.
Command a one-word instruction that is used in the immediate mode to tell the computer to do something to a program.
Commission an amount of money paid in addition to a regular wage for each sale rendered.
Graduated commission commission that increases as sales increase.
Common factors factors that are common among numbers.
Complementary angles two angles whose measures have a sum of 90°.
Composite number a number that has more than one set of factors.
Condition the state of a variable in relation to another variable, a string, or a number. The comparison in an IF-THEN statement is a condition. If the condition is fulfilled (if the comparison is true), then the statement(s) after the THEN is carried out.
Congruent equal in measure.
Constant a numerical value that never changes.
Coordinate one of an ordered number pair that correspond to a point in a coordinate plan.
Counter a set of program instructions that enable the computer to count.
Cube (of a number) a number used as a factor three times.
Cursor the blinking light on a computer that tells you which space you are at on a line and where the next character will be printed.

Diagonal a line connecting two nonconsecutive vertices of a polygon.
Diameter the line that divides a circle exactly in half.
Difference the result of a subtraction.
DIM a computer instruction that tells the computer how many memory locations to reserve for subscripted variables.
Discount an amount taken off the original selling price of an item.

Diskette (disk) a small, flat, flexible magnetic plate that is used to store information.
Disk drive the device that lets the computer read from and write on the diskette.
Distributive property the fundamental principle connecting addition and subtraction. $a \times (b+c) = a \times b + a \times c$.
Dividend the number you are dividing into.
Divisor the number you are dividing by.

Edit to add, correct, or delete lines in an existing program.
END an instruction that tells the computer the program is finished.
Equilateral triangle a triangle with all sides the same length.
Equation a mathematical sentence using the equality symbol.
Exponent a numerical value written above and to the right of a mathematical expression to indicate how many times a number is to be used as a factor.

Factor any one number of a set of numbers to be multiplied.
Fahrenheit the English unit for measuring temperature.
Field a section of the computer's display screen. The display screens of different computers are divided into varying numbers of fields.
Flag a signal in a program used in a bubble sort to indicate to the computer that the sorting has been completed.
Flowchart a graphic representation of a solution to a problem.
Format the way information is arranged on the screen.
FOR-NEXT loop a pair of computer instructions used for counting.
FOR NEXT with steps the step part of a FOR-NEXT loop refers to the increment.

General form a mathematical statement that applies to all examples of the same type.
Geometric sequence a sequence with a common ratio between terms.
Geometric series the sum of a geometric sequence.
GOTO an instruction that sends the computer to another part of the program (sometimes called an unconditional branch).
Graphics pictures, diagrams, and artwork drawn on the computer screen.

Hardware computer machinery. *See also* SOFTWARE.
Hypotenuse the longest side in a right triangle; the side opposite the right angle.
Hundredth the second place value to the right of a decimal point.

IF-THEN a computer instruction that poses a condition in the program. If the condition is true, the THEN part of the instruction sends the computer to another part of the program. If false, the computer simply proceeds to the next program line.
Immediate mode the mode in which statements are used without line numbers or any commands. As soon as RETURN (or ENTER) is pressed, the instructions on the line just typed will be carried out. *See also* PROGRAM MODE.
Increment the amount of increase or decrease necessary to arrive at each new number in a series.
Initialize a term referring to assigning a value to a variable in a program.
INPUT a computer instruction that stops the computer to allow the entry of a value to a variable. This instruction produces a ? on the display. The user responds with the value he or she wishes to assign to the variable while the program is running.
Inscribe a geometric figure drawn inside another figure in which all vertices are touching the outside figure.
INT refers to the computer's integer function. This function gives back the largest whole number that is not greater than the starting number; e.g.,
INT (7.5) gives you 7
INT (−3.36) gives you −4
Integer the set of numbers consisting of all positive natural numbers, all negative natural numbers, and zero.
Interest an amount of money paid in excess of a loan.
Isosceles triangle a triangle with two equal sides.

Keypunch operator the person who types instructions onto cards.

LEN a computer function that returns the length of a string.
Line number a number that preceeds every program line, and shows the order in which the lines are carried out in the program.
LIST a system command that produces an up-to-date list of the program.
List price the original selling price of an item.

Logical operators refers to the computer terms AND, OR which can be used to combine IF-THEN statements in a program.
Loop a group of instructions that are repeated a set number of times until a condition is satisfied.

Median the middle number.
Memory cell a small part of the computer's memory where information is stored.
Meter the principle unit of length in the metric system. It is approximately equal to 39 inches.
**MID$** a computer function that returns a substring of A$ beginning at position P of length L.
Millimeter a metric unit of length equaling 1/1000 of a meter.
Monitor another name for the video screen on which the computer displays the output.

Nested FOR-NEXT when a FOR-NEXT loop is found inside another FOR-NEXT loop.
NEW a system command that clears the memory of the computer.
Numeric data information in number form.

Obtuse triangle a triangle with one obtuse angle (one angle with measure greater than 90°).
Output the computer's response to your program.

Parallel lines lines in the same plane that will never intersect.
Parallelogram a quadrilateral whose opposite sides are parallel.
Pentagon any five-sided polygon.
Perfect number a number that is equal to the sum of all of its proper factors.
Perimeter the sum of the lengths of the sides of any polygon.
Pi (π) the ratio of the circumference of a circle to its diameter. Pi is approximately equal to 3.14.
Plan a strategy that outlines how an algorithm is used to solve a problem.
Planned output the programmer's decision as to what the display should look like before writing the program.
Positive a number that is greater than zero.
Prime number a number with exactly two factors; itself and one.
PRINT the instruction that displays the output of a program.
Product the result in multiplication.
Program a set of instructions written in a computer code that is used to solve a problem.
Programmer a person who translates the plan into computer code.
Proper factors factors that are less than the number.

Quadrilateral any four-sided polygon.
Quotient the result in division.

Radius a line segment drawn to a circle from its center.
Rate of discount the percent to be deducted from the price.
Ratio the fractional relationship that exists between two quantities.
READ-DATA a pair of instructions used to assign values to variables without having to use an INPUT statement.
Rectangle a quadrilateral with opposite sides equal and opposite angles equal.
Regular polygon any polygon with all sides equal and all angles equal.
Relation operator a mathematical symbol used in a conditional statement to show a relationship between two expressions. The operators are ($<$) less than, ($>$) greater than, ($\geq$) greater than or equal to, ($\leq$) less than or equal to, ($\neq$) not equal and ($=$) equal.
REM refers to the word *remark*. It is an instruction used to make a comment in a program. The comment will not appear in the output.
Return or enter key key that, by pressing, passes control to the computer.
Right triangle a triangle that contains one 90° angle.
RND a computer function used to generate random numbers from within the computer.
RUN tells the computer to carry out the program instructions.

Sale price the amount payable after the discount is subtracted from the original price.
Scalene triangle a triangle with no equal sides and angles.
Semi-circle half of a circle.
Software anything related to computers that is not physically part of the computer's machinery, such as programs. *See also* HARDWARE.
SQR the computer function for finding the square root of a number.
Square a four-sided polygon (quadrilateral) with all equal sides and angles.
Square (of a number) a number multiplied by itself.
Square root (of a number) the number that, when multiplied by itself, gives back the number in the square root symbol; e.g., $\sqrt{16}=4$.

Straight angle an angle whose rays form a straight line; it has a measure of 180°.

String variable a variable used to store alphanumeric characters.

Subscripted variable a variable written as X(0), X(1), X(2), and read X sub 0, X sub 1, etc.

Sum the result of addition.

Supplementary angles two angles whose measures have a sum of 90°.

Syntax the grammar for a statement; the way it must be written.

System command an instruction without a line number. These instructions are carried out immediately by the computer.

Systems analyst a person who writes a plan for solving a problem.

TAB a computer function used in a print statement to allow the computer to print at a specific position on the line.

Tenth the first place value to the right of the decimal point.

Terminal operator a person who operates the computer.

Thousandth the third place value to the right of the decimal point.

Tracing going through a program in a step-by-step procedure to identify any errors that may exist in a program.

Truncate to cut off a number at a certain decimal place.

User a person who interacts with the computer.

Value a numeric or string quantity.

Variable a letter of the alphabet that stands for a number. In addition, variables can also be written with a letter followed by a number or two letters together; e.g.,
A, V2, LM.

Vertex the point where the two rays in an angle meet.

Index

Photo Acknowledgments

A Little Gallery of Hardware: A Look at the Main Parts of a Computer System

1. Ramtek Corporation
2. Cray Research, Inc.
3. © Michal Heron/Woodfin Camp & Associates
4. Smith College/courtesy Chuck Kidd Photography
5. Lawrence Livermore National Laboratory
6. IBM Corporation
7. Anacomp, Inc.
8. Inforex, Inc.
9. Bell Helicopter/Textron
10. IBM Corporation
11. © TRW Inc.
12. Honeywell Incorporated/Honeywell Information Systems
13. © Glenn R. Steiner

The "All-Purpose Machine": The Many, Often Hidden, Uses of Computing

1. IBM Corporation
2. Anacomp, Inc.
3. U.S. Census Bureau/Sperry Univac, A Division of Sperry Corporation
4. © Peter Menzel, 1981

5, 6. Sperry Univac, A Division of Sperry Corporation

7, 8. Ramtek Corporation

9. Smith College/courtesy Chuck Kidd Photography
10. Seabrook Training Center/PSNH (Public Service Company of New Hampshire)
11. Ramtek Corporation

12, 13. IBM Corporation

14, 15. Hal Anderson/courtesy Hewlett Packard Company/courtesy Dr. Roy Ing/The Center for Environmental Health at Centers for Disease Control, Atlanta, Ga.

16. Ramtek Corporation
17. Anacomp, Inc.
18. R. L. Jairell/courtesy Wyoming Department of Highways/courtesy Keyboard Magazine
19. © Peter Menzel
20. © John Zainer/Peter Arnold
21. © TRW Inc.